Biological and Medical Physics, Biomedical Engineering

BIOLOGICAL AND MEDICAL PHYSICS, BIOMEDICAL ENGINEERING

This series is intended to be comprehensive, covering a broad range of topics important to the study of the physical, chemical and biological sciences. Its goal is to provide scientists and engineers with textbooks, monographs, and reference works to address the growing need for information. The fields of biological and medical physics and biomedical engineering are broad, multidisciplinary and dynamic. They lie at the crossroads of frontier research in physics, biology, chemistry, and medicine.

Books in the series emphasize established and emergent areas of science including molecular, membrane, and mathematical biophysics; photosynthetic energy harvesting and conversion; information processing; physical principles of genetics; sensory communications; automata networks, neural networks, and cellular automata. Equally important is coverage of applied aspects of biological and medical physics and biomedical engineering such as molecular electronic components and devices, biosensors, medicine, imaging, physical principles of renewable energy production, advanced prostheses, and environmental control and engineering.

More information about this series at http://www.springer.com/series/3740

Kiao Inthavong · Narinder Singh ·
Eugene Wong · Jiyuan Tu
Editors

Clinical and Biomedical Engineering in the Human Nose

A Computational Fluid Dynamics Approach

Springer

Editors
Kiao Inthavong
Mechanical and Automotive Engineering
School of Engineering
RMIT University
Bundoora, VIC, Australia

Eugene Wong
Mechanical and Automotive Engineering
School of Engineering
RMIT University
Bundoora, VIC, Australia

Narinder Singh
Sydney Medical School
University of Sydney
Sydney, NSW, Australia

Jiyuan Tu
Mechanical and Automotive Engineering
School of Engineering
RMIT University
Bundoora, VIC, Australia

ISSN 1618-7210 ISSN 2197-5647 (electronic)
Biological and Medical Physics, Biomedical Engineering
ISBN 978-981-15-6715-5 ISBN 978-981-15-6716-2 (eBook)
https://doi.org/10.1007/978-981-15-6716-2

This Springer imprint is published by the registered company Springer Nature Singapore Pte Ltd.
The registered company address is: 152 Beach Road, #21-01/04 Gateway East, Singapore 189721, Singapore

Preface

In recent years, there has been an explosion in interest in the field of Computational Fluid Dynamics (CFD) of the Nose and Airway. Historically, this period of growth can be traced back to the early 1990s when rapid developments in computing technology occurred. Researchers began exploring the use of CFD in the nose and airway, firstly by modelling airflow patterns, calculating changes in temperature and pressure, and measuring wall shear stress. Next, researchers studied common abnormalities, such as septal deviation, septal perforations, and inferior turbinate hypertrophy. Recently, CFD has been used to model and predict the effects of surgery and therapeutic interventions.

It was evident that the growth in computational capability and uptake of CFD technology in clinical applications was gaining significant traction. The international Society for CFD Of the Nose and Airway (SCONA, www.scona.org) held its inaugural world scientific congress in London, UK, in 2018. This brought together leading CFD clinicians and engineers with the aim of fostering collaboration and collegiality, increasing the impact of the work being performed and exploring the technology's significant potential to reveal the biomechanics of nasal physiology. In 2019, the second SCONA world scientific congress was held in Chicago, USA, bringing CFD's insights to a newer and larger audience.

The meetings demonstrated a strong need to bridge the gap between engineering knowledge and clinical experience, culminating in the creation of this book. In the spirit of SCONA—to bring the diverse mix of expertise together—a call out for contributions to this text was made to SCONA participants and members. The feedback and enthusiasm from all contributors made the editorial process a wonderful experience and the book was a pleasure to compile.

Our goal for this book was to showcase the wide variety of work being undertaken worldwide in this field and to provide foundational knowledge to fill the gaps that students entering this field may encounter. We sincerely thank all the authors for their fascinating and insightful contributions in creating this state-of-the-art work. To you, the reader of this book, we trust that the experience and innovation contained within each chapter will inspire new research ideas and effective clinical outcomes.

Bundoora, VIC, Australia

Kiao Inthavong
Narinder Singh
Eugene Wong
Jiyuan Tu

Acknowledgements

The editors are grateful to Springer's publishing team, in particular, Ramesh Premnath for supporting and endorsing this book and Ashok Arumairaj for the patience and support shown in putting the book together. We would like to thank all those who participated in the SCONA events. The enthusiasm and encouragement for the creation of SCONA was immense, which led to successful meetings and the creation of this book.

We are also grateful for the research grant from Garnett Passe Rodney Williams Memorial Foundation (Conjoint Grant 2019), which supported the collaborative efforts of clinicians and engineers to compile this book and to make progress in research towards improving clinical practise of otolaryngology, head, and neck surgery (OHNS).

Contents

Contributors

Jim Bartley University of Auckland, Auckland, New Zealand

Alister J. Bates Cincinnati Children's Hospital, Cincinnati, USA

Kimberley Bradshaw University of Sydney, Sydney, Australia

Hadrien Calmet Barcelona Supercomputing Center, Barcelona, Spain

Jose Luis Cercos-Pita Postdoctoral researcher, Department of Surgical Sciences, Hedenstierna laboratory, Uppsala Universitet, Uppsala, Sweden

Richard Douglas University of Auckland, Auckland, New Zealand

Yu Feng Oklahoma State University, Stillwater, USA

Dennis Onyeka Frank-Ito Duke University, Durham, NC, USA

Guilherme Garcia Medical College of Wisconsin, Wauwatosa, WI, USA

Koch Gerda AIT Angewandte Informationstechnik Forschungsgesellschaft mbH, Graz, Austria

Hamideh Hayati Oklahoma State University, Stillwater, USA

Kiao Inthavong Mechanical & Automotive Engineering, School of Engineering, RMIT University, Melbourne, Australia

Chengyu Li Villanova University, Villanova, USA

Andreas Lintermann Forschungszentrum Jülich GmbH, Jülich, Germany

Lehner Matthias AIT Angewandte Informationstechnik Forschungsgesellschaft mbH, Graz, Austria

Benda Odo AIT Angewandte Informationstechnik Forschungsgesellschaft mbH, Graz, Austria

Ortiz Ramiro AIT Angewandte Informationstechnik Forschungsgesellschaft mbH, Graz, Austria

Hana Salati Auckland University of Technology, Auckland, New Zealand

Yidan Shang RMIT University, Melbourne, Australia

Kendra Shrestha RMIT University, Melbourne, Australia

Dennis Shusterman University of California, San Francisco, USA

Narinder Singh University of Sydney, Sydney, Australia

Joey Siu University of Auckland, Auckland, New Zealand

Jiyuan Tu RMIT University, Melbourne, Australia

Ross Walenga U.S. Food and Drug Administration, Silver Spring, USA

Koch Walter AIT Angewandte Informationstechnik Forschungsgesellschaft mbH, Graz, Austria

David White Auckland University of Technology, Auckland, New Zealand

Eugene Wong RMIT University, Melbourne, Australia

Jinxiang Xi University of Massachusetts Lowell, Lowell, England

Kai Zhao Ohio State University, Columbus, USA

List of Figures

List of Tables

Chapter 1
Clinical and Biomedical Engineering in the Human Nose

Kiao Inthavong, Eugene Wong, Jiyuan Tu, and Narinder Singh

Abstract The convergence of Computational Fluid Dynamics (CFD) with otorhinolaryngology has provided an avenue for a multidisciplinary approach to observe nasal physiology. This chapter provides a history of its beginnings to its advances that parallel with increased computational power, and a discussion on how it could be applied to specific clinical applications. Furthermore, the challenges in bringing the techniques to be used as a diagnostic for clinical practice is discussed.

1.1 Introduction

Computational fluid dynamics (CFD) has emerged as a powerful objective tool in otorhinolaryngology as it provides detailed and quantified measures of the upper airway physio- and pathophysiological and functions. This includes airflow conductance, nasal air conditioning (heating and humidification of inspired air), delivery of odorant molecules to the olfactory epithelium, flow-induced phonation, and delivery of aerosol medications to their target sites. It has also been used to investigate a wide range of upper airway diseases, including nasal airway obstruction, chronic rhinosinusitis, empty nose syndrome, nasal septal perforations, and laryngotracheal stenosis. As it evolves, we expect advanced topics will deal with rapid virtual surgery applications through coupling with machine learning, leading to the next revolution in health care in the so-called Digital Twin.

The field of CFD is well established, having a long history and development from the field of aerodynamics, where the study of fluid flow is represented by mathematical equations, and solved using computer programs. The term "Fluid Dynamics" encompasses both the study of fluids either in motion (fluid in dynamic mode) or at rest (fluid in stationary mode). However, it is mainly dedicated to the former, fluids

K. Inthavong (✉) · E. Wong · J. Tu
RMIT University, Melbourne, Australia
e-mail: kiao.inthavong@rmit.edu.au

N. Singh
University of Sydney, Sydney, Australia

K. Inthavong et al. (eds.), *Clinical and Biomedical Engineering in the Human Nose*, Biological and Medical Physics, Biomedical Engineering,
https://doi.org/10.1007/978-981-15-6716-2_1

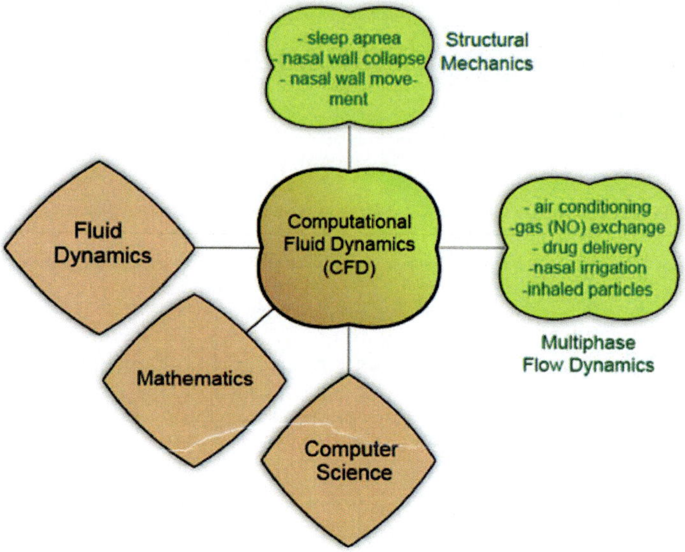

Fig. 1.1 CFD and its applications in the human nasal airway when coupled with additional physics

that are in motion–in their dynamic state. Its capability is analogous to wind-tunnel and laser-photography visualisation can be employed as a research tool to perform numerical experiments to better understand the physical nature of the fluid-particle dynamics. By performing these numerical experiments, advanced models can then be developed to increase the capabilities of the computational modelling

When additional physics are coupled with the CFD, new applications can be explored, which has provided new opportunities in otorhinolaryngology (Fig. 1.1). For example, including mathematical descriptions of structural mechanics allows studies of sleep apnea and nasal wall collapse by coupling the pressure, and forces generated by the fluid flow onto an elastic wall. While multiphase flows enable studies of heating and conditioning the air, nitric oxide gas (from the sinuses) exchange, drug delivery or inhaled particles (using Lagrangian particle tracking), and nasal irrigation (using liquid in air). Continued development involving multiphysics will indeed advance the current capability to allow greater exploratory research and account for more complex physiology.

1.2 Historical Perspectives

Among the earliest computational studies of airflow through a human nasal cavity are: Elad et al. [1], which used an ideal nose-like model where the turbinates were rectangular blocks connected to an orthogonal main nasal passage; and Keyhani et al.

Table 1.1 Comparison of computational model capability from 1995 to present day, 2019

	Keyhani et al. [2]	Current practices
CT slices	42	100–200
Slice thickness	2 mm	0.5 mm
Mesh elements	0.077 mil	3.0–10.0 mil
Model	Unilateral cavity	Both left, right cavities, and multiple patients/models
Computing power	Sun Sparcstation, 36–40 MHz processor, maximum of 512 MB RAM in eight slots	Workstations up to 64 Cores at 3–4 GHz, and supercomputers (IBM Blue Gene/P) with up to 164,000 processor cores

[2]'s which reconstructed a unilateral nasal cavity truncated at the choanae, from CT-scans. Both works applied structured meshing which was the most straight-forward algorithm at the time. To appreciate how far the research has progressed a comparison of Keyhani et al. [2]'s method with current practices (over thirty years apart) is given in Table 1.1.

The astounding technological advances have seen this field become more accessible and greater opportunities for more complex analysis. The advancement of the discipline is attributed to availability of Computational Fluid Dynamics (CFD), and advances in computer science and clinical imaging modalities. This is evident through the number of published studies using computational modelling of nasal airflow. A review of the number of articles for CFD in the nasal airway published annually between 1993 to 2018 was found by using the keywords "nasal CFD" in the search engine Google Scholar. This was compared with results from the Scopus (Web of Science) database which used the keywords "nasal" + "Computational Fluid Dynamics", shown in Fig. 1.2a. Between 2004 and 2009, there was a surge in the number of articles published, and overall there is an exponential growth to 2018, with nearly 80 related publications in 2017 and 2018 alone.

A significant contributor to the growing trend is undoubtedly the availability of computational resources, and advances in model algorithms to automate many processes that were once performed by hand. As an example, Fig. 1.2b shows a positive correlation in the number of mesh elements with the year, where the earliest reported computational studies contained around 0.01 [1] to 0.07 [2] million cells, and later studies had up to 36 [3] to 44 million cells [5]. These advancements have opened new opportunities of research and there is a need for clinicians to work with engineers to produce meaningful outcomes, which is often lacking from the technical modelling studies. An aim of this book is to bridge the knowledge gap between current clinicians and fluid dynamics engineers.

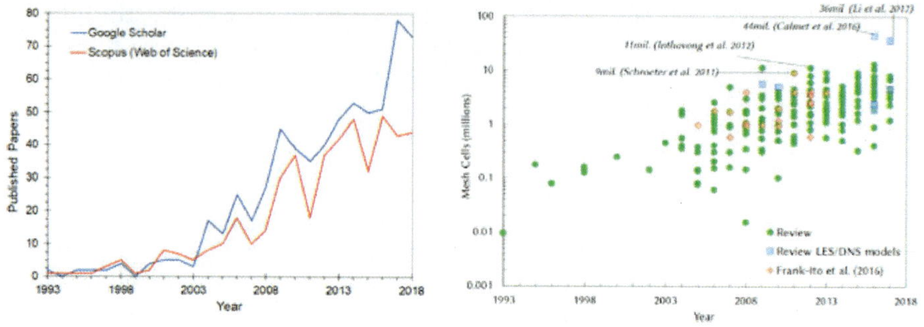

(a) articles per year (b) mesh elements

Fig. 1.2 **a** Number of published articles relevant to computational fluid dynamics in the human nasal cavity, obtained in the literature using the keyword search "nasal CFD" and a search on Scopus (web of Science) using the keyword search "nasal" and "Computational Fluid Dynamics". **b** Mesh cell elements used per nasal cavity model study, reported in the literature between 1993 and 2017, taken from Inthavong et al. [4]

1.3 CFD for Clinical Practice in the Human Nose

In clinical practice patients, most commonly present with disordered nasal airflow can lead to significant impairment in quality-of-life, secondary to a myriad of potential sequelae, including chronic nasal obstruction, mouth breathing, chronic rhinosinusitis, sleep-disordered breathing and obstructive sleep apnoea. Objective tests in otolaryngology practice include anterior rhinomanometry, acoustic rhinometry and nasal peak inspiratory flow which suffer from limited sensitivity, specificity and reliability and provide limited insight into the specific cause or site of obstruction. As a result, they provide little guidance as to the appropriate choice of surgical interventions.

Subjective measures of patient-reported nasal function, such as the NOSE (Nasal Obstruction and Septoplasty Effectiveness) scale and SNOT-22 (Sino-Nasal Outcome Test) similarly have limited usefulness in guiding surgical manoeuvres. As a result, most surgeons rely on nasendoscopy, imaging (such as CT, or MRI) and clinical experience to determine the precise site of obstruction and the appropriate surgical interventions that should be considered. This surgeon-specific interpretation of clinical findings leads to subjective opinion, inter-observer variability and an inability to accurately and scientifically analyse medical treatments.

The resulting treatment strategy is based principally on surgeons' anecdotal experience or low-level evidence, along with patient self-reporting of improved outcomes. To bring the "art" of nasal airway obstruction treatment into the modern scientific domain, CFD analysis coupled with additional physics provides new methods of testing, diagnosis and outcome reporting. Additionally, one of the most attractive features of the CFD technology is the ability to provide visualisation and analysis of flow properties such as localised velocity, pressure, and wall shear stress, dynamically during the respiratory cycle to support the quantitative data.

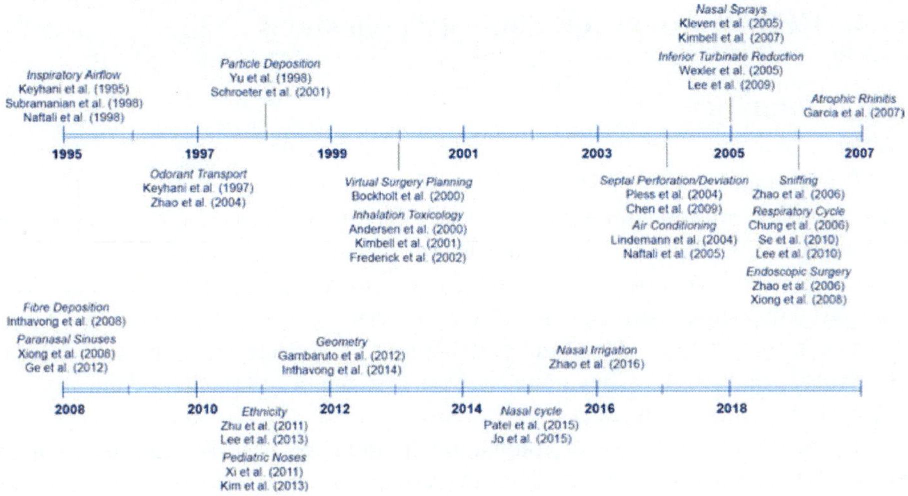

Fig. 1.3 A timeline showing some of the earliest representative studies for different research topics is given to indicate the growth in research interest in CFD for clinical applications of the human nose with example works [23–41] of different nasal topics related to clinical and biomedical engineering in the human nose

Figure 1.3 presents a timeline that shows two early studies representative of different research topics (although it is not an exhaustive list). Early studies involved inspiratory airflow [2, 6], odorant transport [7] and particle deposition [8]. From the year 2000, studies with focussed applications began emerging with topics involving, virtual surgery [9], and inhalation toxicology [10–12]. This was followed by even more complex multiphysics application which involved effects of surgery [13, 14], air conditioning [15, 16], nasal spray drug delivery [17, 18], flow dynamics of sniffing [19], and unsteady modelling of the respiratory cycle [20]. Indeed from the mid 2000s the research had gathered critical mass and developed into its own specialised field, and the wide variety of studies correspond with the increased published articles found in Fig. 1.2a.

CFD studies have demonstrated its ability to show the effects of surgery including the effects of turbinate surgery [22] and sinus surgery [21]. However, there is quite a delay in obtaining the results and future uptake of CFD for practical clinical use will depend on the turnaround time from imaging/diagnosis to an outcome, with results in produced close to real-time. This may involve the integration of workflow components, e.g. imaging, segmentation, reconstruction, meshing, analysis, and post-processing. Additionally, the graphical interface that becomes the bridging communication between the surgeon's needs and knowledge to the engineer's report and outcomes needs to align and easy for both groups to understand. CFD results and in particular fluid dynamics concepts can be perplexing to surgeons and physicians, and therefore how the information is presented to surgeons and physicians need to be uncomplicated but still be true to what the answers are, is user-friendly, fast, interactive and intuitive.

1.4 CFD Workflow for Clinical Applications

1.5 Summary

CFD has a long-standing history in the fields of aerodynamics and aerospace, but its application is not just limited to these traditional fields. From early beginnings, its use in the field of otorhinolaryngology was limited by the time-consuming step of re-creating the computational models of the nasal airway. Its use rapidly increased over the past three decades driven mainly by the exponential growth in computing power, advancements in technology, and collaborative interdisciplinary. It is an exciting technology that provides an alternative objective tool to complement clinical testing of the human airway which is often invasive.

To yield clinical impact, a computational approach is truly multi-disciplinary, incorporating the fundamentals of CFD (computational science, mathematics, fluid dynamics) with human anatomy, and physiology—i.e. expertise from an engineer and a clinician. This suggests that a researcher working in this field and using a computational approach should obtain some knowledge subsets from each discipline. Therefore, this book is compiled by a diverse mix of researchers, engineers, and clinicians to provide a range of topics that informs on the current state of the art and a look forward to new research directions.

References

1. M. Andersen, R. Sarangapani, R. Gentry, H. Clewell, T. Covington, C.B. Frederick, Application of a hybrid CFD-PBPK nasal dosimetry model in an inhalation risk assessment: an example with acrylic acid. Toxicol. Sci. **57**(2), 312–325 (2000)
2. U. Bockholt, W. Muller, G. Voss, U. Ecke, L. Klimek, Real-time simulation of tissue deformation for the nasal endoscopy simulator (NES). Comput. Aided Surg. **4**(5), 281–285 (1999)
3. H. Calmet, A.M. Gambaruto, A.J. Bates, M. Vazquez, G. Houzeaux, D.J. Doorly, Large-scale CFD simulations of the transitional and turbulent regime for the large human airways during rapid inhalation. Comput. Biol. Med. **69**, 166–180 (2016)
4. X.B. Chen, H.P. Lee, V.F.H. Chong, D.Y. Wang, Assessment of septal deviation effects on nasal air flow: a computational fluid dynamics model. The Laryngoscope **119**(9), 1730–1736 (2009)
5. X.B. Chen, H.P. Lee, V.F. Chong, D.Y. Wang, Impact of inferior turbinate hypertrophy on the aerodynamic pattern and physiological functions of the turbulent airflow - a CFD simulation model. Rhinology **48**, 163–168 (2010)
6. S.-K. Chung, Y.R. Son, S.J. Shin, S.-K. Kim, Nasal airflow during respiratory cycle. Am. J. Rhinol. **20**(4), 379–384 (2006)
7. D. Elad, R. Liebenthal, B.L. Wenig, S. Einav, Analysis of air flow patterns in the human nose. Med. Biol. Eng. Comput. **31**(6), 585–592 (1993)
8. C.B. Frederick, L.G. Lomax, K.A. Black, L. Finch, H.E. Scribner, J.S. Kimbell, K.T. Morgan, R.P. Subramaniam, J.B. Morris, Use of a hybrid computational fluid dynamics and physiologically based inhalation model for interspecies dosimetry comparisons of ester vapors. Toxicol. Appl. Pharmacol. **183**(1), 23–40 (2002)
9. A.M. Gambaruto, D.J. Taylor, D.J. Doorly, Decomposition and description of the nasal cavity form. Ann. Biomed. Eng. **40**(5), 1142–1159 (2012)

10. G.J.M. Garcia, N. Bailie, D.A. Martins, J.S. Kimbell, Atrophic rhinitis: a CFD study of air conditioning in the nasal cavity. J. Appl. Physiol. **103**(3), 1082–1092 (2007)
11. K. Inthavong, J. Wen, Z. Tian, J. Tu, Numerical study of fibre deposition in a human nasal cavity. J. Aerosol Sci. **39**(3), 253–265 (2008)
12. K. Inthavong, Y.D. Shang, J.Y. Tu, Surface mapping for visualization of wall stresses during inhalation in a human nasal cavity. Respir. Physiol. Neurobiol. **190**(1), 54–61 (2014)
13. K. Inthavong, A. Chetty, Y. Shang, J. Tu, Examining mesh independence for flow dynamics in the human nasal cavity. Comput. Biol. Med. **102**, 40–50 (2018)
14. G. Jo, S.-K. Chung, Y. Na, Numerical study of the effect of the nasal cycle on unilateral nasal resistance. Respir. Physiol. Neurobiol. **219**, 58–68 (2015)
15. K. Keyhani, P.W. Scherer, M.M. Mozell, Numerical simulation of airflow in the human nasal cavity. J. Biomech. Eng. **117**(4), 429–441 (1995)
16. K. Keyhani, P. Scherer, M. Mozell, A numerical model of nasal odorant transport for the analysis of human olfaction. J. Theor. Biol. **186**, 279–301 (1997)
17. J.W. Kim, J. Xi, X.A. Si, Dynamic growth and deposition of hygroscopic aerosols in the nasal airway of a 5-year-old child. Int. J. Numer. Methods Biomed. Eng. **29**(1), 17–39 (2013)
18. J. Kimbell, R. Subramaniam, Use of computational fluid dynamics models for dosimetry of inhaled gases in the nasal passages. Inhal. Toxicol. **13**(5), 325–334 (2001)
19. J. Kimbell, R. Segal, B. Asgharian, B. Wong, J. Schroeter, J. Southall, C. Dickens, G. Brace, F. Miller, Characterization of deposition from nasal spray devices using a computational fluid dynamics model of the human nasal passages. J. Aerosol Med. **20**(1), 59–74 (2007)
20. M. Kleven, M.C. Melaaen, M. Reimers, J.S. Røtnes, L. Aurdal, P.G. Djupesland, Using computational fluid dynamics (CFD) to improve the bi-directional nasal drug delivery concept. Food Bioprod. Process. **83**(2), 107–117 (2005)
21. C.F. Lee, M.Z. Abdullah, K.A. Ahmad, I. Lutfi Shuaib, Standardization of Malaysian adult female nasal cavity. Comput. Math. Methods Med. **2013**, 519071 (2013)
22. H.P. Lee, H.J. Poh, F.H. Chong, D.Y. Wang, Changes of airflow pattern in inferior turbinate hypertrophy: a computational fluid dynamics model. Am. J. Rhinol. Allergy **23**(2), 153–158 (2009)
23. J.-H. Lee, Y. Na, S.-K. Kim, S.-K. Chung, Unsteady flow characteristics through a human nasal airway. Respir. Physiol. Neurobiol. **172**, 136–146 (2010)
24. C. Li, J. Jiang, H. Dong, K. Zhao, Computational modeling and validation of human nasal airflow under various breathing conditions. J. Biomech. **64**, 59–68 (2017)
25. J. Lindemann, T. Keck, K. Wiesmiller, B. Sander, H. Brambs, G. Rettinger, D. Pless, Nasal air temperature and airflow during respiration in numerical simulation based on multislice computed tomography scan. Am. J. Rhinol. **20**, 219–223 (2006)
26. S. Naftali, R. Schroter, J. Shiner, D. Elad, Transport phenomena in the human nasal cavity: a computational model. Ann. Biomed. Eng. **26**, 831–839 (1998)
27. S. Naftali, M. Rosenfeld, M. Wolf, D. Elad, The air-conditioning capacity of the human nose. Ann. Biomed. Eng. **33**, 545–553 (2005)
28. R.G. Patel, G.J.M. Garcia, D.O. Frank-Ito, J.S. Kimbell, J.S. Rhee, Simulating the nasal cycle with computational fluid dynamics. Otolaryngol.-Head Neck Surg. **152**, 353–360 (2014)
29. D. Pless, T. Keck, K. Wiesmiller, G. Rettinger, A. Aschoff, T. Fleiter, J. Lindemann, Numerical simulation of air temperature and airflow patterns in the human nose during expiration. Clin. Otolaryngol. **29**(6), 642–647 (2004)
30. C.M. Se, K. Inthavong, J. Tu, Unsteady particle deposition in a human nasal cavity during inhalation. J. Comput. Multiph. Flows **2**(4), 207–218 (2010)
31. R. Subramaniam, R. Richardson, K. Morgan, J.S. Kimbell, R. Guilmette, Computational fluid dynamics simulations of inspiratory airflow in the human nose and nasopharynx. Inhal. Toxicol. **10**, 91–120 (1998)
32. D. Wexler, R. Segal, J. Kimbell, Aerodynamic effects of inferior turbinate reduction. Arch. Otolaryngol. Head Neck Surg. **131**, 1102–1107 (2005)
33. J. Xi, X. Si, J.W. Kim, A. Berlinski, Simulation of airflow and aerosol deposition in the nasal cavity of a 5-year-old child. J. Aerosol Sci. **42**(3), 156–173 (2011)

34. G. Xiong, J. Zhan, H. Jiang, J. Li, L. Rong, G. Xu, Computational fluid dynamics simulation of airflow in the normal nasal cavity and paranasal sinuses. Am. J. Rhinol. **22**, 477–482 (2008)
35. G. Xiong, J. Zhan, K. Zuo, J. Li, L. Rong, G. Xu, Numerical flow simulation in the post-endoscopic sinus surgery nasal cavity. Med. Biol. Eng. Comput. **46**(11), 1161–1167 (2008)
36. G. Yu, Z. Zhang, R. Lessmann, Fluid flow and particle diffusion in the human upper respiratory system. Aerosol Sci. Technol. **28**(2), 146–158 (1998)
37. K. Zhao, P.W. Scherer, S.A. Hajiloo, P. Dalton, Effect of anatomy on human nasal air flow and odorant transport patterns: implications for olfaction. Chem. Senses **29**(5), 365–379 (2004)
38. K. Zhao, P. Dalton, G.C. Yang, P.W. Scherer, Numerical modeling of turbulent and laminar airflow and odorant transport during sniffing in the human and rat nose. Chem. Senses **31**(2), 107–118 (2006)
39. K. Zhao, E.A. Pribitkin, B.J. Cowart, D. Rosen, P.W. Scherer, P. Dalton, Numerical modeling of nasal obstruction and endoscopic surgical intervention: outcome to airflow and olfaction. Am. J. Rhinol. **20**(3), 308–316 (2006)
40. K. Zhao, J. Craig, N. Cohen, N. Adappa, S. Khalili, J. Palmer, Sinus irrigations before and after surgery-visualization through computational fluid dynamics simulations. Laryngoscope **126**, 90–96 (2016)
41. J.H. Zhu, H.P. Lee, K.M. Lim, S.J. Lee, D.Y. Wang, Evaluation and comparison of nasal airway flow patterns among three subjects from Caucasian, Chinese and Indian ethnic groups using computational fluid dynamics simulation. Respir. Physiol. Neurobiol. **175**(1), 62–69 (2011)

Chapter 2
Anatomy and Physiology of the Human Nose

Eugene Wong, Joey Siu, Richard Douglas, and Narinder Singh

Abstract A good understanding of the normal human nose anatomy and function is critical to perform human respiratory system computer modelling and computational fluid dynamic (CFD) analysis. This chapter provides a broad overview of the function and organisation of the human respiratory system, including essential anatomical definitions, anatomy and naming conventions, as well as particular areas of surgical interest. The topics include Functions and Organisation of the Respiratory System, and the nasal cavity anatomy, function, physiology, and histology.

2.1 Functions and Organisation of the Respiratory System

The primary role of the human respiratory system is gas exchange; inhaling oxygen (O_2) and exhaling carbon dioxide (CO_2). O_2 is transferred from inhaled air, via the arterial blood stream, to the body's cells where it is consumed during normal cell function. CO_2 is a by-product of such activity and is carried from the cells, back via the veins, to the lungs, where it is exhaled. Gas exchange occurs in the lungs at the alveoli, the final component of the respiratory system. However, inhaled air needs to first pass through the upper respiratory tract, starting from either the nose or the mouth. From here, air passes through the pharynx (throat), larynx (voicebox) and trachea (windpipe), before passing through the bronchi, the primary, secondary and tertiary bronchioles and, finally, the alveoli, where the large surface area and specialised tissue facilitates effective gas exchange [21] (Fig. 2.1).

The respiratory system can therefore be separated into two broad regions based on function, including the conducting zone and the respiratory zone. The conducting

E. Wong (✉)
RMIT University, Melbourne, Australia
e-mail: eugene.hl.wong@gmail.com

J. Siu · R. Douglas
University of Auckland, Auckland, New Zealand

N. Singh
University of Sydney, Sydney, Australia

K. Inthavong et al. (eds.), *Clinical and Biomedical Engineering in the Human Nose*, Biological and Medical Physics, Biomedical Engineering,
https://doi.org/10.1007/978-981-15-6716-2_2

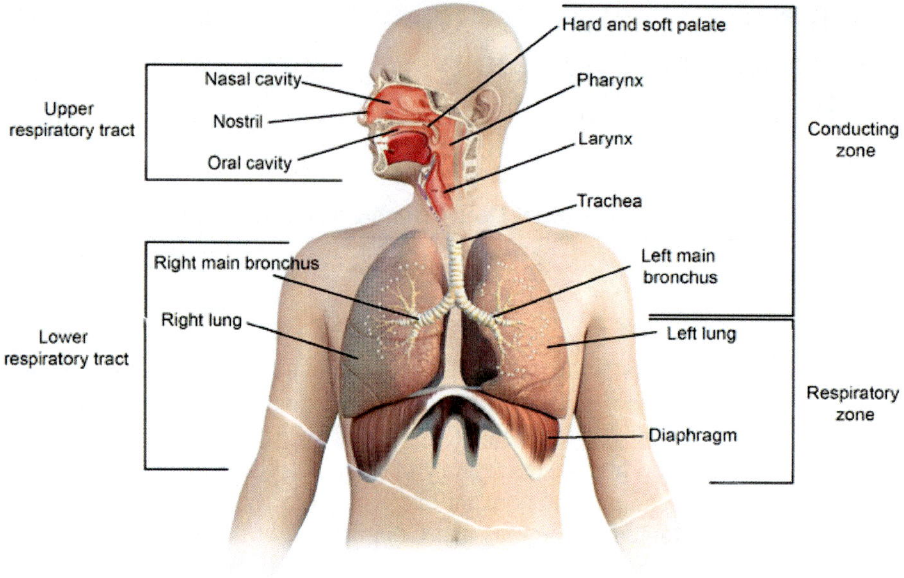

Fig. 2.1 Diagram of the respiratory system and it's organs

zone begins at the nares (nostril openings) and ends at the bronchioles. Its primary role is to conduct air to the respiratory zone. The respiratory zone consists of alveolar ducts and alveoli, where gas exchange takes place [2, 20].

2.1.1 Secondary Functions of the Conducting Zone

In addition to the transmission of inhaled and exhaled air, the conducting zone of the respiratory system has several secondary functions, some of which will be described in further detail later in the chapter. The nasal cavity is highly effective at regulating the temperature of inhaled air, with the ability to create a consistent 34 degrees output, given a range of inputs from 24.4 to 34.7 degrees [10, 13]. The nasal cavity humidifies inhaled air by adding moisture. Inhaled particles are trapped and prevented from entering the lungs.

The sensitive olfactory neuroepithelium, located high in the nasal cavity, is directly involved in the sense of smell [15, 24]. The nasal cavity contains several ostia (openings) that communicate with the paranasal sinuses and the nasolacrimal apparatus. As a summary, the functions of the nose include [8]:

- Transmitting air from nostrils to the lungs
- Thermoregulation (warming inhaled air)
- Humidification of inhaled air
- Filtration (capture of inhaled particles)

- Olfaction
- Immune defence (capture and destruction of microbes)
- Phonation—modifies sound of speech
- Communication with sinuses and nasolacrimal apparatus
  - Transmit mucous from sinuses to nose
  - Transmit nitrous oxide from sinuses to lungs
  - Transmit lacrimal fluids from eye to nose

Further down, the vocal cords of the larynx (voicebox) aid in phonation (speech) and generation of intrathoracic pressures (such as in a Valsalva manoeuvre) [18]. The conducting zone also acts as a regulator between the oral cavity (mouth) and the oesophagus (food pipe), to facilitate the transfer of food and liquids into the stomach and digestive system, while protecting the lungs from inadvertent entry of these substances (aspiration) [8].

2.2 Nasal Cavity Anatomy

Anatomical Terminology

An overview of common anatomical terms is provided. Readers seeking more in-depth information are referred to the *Terminologica Anatomica*, the international standard on human anatomic terminology [19].

Standard Anatomical Position

Just as cartographers draw maps with north standardised at the top of the map, anatomists view the human body in a standardised position, standing upright with the arms at the side and the palms of the hands facing forwards. This is defined as the *anatomical position*. In this way, the same descriptive terminology is used, irrespective of the position of the body being described. For example, the tip of the nose is always described as being "anterior", regardless of whether the body is standing or lying face down.

Directional Terms (Figs. 2.2 and 2.3)

Standard terms are used in anatomical descriptions in this and all biomedical textbooks. The reader is advised to learn and use the following terms when describing human anatomy.

- Anterior (or ventral): The front or towards the front of the body e.g.. the nose is anterior to the face.
- Posterior (or dorsal): The back or towards the back of the body e.g.. the brain is posterior to the nose.
- Superior (or cranial): The top or towards the head e.g.. the head is superior to the neck.

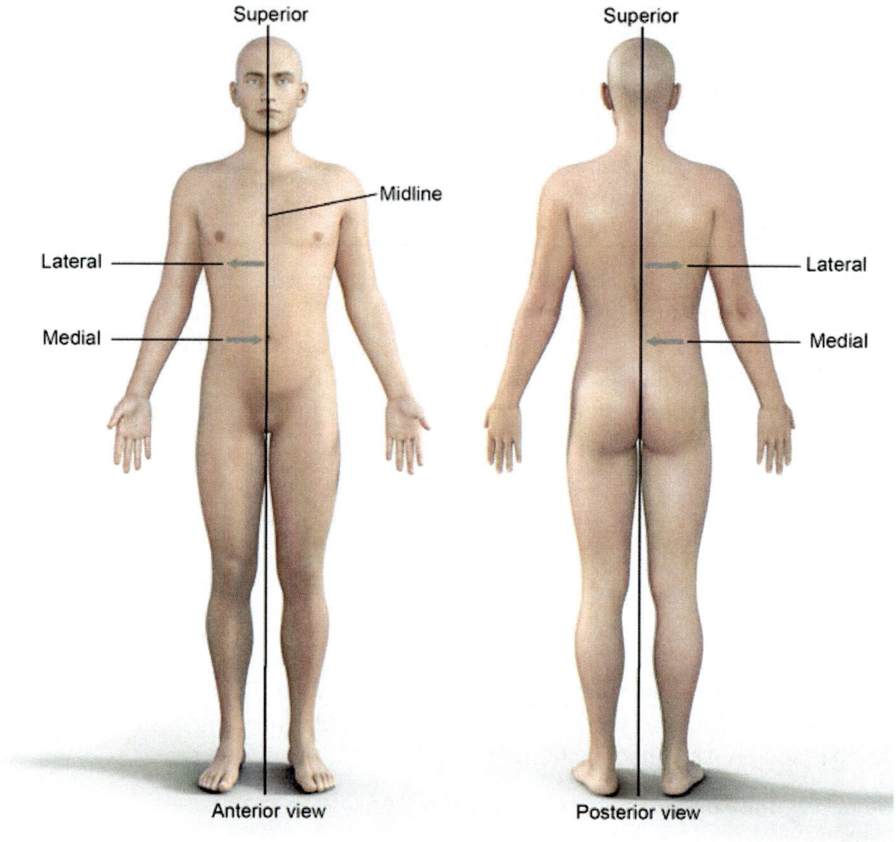

Fig. 2.2 Descriptions of anatomical positions

- Inferior (or caudal): The bottom (lower) or towards the fee e.g..The mouth is inferior to the nose.
- Medial: The middle or towards the middle e.g.. the nose is medial to the ear.
- Lateral: The side or towards the side e.g.. the ear is lateral to the nose.
- Proximal: A point that is closer to where a body part attaches to the trunk of the body e.g.. the neck is proximal to the head.
- Distal: A point that is farther away from where a body part attaches to the trunk of the body e.g.. the hand is distal to the shoulder.
- Superficial: Closer to the surface e.g.. the skin is superficial to the muscles.
- Deep: Further away from the surface e.g.. the brain is deep to the scalp.

Anatomical Planes (Fig. 2.4)

The body can be cut virtually into imaginary two-dimensional slices or planes:

- Coronal (or frontal) plane: A vertical plane running from side to side that splits the body into anterior (front) and posterior (back) sides

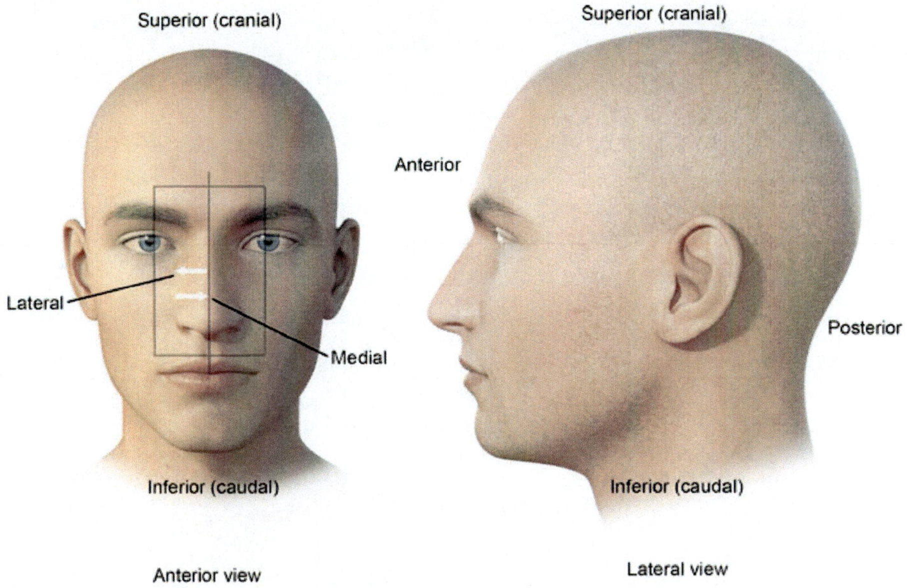

Fig. 2.3 Descriptions of anatomical positions relative to the face and nose

- Transverse (or axial) plane: A horizontal plane running from side to side that splits the body into superior (top) and inferior (bottom) sides
- Sagittal plane: A vertical plane running from front to back that splits the body into right and left sides

2.3 Nasal Anatomy

This section contains a detailed discussion of the external nose, nasal cavity and sinuses, which are the primary areas of interest of this book. Understanding sinonasal anatomy is often a challenging experience, particularly for those not in the medical field. Anatomical textbooks often describe minutiae in great detail rather than focusing on the key areas that hold greater relevance to the clinician.

For most students of anatomy, sinonasal anatomy is learnt in the context of clinical medicine and surgery. Therefore, the nose is conceptualised as a series of three-dimensional structures and their relationships—a difficult task to replicate in two dimensional illustrations and wordy descriptions in a textbook. Surgical trainees are also exposed to repeated clinical examination of the nose, externally and internally, along with increasing surgical experience to cement their learning. This visual and tactile input is missed when approaching anatomy from a non-medical background.

As a result, the aim of this section is to describe the anatomy of the nasal cavity in a clinically relevant manner similar to the approach taken by clinicians. In further

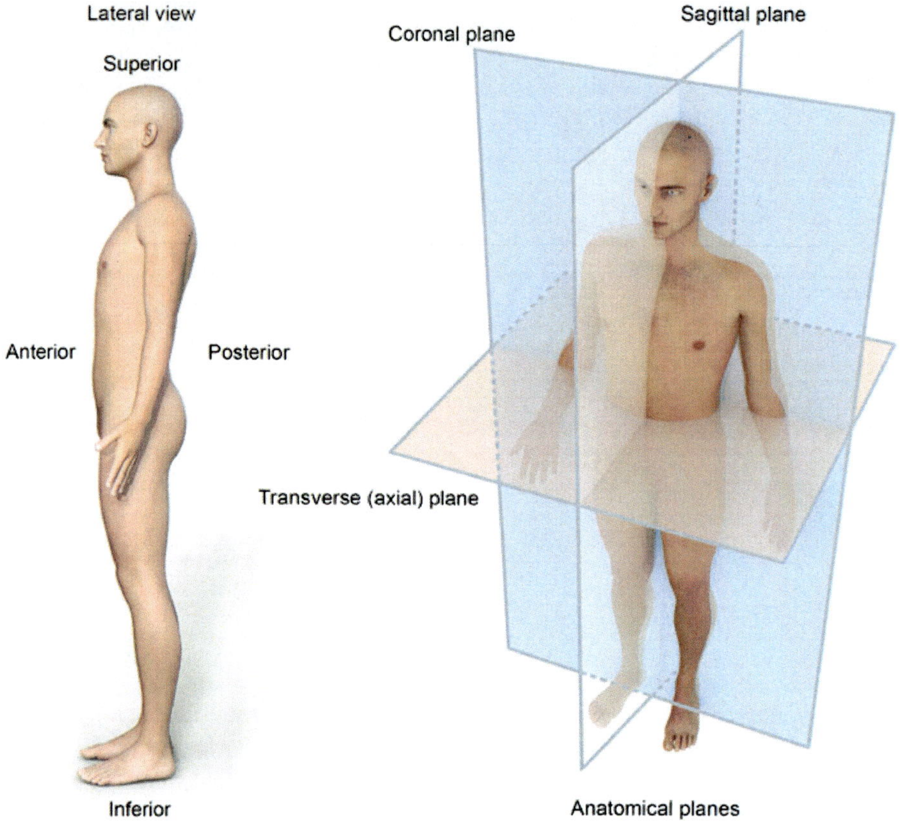

Fig. 2.4 Descriptions of anatomical planes

sections of this chapter, the function of nasal structures will be discussed, in normal conditions. In the next chapter, variations and disease will be discussed, as will potential areas where CFD may bridge a gap in current understanding.

2.3.1 *The External Nose*

The external nose describes all aspects of the visible nasal structure projecting anteriorly from the face (see Fig. 2.5). The upper third of the nasal framework consists of bone and the lower two thirds is made of cartilage.

The root or base of the nose begins with a notch formed by the frontal bones. The paired nasal bones insert in to this notch to form the nasal bridge. Laterally (to the sides), the nasal bones articulate with the frontal process of the maxilla. Centrally, the nasal bones articulate with the nasal septum.

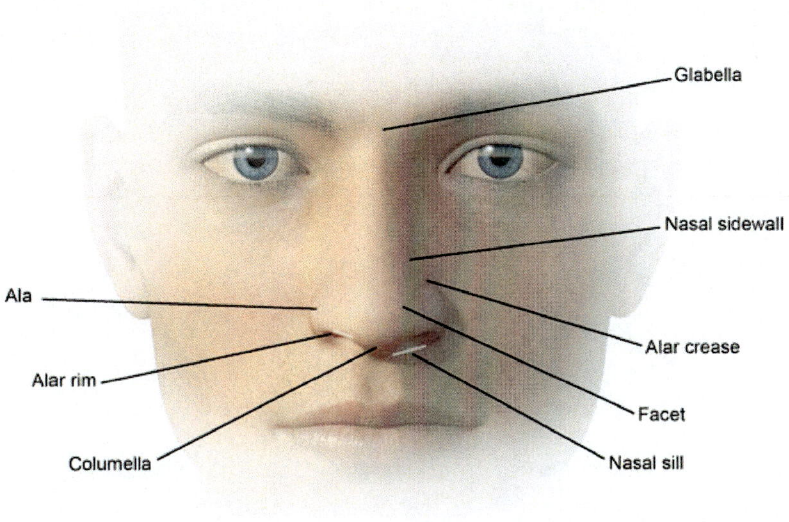

Fig. 2.5 Topographic features of the external nose

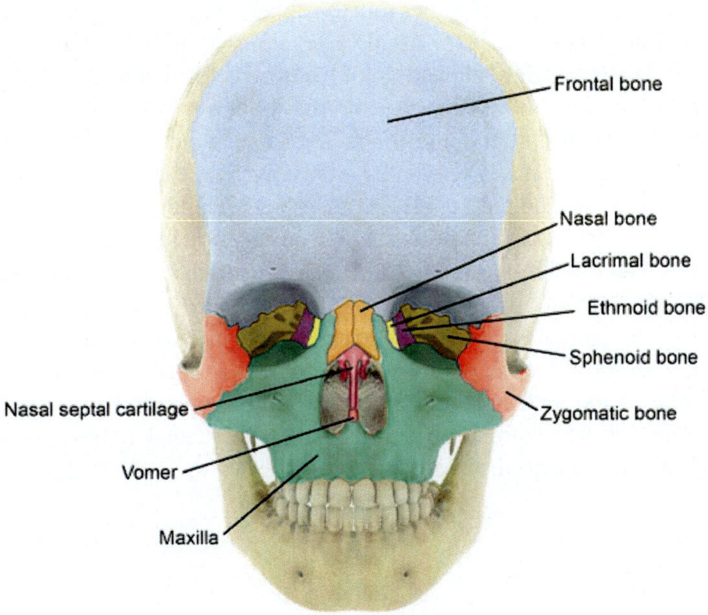

Fig. 2.6 Anterio-Posterior view of the sinonasal bones

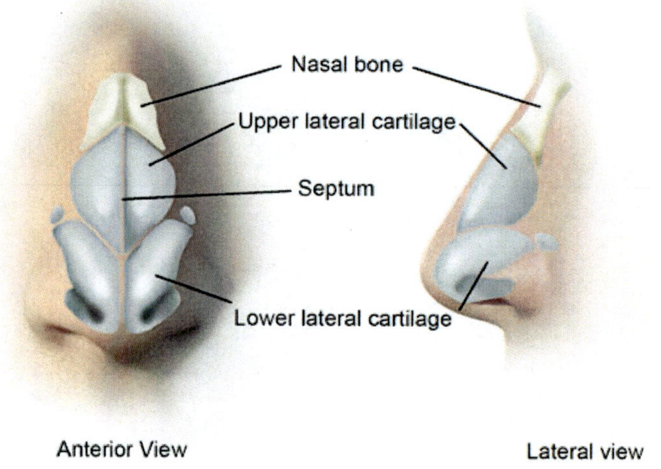

Fig. 2.7 Topographic view of nasal cartilages

Inferiorly (below), the nasal bones join the upper lateral cartilages which, together, form the mid-third of the nose, or nasal vault, along with the septum in the midline (see Fig. 2.7. The lower third ala and tip of the nose is formed by the U-shaped lower lateral cartilages with central support from the septum (see Fig. 2.8).

The nostrils or nares are the inferior-facing (downward) openings of the external nose, which lead to the nasal vestibules internally. The external nose is covered by skin, which wraps around the nostril entrances and extends into the nasal vestibules by a few millimeters , at which point there is a transition to the nasal mucosa (pseudostratified ciliated columnar epithelium). The vestibular skin often contains nasal hairs, although the proportion and distribution of these hairs can be variable based on ethnicity and gender. Between the nostrils, the central structure is known as the columella, with a framework composed of the medial portions (crura) of the lower lateral cartilages and the caudal continuation of the septum, all covered by skin.

Small muscles attach to the nasal cartilages and facial bones, allowing some control of airflow through the nostrils (see Fig. 2.9).

2.3.2 *Nasal Valves*

Figure 2.10 depicts the anterior and oblique views of the internal nasal valve. This region has the following features

- External nasal valve: The external nasal valve is the name given to the most antero-inferior region of the external nose, where the nostrils open to the external environment.

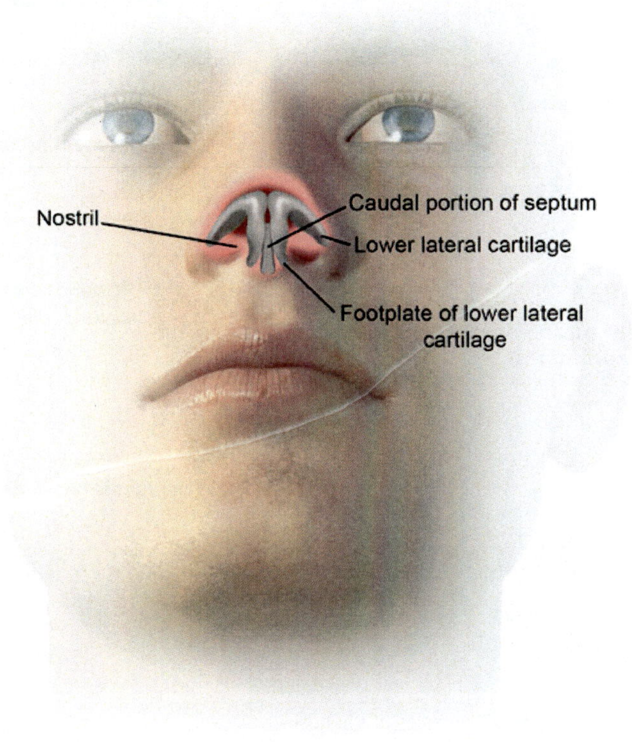

Fig. 2.8 Anatomy of the nostril from below the face

- Internal nasal valve: The internal nasal valve describes a region approximately 1.3 cm from the naris. The boundaries of the internal nasal valve include the junction between the upper lateral and lower lateral cartilages, the septum, the floor of the nose and the head (most anterior portion) of the inferior turbinate

2.3.3 The Nasal Cavity

The nasal cavity is defined as the space between the nares anteriorly to the choanae posteriorly, finally exiting into the nasopharynx. The nasal cavity is divided into left and right in midline by the nasal septum.

The *nasal septum* is the central dividing structure of the nose (see Fig. 2.11. Anteriorly, the septum is composed of cartilage, which allows some side to side movement of the nasal tip. Posteriorly, the septum is composed of rigid bone. The

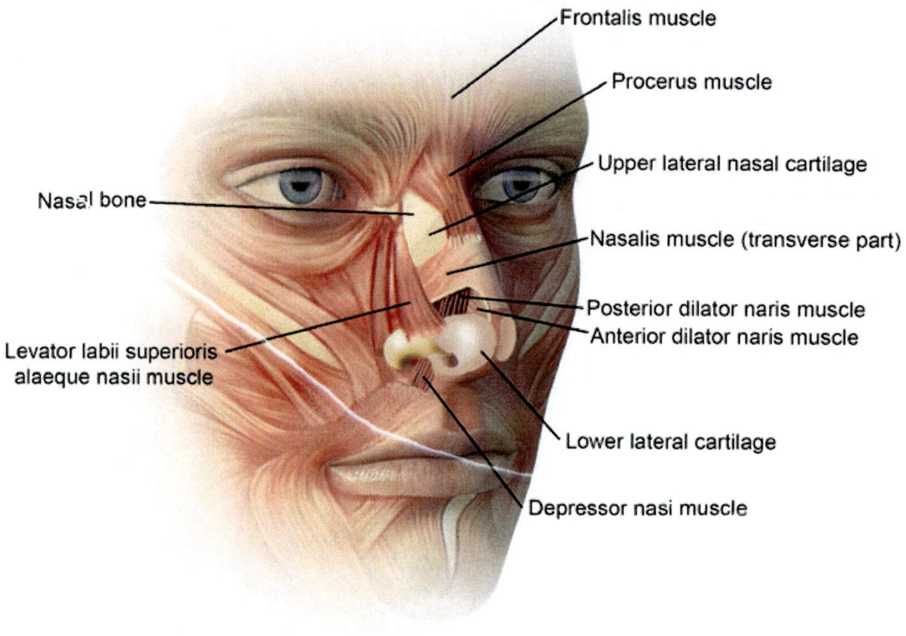

Fig. 2.9 Diagram of small muscles of the nose

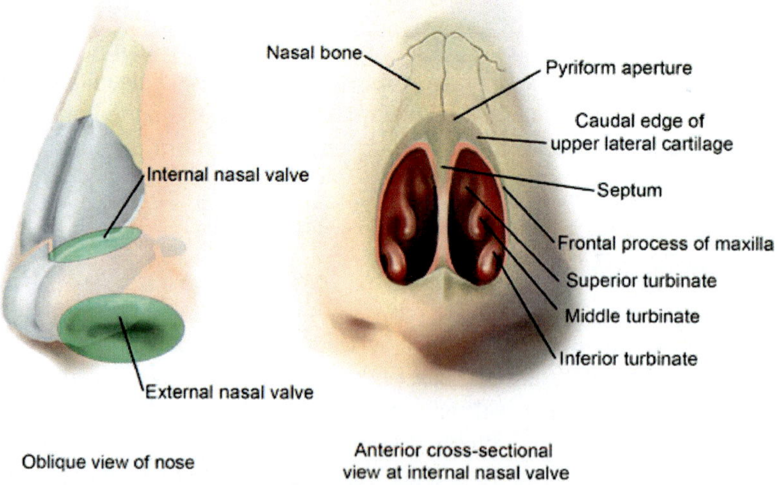

Fig. 2.10 Anterior and oblique views of the internal nasal valve

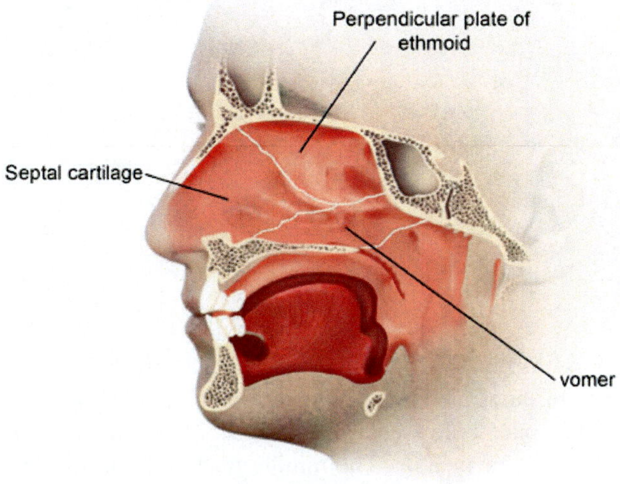

Fig. 2.11 Sagittal cross-section slice of the nasal cavity showing bones and cartilage

main bones in the septum are the vomer, the perpendicular plate of the ethmoid and the maxillary crest, but several other bones provide small contributions to the septum including the nasal bones, frontal bones, palatine bones and sphenoid.

The septum is covered by mucosa on both sides with a total thickness of around 2 mm. The septal swell body is an area of increased thickness located anterior to the middle turbinate approximately 2.5 cm above the nasal floor. This area contains submucosal venous sinusoids and seromucinous glands, similar histologically to the inferior turbinates, with the capacity to vary in size in order to regulate nasal airflow. The precise role of the septal swell body in normal physiol- ogy and disease is an area requiring further study and may constitute a potential therapeutic target [4, 22].

The *lateral walls* of the nasal cavity are formed by various bones (see Fig. 2.6. Anteriorly, they are formed by the nasal bones. The maxilla forms the bulk of the lateral wall antero- inferiorly, the lacrimal bone antero-superiorly, and the ethmoid contributes postero- superiorly. The perpendicular plate of the palatine bone is located postero-inferiorly. At the very posterior end of the lateral nasal wall lies the medial pterygoid plate of the sphenoid bone.

The most prominent feature of the lateral walls are the turbinates (conchae)—inferior, medial and superior. The middle and superior turbinates are extensions of the ethmoid bone. By contrast, the inferior turbinate is a separate bone, articulating anteriorly with the conchal crest of the maxillary bone and posteriorly with the palatine bones.

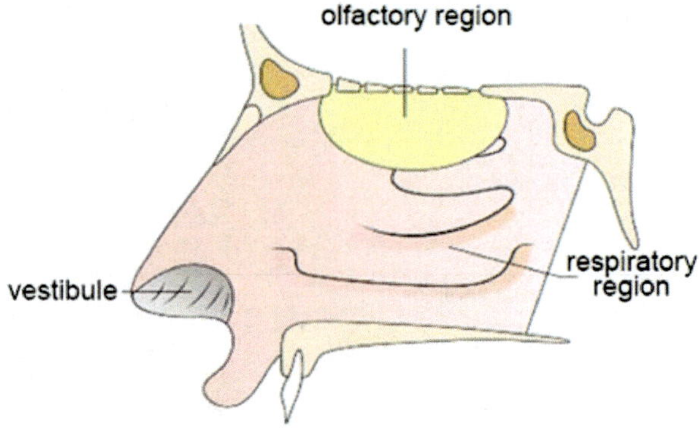

Fig. 2.12 The olfactory region is located in the nasal cavity roof

Of the three paired turbinates, the inferior turbinates are believed to carry the greatest significance in nasal airflow and conditioning. This is described in more detail in Chap. 9. The space below each turbinate is known as the corresponding inferior, middle and superior meatus. The middle meatus is an important junction point where the maxillary, anterior ethmoid and frontal sinuses drain, collectively known as the osteo-meatial complex. The posterior ethmoidal air cells open into the superior meatus and the nasolacrimal duct (draining tears from the lacrimal gland and eye) into the inferior meatus. The sphenoid sinus opens on the posterior wall of the nasal cavity, infero-medially.

The *roof of the nasal cavity* is comprised primarily of the cribriform plate of the ethmoid bone and, laterally, the fovea ethmoidalis of the ethmoid bone. Anterior to this is the nasal spine of the frontal and nasal bones and, posteriorly, the sphenoid bone. The superior aspect of the nose corresponding to the cribriform plate, superior nasal septum and superior lateral nasal wall is often called the "olfactory area" of the nose, as it houses delicate neuroepithelium responsible for the sense of smell (see Fig. 2.12). The thin cribriform plate contains perforations allowing nerve fibres to enter the nasal cavity from the olfactory bulb within the cranial cavity.

The *floor of the nasal cavity* is a bony septation separating the nose from the roof of the mouth. The floor is the widest part of the nasal cavity, which narrows as the lateral walls slant superiorly. At the posterior end of the nasal cavity are the *choanae*, which join to form a single passageway at the nasopharynx.

2.3.4 The Paranasal Sinuses

The paranasal sinuses are a group of four, paired, air-filled spaces surrounding the nasal cavity, lined by pseudostratified columnar epithelium. They comprise the maxillary, ethmoid, sphenoid and frontal sinuses, with the roof formed by the frontal bone lateral to the cribriform plate and the crista galli in the midline [1, 14, 17]. The exact function of the paranasal sinuses has not been fully determined, however, they have been implicated in several functions [3], including:

- Heating and humidifying inspired air
- Insulating structures from rapid and frequent temperature fluctuations in the nose
- Increasing vocal resonance
- Decreasing the relative weight of the skull
- Acting as a buffer against facial trauma
- Providing immunological defense against pathogens
- Creating mucous, and Creating nitric oxide

Development of the paranasal sinuses begins through pneumatic diverticulae (excavation of bone by air-filled sacs) from the nasal cavity, which starts in-utero and continues through the course of a lifetime. Each paired paranasal sinus develops in different time frames. The maxillary sinuses are the first to develop in-utero, followed by ethmoid sinuses with full pneumatisation of both only occurring around the age of seven. The sphenoid sinuses develop during the third month of gestation and will only appear at the age of three, with full pneumatisation at the age of twelve. The frontal sinuses develop last and do not appear until at least five years of age, and completes around 14–15 years of age [11, 23].

2.3.5 *Maxillary Sinuses*

The maxillary sinus is the largest paranasal sinus in the body, and is located in the cheek, under the eye. It is situated in the body of the maxilla with the roots of the molars and upper premolars projecting into its floor. It has an open medial wall, contained by the inferior turbinate, uncinate bone superiorly and the ethmoid posteriorly. The maxillary ostium drains into a slit-like opening into the middle meatus of the nasal airway which also allows aeration of the sinus. The maxillary ostium is shielded medially by the uncinate process, and therefore does not open directly into the nasal airway. Typically, there is only one ostium per maxillary sinus, however cadaver studies have shown that additional “accessory” maxillary ostia (opening into a cavity of the body) can be present in 10–30% of people [7].

2.3.6 *Ethmoid Sinuses*

The ethmoid sinuses are located between the eyes. There are approximately eight to fifteen ethmoidal air cells, forming a bony labyrinth in the superior and lateral aspect of the nasal cavity. The ground lamella, which is a condensation of the bony partitions separating the ethmoidal air cells, divides the ethmoid sinuses into anterior and posterior air cells. The anterior ethmoid air cells drain into the ethmoid infundibulum in the middle meatus, whereas the posterior ethmoid air cells drain into the sphenoethmoidal recess in the superior meatus.

2.3.7 *Sphenoid Sinuses*

The sphenoid sinuses are located posteriorly within the sphenoid bone. They drain into the sphenoethmoidal recess, together with the posterior ethmoid sinuses. The sphenoid sinuses have a close anatomical relationship to the carotid arteries and optic nerve. The sphenoid ostium lies in the posterior wall of the nose, drains into the sphenoethmoidal recess, the highest point of which is about the center between the choanae and the roof of the nasal cavity.

2.3.8 *Frontal Sinuses*

The frontal sinus is located above the orbit and within the frontal bone. They drains into the middle meatus. The drainage can be medial or lateral to the uncinate process, depending on where the uncinate attaches to the skull base.

2.3.9 *The Ostiomeatal Complex*

The ostiomeatal complex refers to the region in the anterior ethmoid sinuses, bounded by the middle turbinate medially, the lamina papyracea laterally, and the basal ground lamella superiorly and posteriorly. The region contains the ostia of the maxillary, frontal, and anterior ethmoid sinuses, where it serves as the final common pathway for the drainage and ventilation of the aforementioned sinuses. A diagram of the paranasal sinuses and ostiomeatal complex is demonstrated in Figs. 2.13 and 2.14.

2.4 Function and Physiology of the Nose

The various functions of the sinonasal cavity have previously been outlined. Essentially, the main role is to deliver filtered, warmed, humidified, oxygenated air to the lungs and return CO_2-laden air to the external environment. It also facilitates the perception of smell (olfaction). Various anatomical features contribute to these functions and the physiological processes are described below. Detailed discussion of airflow is explored in Chap. 9.

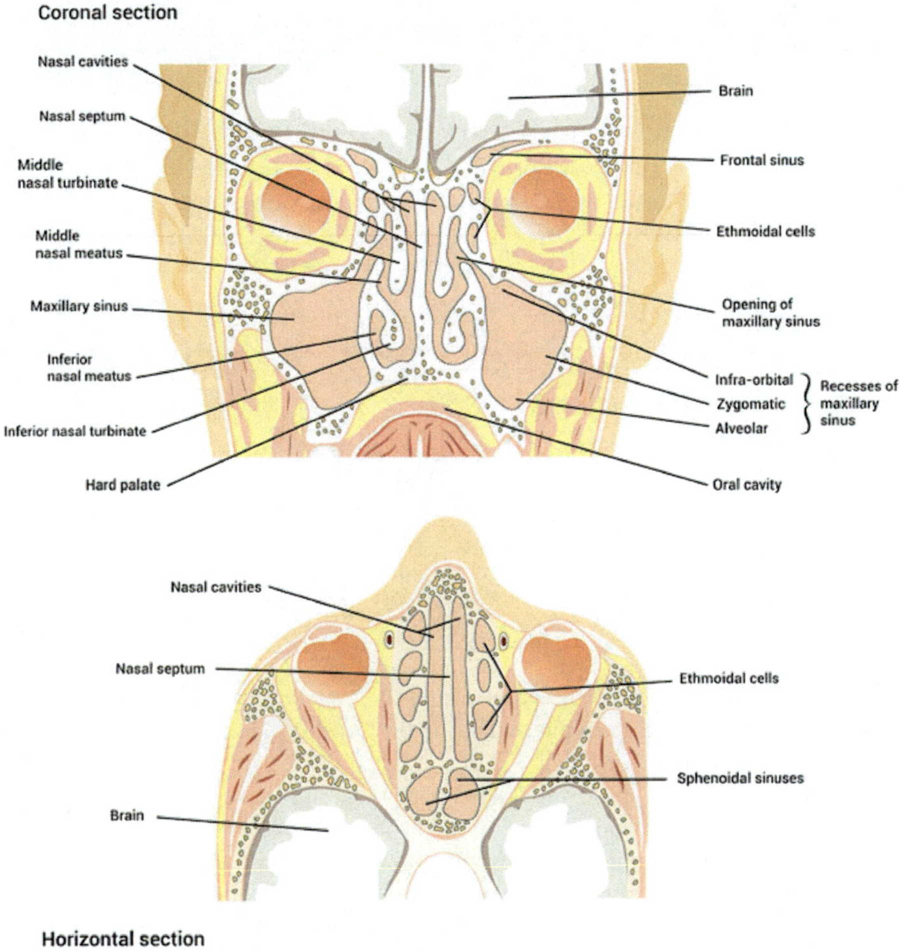

Fig. 2.13 Locations of the paranasal sinuses depicted in the coronal and axial (horizontal) sections

2.4.1 The Nasal Cycle

The nasal cycle is defined as alternating physiological congestion and decongestion between the left and right nasal airways, which leads to unilateral increase in resistance and decrease in airflow and turbulence, while the total combined nasal airflow remains constant [6]. The nasal cycle is considered to be an ultradian rhythm of side-to-side nasal mucosal engorgement which can be detected in 70–90% of humans [16], with a mean duration of 2.5 h (range, 30 min–6 h) [5]. The physiological congestion of the inferior and middle turbinates by dilation of the venous cavernous tissue in the submucosa is due to selective activation of one half of the autonomic nervous system, regulated by the hypothalamus [9, 12] .

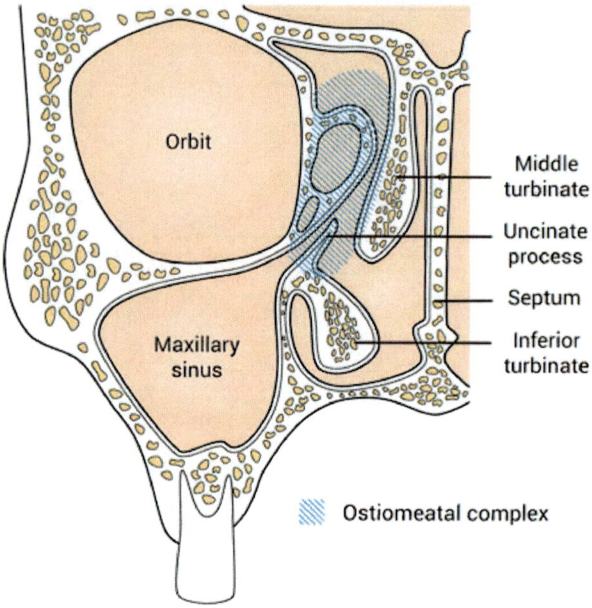

Fig. 2.14 Schematic of ostiometal complex in coronal plane

There are two phases of the nasal cycle, the 'working' or decongested phase, and the 'resting' or congested phase. During the 'working' phase, the diameter, passage of nasal airflow and turbulence is increased, while resistance is decreased. Conversely, in the 'resting' phase, the diameter, passage of nasal airflow and turbulence is decreased, while resistance is increased. The nasal cycle is postulated to be an evolutionary mechanism that enables moisturising, cleaning and humidification of the nasal mucosa on the 'resting' side of the nasal cavity, although the exact functions of the cycle have yet to be determined. The cycle may provide possible benefits to olfaction by allowing slower flow in the olfactory region on the resting side and faster flow on the opposite working side, thus increasing the range of odours detected.

Three patterns of nasal cycle have been described through studies using acoustic rhinometry [9]. These include: 1. Classic nasal cycling, where there is reciprocal congestion and decongestion alterations while maintaining the total combined nasal airflow 2. Parallel nasal cycling, where there is congestion or decongestion appearing in both nasal cavities at the same time 3. Irregular nasal cycling, where there is mutual alteration in nasal volume with no constant total combined nasal airflow and defined pattern of congestion and decongestion It has been shown that different nasal cycle patterns may alternate in the same person, with shifts in nasal cycling that may potentially be affected by environmental or physical factors.

Furthermore, studies using acoustic rhinometry have demonstrated that the subjective sensation of nasal patency has no correlation with the changing nostril volumes or cross-sectional areas during the nasal cycle. This is because the total nasal resistance usually remains constant and is less than the resistance of the individual left or right nasal passages. It should be noted that the nasal cycle is a normal physiologic phenomenon and largely unnoticed in healthy people with no sinonasal disease. Conscious awareness of reduction in nasal patency may be noted when the total nasal airflow suddenly changes, such as during the lateral recumbent position or when systemic stimuli lead to reflexes that dilate or obstruct the nostrils. In patients with nasal septal deviation or other pathological causes of obstruction, the changes during the nasal cycle become evident, and the overall sensation of reduced nasal patency frequently mirrors the unilateral 'resting' or congestion phase.

Unfortunately, the nasal cycle does complicate modelling and analysis. Computer models are typically derived from radiological images (CT or MRI). Such images are taken from a single snapshot in time and will often show the nose part-way through a cycle. This will result in one side being congested while the other will appear decongested, potentially distorting the computer model and subsequent CFD analysis. To avoid this problem, decongestants may be used prior to scanning which will result in bilateral mucosal decongestion. While this will result in mucosal symmetry in the model, it may not accurately represent true physiology.

Postural Changes

Lying on one side (lateral recumbent position) results in increased congestion on that side of the nose. Similarly, lying on the back (supine position) increases overall nasal resistance. This may have significance for CFD-related CT scans which are typically taken in the supine position.

2.4.2 Filtration

Initial filtration of particles and foreign bodies occurs at the nasal hairs in the vestibule. Further filtration is enhanced by the labyrinthine flow of air, turning from vertical to horizontal at the vestibule and internal valve, then passing through the complex anatomy of the turbinates, resulting in deposition of particles in the mucous blanket, where they are trapped and engage with components of the immune system, before being passed posteriorly to be swallowed into the gastrointestinal system.

2.4.3 Thermoregulation and Humidification

The shape of the inferior turbinates results in increased contact of inspired air with the mucosal surface of the nose. However, the efficiency of the turbinates in humidifying and warming air likely relies more on how they influence airflow than surface area contacted by inspired air.

2.4.4 *Olfaction*

The smooth and even flow of air through most of regions of the nasal cavity assists the diffusion of volatile odour molecules to the olfactory region. The sense of smell begins when these molecules activate olfactory receptors proteins in the olfactory mucosa. The olfactory receptor proteins are located in hair-like projections of the olfactory sensory nerves. Activation of the olfactory receptor proteins sets in chain a complex sequence of biochemical reactions which produces signals that are picked up by the olfactory bulb. The signal patterns leaving the olfactory bulb then travel to different regions in the brain and are then combined to be interpreted as an odour perception.

2.5 Histology

2.5.1 *Nasal Mucosal Surface*

At the entrance to the nose, the external skin, which contains keratinising stratified squamous epithelium, carries into the nostrils then transitions to respiratory mucosa, which contains pseudo stratified ciliated columnar epithelium. Interspersed among the columnar cells are mucous-producing goblet cells. Respiratory epithelium (Fig. 2.15) lines all of the nasal cavity and sinuses and is replaced by olfactory epithelium superiorly (Fig. 2.16)

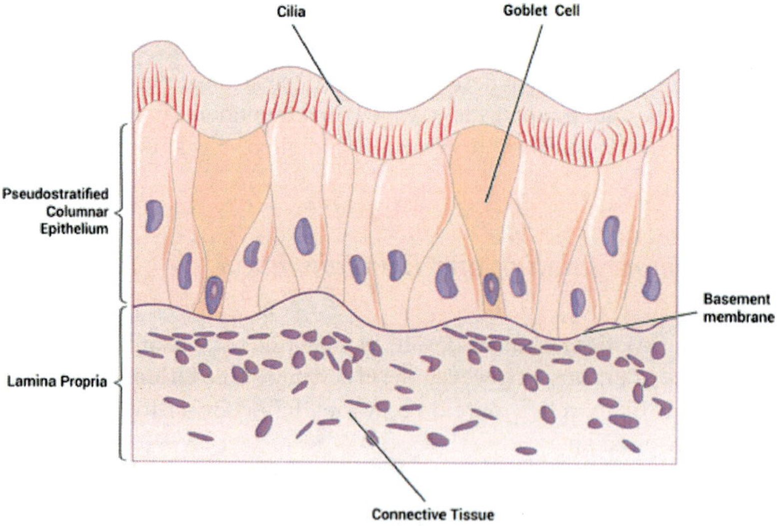

Fig. 2.15 Schematic of respiratory epithelium

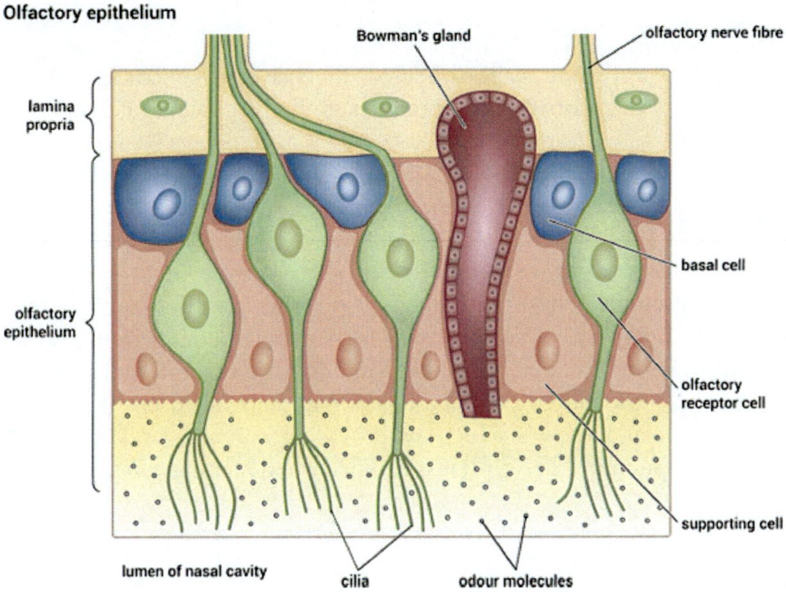

Fig. 2.16 Schematic of olfactory epithelium

Epithelial cells provide a physical barrier within the nose and play a role in the inflammatory response by releasing cytokines, chemicals that attack microbes. The columnar cells of the respiratory epithelium overlie the basement membrane, which itself overlies the lamina propria (loose connective tissue). The epithelium, basement membrane and lamina propria are together referred to as the mucosa. Within the lamina propria are seromucosal glands which produce secretions that form a mucous blanket over the nasal surface. Around 125 mL of fluid is produced daily. The cilia on the surface of the cells beat around 10–15 times per minute, moving the continuously-produced mucous blanket around 1–2 cm/h resulting in renewal every 10–15 min. The mucous blanket traps inhaled particles, which are then carried posteriorly and eventually swallowed. Antimicrobial agents secreted into the mucous blanket aid in protecting the host. The lamina propria contains thin-walled blood vessels that are effective in heating inspired air. It also contains venous spaces (cavernous sinusoids) that have the capacity to enlarge in response to inflammation or allergy. These are particularly found in the inferior and middle turbinates [22] along with the septal swell body [4].

The lamina propria contains thin-walled blood vessels that are effective in heating inspired air. It also contains venous spaces (cavernous sinusoids) that have the capacity to enlarge in response to inflammation or allergy. These are particularly found in the inferior and middle turbinates, along with the septal swell body. In the olfactory mucosa the olfactory receptors also have hair-like projections that contain olfactory receptor proteins that are activated by binding odorant molecules.

Olfactory Mucosa

Unlike the respiratory mucosa, the olfactory mucosa does not contain cilia, but does contain olfactory receptors that combine to supply the olfactory nerve.

2.6 Summary

This chapter presents a basic introduction to nasal anatomy and physiology that is aimed at bridging the knowledge gap for students coming from a non-clinical background. The information presented had a focus on its relevance to performing computer modelling and CFD of the physiology and functions of the respiratory system In particular, there was a focus on providing a broad overview of the function the important anatomical descriptions, geometry and naming conventions.

References

1. W.E. Bolger, D.S. Parsons, C.A. Butzin, Paranasal sinus bony anatomic variations and mucosal abnormalities: Ct analysis for endoscopic sinus surgery. The Laryngoscope **101**(1), 56–64 (1991)
2. K. BéruBé, Z. Prytherch, C. Job, T. Hughes, Human primary bronchial lung cell constructs: the new respiratory models. Toxicology **278**(3), 311–318 (2010)
3. Z.J. Cappello, A.B. Dublin, *Anatomy, Head and Neck, Nose Paranasal Sinuses* (StatPearls Publishing, 2018)
4. D.J. Costa, T. Sanford, C. Janney, M. Cooper, R. Sindwani, Radiographic and anatomic characterization of the nasal septal swell body. Arch. Otolaryngol.-Head Neck Surg. **136**(11), 1107–1110 (2010)
5. A. Gungor, R. Moinuddin, R.H. Nelson, J.P. Corey, Detection of the nasal cycle with acoustic rhinometry: Techniques and applications. Otolaryngology - Head and Neck Surgery (1999)
6. M. Hasegawa, E. Kern, The human nasal cycle, in *Mayo Clinic Proceedings*, vol. 52, pp. 28–34
7. M. Jog, G. McGarry, How frequent are accessory sinus ostia? J. Laryngol. Otol. **117**(4), 270–272 (2003)
8. N. Jones, The nose and paranasal sinuses physiology and anatomy. Adv. Drug Deliv. Rev. **51**(1–3), 5–19 (2001)
9. R. Kahana-Zweig, M. Geva-Sagiv, A. Weissbrod, L. Secundo, N. Soroker, N. Sobel, Measuring and characterizing the human nasal cycle. PLoS ONE (2016)
10. T. Keck, R. Leiacker, H. Riechelmann, G. Rettinger, Temperature profile in the nasal cavity. The Laryngoscope **110**(4), 651–654 (2000)
11. J. Medina, H. Hernandez, L.W. Tom, L. Bilaniuk, Development of the paranasal sinuses in children. Am. J. Rhinol. **11**(3), 203–210 (1997)
12. N. Mirza, H. Kroger, R.L. Doty, Influence of age on the 'nasal cycle'. Laryngoscope (1997)
13. N. Mygind, R. Dahl, Anatomy, physiology and function of the nasal cavities in health and disease. Adv. Drug Del. Rev. **29**(1–2), 3–12 (1998)
14. S. Nouraei, A. Elisay, A. Dimarco, R. Abdi, H. Majidi, S. Madani, P. Andrews, Variations in paranasal sinus anatomy: implications for the pathophysiology of chronic rhinosinusitis and safety of endoscopic sinus surgery. J. Otolaryngol.-Head Neck Surg. **38**(1), 32 (2009)
15. R.M. Patel, J.M. Pinto, Olfaction: anatomy, physiology, and disease. Clin. Anat. **27**(1), 54–60 (2014)

16. A.L. Pendolino, V. Lund, E. Nardello, G. Ottaviano, The nasal cycle: a comprehensive review, in *Rhinology Online* (2018)
17. I. Perez-Pinas, J. Sabate, A. Carmona, C. Catalina-Herrera, J. Jimenez-Castellanos, Anatomical variations in the human paranasal sinus region studied by ct. J. Anat. **197**(2), 221–227 (2000)
18. N.P. Solomon, S.J. Garlitz, R.L. Milbrath, Respiratory and laryngeal contributions to maximum phonation duration. J. Voice **14**(3), 331–340 (2000)
19. F.C.O.C. Terminology, *Terminologia Anatomica: International Anatomical Terminology* (Georg Thieme Verlag, 1998)
20. W.A. Wall, L. Wiechert, A. Comerford, S. Rausch, Towards a comprehensive computational model for the respiratory system. Int. J. Numer. Methods Biomed. Eng. **26**(7), 807–827 (2010)
21. E. Weiss, Ventilation/blood flow and gas exchange. New England J. Med. **325**, 4 (1991)
22. D. Wexler, I. Braverman, M. Amar, Histology of the nasal septal swell body (septal turbinate). Otolaryngol. - Head Neck Surg. **134**(4), 596–600 (2006)
23. G. Wolf, W. Anderhuber, F. Kuhn, Development of the paranasal sinuses in children: implications for paranasal sinus surgery. Ann. Otol., Rhinol. Laryngol. **102**(9), 705–711 (1993)
24. B.B. Wrobel, D.A. Leopold, Olfactory and sensory attributes of the nose. Otolaryngol. Clin. North Am. **38**(6), 1163–1170 (2005)

Chapter 3
Sinonasal Anatomic Variants, Diseases and their Management

Joey Siu and Richard Douglas

Abstract There is an enormous range of anatomical variants of the nasal cavity and paranasal sinuses, and these can be influenced by gender, age and ethnic adaptations to local climate. A further factor is pneumatization of the paranasal sinuses which is highly variable between individuals. When significant deviations occur from typical airway geometries, this can lead to nasal airway obstruction having the potential to significantly reduce quality of life in patients. This chapter presents the main factors affecting anatomy geometry, and common disorders, diseases, and its management.

3.1 Nasal Cavity Variants

3.1.1 Gender Difference

The male nose is typically larger and more capacious than the female nose, with thicker cartilages and nasal bones. It is characterized by a longer, narrower and higher nasal floor. Bastir et al. [2] measured twenty-five, 3-dimensional landmarks in five populations of 212 adult humans of various geographic regions. Multivariate regression analysis and residual analysis demonstrated that even adjusting for body size, males had relatively larger nasal passages than females. Male nasal airways are larger due to taller piriform apertures, internal nasal cavities and choanae than females of the same body size. The tip of the female nose is typically rotated more superiorly and the nasolabial angle, formed by the plane of columella meeting the upper lip, is usually greater.

J. Siu (✉) · R. Douglas
University of Auckland, Auckland, New Zealand
e-mail: joeysiu.nz@gmail.com

K. Inthavong et al. (eds.), *Clinical and Biomedical Engineering in the Human Nose*, Biological and Medical Physics, Biomedical Engineering,
https://doi.org/10.1007/978-981-15-6716-2_3

3.1.2 Age Variation

Craniofacial measurements also change substantially with age. While the bony structures of the head typically reach their definitive lengths by early adulthood, the nose contains cartilages which may continue to grow throughout life. Zankl et al. [25] examined 2,500 healthy individuals living in Central Europe, from newborn to nonagenarians.

Growth charts for nose length, nasal protrusion and philtrum length were constructed based on these measurements. These showed that the nose, unlike other bony structures of the head and body, continues to grow throughout life, a reflection of its being formed from cartilage. The relationship between age and nasal protrusion was less pronounced. Philtrum length was shown to reach its first peak in adolescence, followed by a decline in early adulthood and then regain length after the age of thirty.

3.1.3 Ethnic Climatic Variation

Some anatomical variations appear to have developed as adaptations to local climatic conditions. The platyrrhine nose has a small external protrusion, large flaring nostrils and is short with a broad base. Its broad cross-sectional area results in reduced heat transfer during exhalation [19]. This shape typically occurs in populations located in hot or moist environments. The leptorrhine nose is tall and narrow, with a large external protrusion and relatively constricted nostrils. It has a narrow cross-sectional area to facilitate heat and moisture exchange [4]. This shape is typically seen in populations located in cold or dry environments Fig. 3.1. The mesorrhine nose is considered to be intermediate, with features of both the leptorrhine nose and the platyrrhine nose. The nasal index, which is the ratio of nasal width to height (width $\times$ 100/height) is used as a way to determine the nose type ($< 70 =$ leptorrhine; $70 - 85 =$ mesorrhine; $> 85 =$ platyrrhine).

A systematic review and meta-analysis of 38 studies [24] described the important nasal angles in the nasal profile. This included the nasofrontal angle, where the plane of the dorsum of the nose meets the forehead and brow, the nasofacial angle, which defines nasal projection from the face and the nasolabial angle. African males were noted to have a smaller nasofrontal angle compared to Caucasian males but a larger nasofacial angle compared to Asian males ($p < 0.05$). The nasolabial angle was more obtuse in Caucasian females than in African and Asian females ($p < 0.05$). Caucasian females had a larger nasofrontal angle than African females, and the nasolabial angle in Caucasian males were larger than in African males (<0.10). Caucasian females were also noted to have a smaller width of the face, shorter height of forehead and shorter physiognomic height of the face compared to Asian females (<0.10). These ethnic variations may influence nasal airflow and drug delivery as reported in a computational study of 16 normal patients from 4 different ethnic groups: African-American, Asian, Caucasian and Latin American [9].

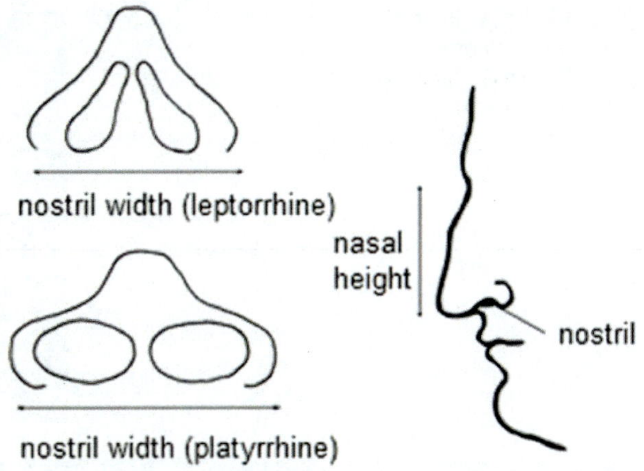

Fig. 3.1 Ethnic Climatic Variation in Nasal Anatomy

3.2 Pneumatization of the Paranasal Sinuses

The paranasal sinuses arise from a slow excavation of bone by pneumatic diverticula (air-filled sacs) growing from the nasal cavity, a process that starts in utero and continues throughout the course of a lifetime [16]. The extent of pneumatization is highly variable between individuals. Sphenoid sinuses arise from presphenoid bone centers. Pneumatization can extend toward the lateral recess of the sphenoid bone and towards the anterior clinoid processes and nasal septum. Extension of pneumatization of the maxillary sinus can also occur for example towards the palatine recess and alveolar recess.

3.3 Ethmoid Cell Variations

There are between eight and fifteen ethmoidal air cells, which form a bony labyrinth in the superior and lateral aspect of the nasal cavity between the eyes. Many ethmoid cell variations exist and these and influence the patency and boundaries of the sinus drainage pathways and so are of significant surgical importances [17]. Infraorbital ethmoid cells or Haller cells are ethmoid air cells located anteriorly to the ethmoid bulla along the orbital floor, adjacent to the natural ostium of the maxillary sinus. Agger nasi cells are the most anterior ethmoid cells and they are located anteriorly to the upper margin of the nasolacrimal duct and anteroinferior to the frontal sinus drainage pathway. Agger nasi cells are the most anterior ethmoid cells and they are located anteriorly to the upper margin of the nasolacrimal duct and anteroinferior to the frontal sinus drainage pathway. Their surgical importance is that upon entry they may be incorrectly identified as sphenoid cells (Figs. 3.2 and 3.3). This section introduces some of the variations to the sinonasal cavity that influence air and particle respiration. This is clinically relevant because they may also influence the pattern

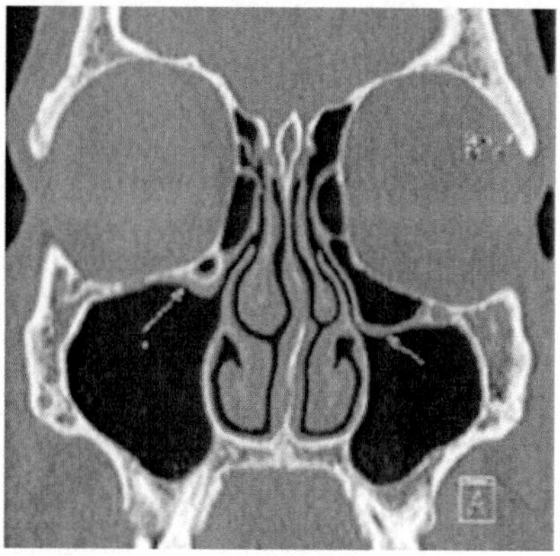

Fig. 3.2 Location of infraorbital ethmoid cells (also known as Haller cells) labelled by the arrows

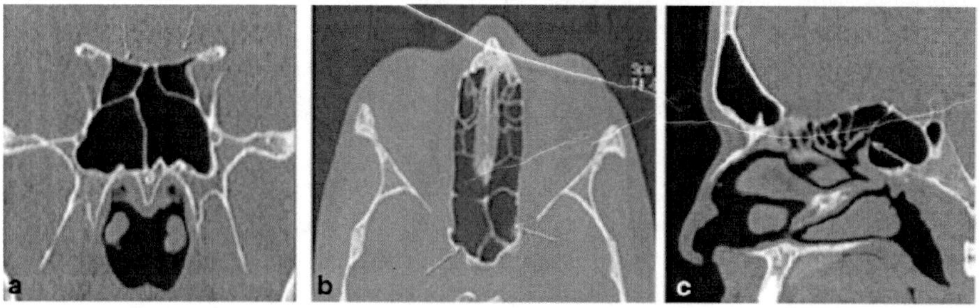

Fig. 3.3 Bilateral Onodi cells (arrows). **a** Coronal **b** Axial **c** Sagittal plane shows no connection to the sphenoid sinus ostium—Images from: Miranda et al. Anatomical variations of paranasal sinuses at multislice computed tomography: what to look for. Radiologia Brasileira, vol. 44, no. 4, 2011, pp. 256–262

of inflammatory disease, nasal drug delivery and required surgery. This approach is replicated through the subsequent sections on variations and diseases of the other parts of the respiratory pathway (pharynx, larynx, and tracheobronchial and lung airway).

3.4 Common Disorders of the Nose, Sinuses and Their Treatment

Most disorders of the nose are not life threatening but have the potential to very significantly reduce quality of life. Patients suffering from common disorders of the nose may present with various sinonasal symptoms.

3.4.1 Sinonasal Symptoms

Common sinonasal symptoms include nasal airway obstruction, congestion, discharge from the nostrils or a sensation of discharge backwards into the throat (post-nasal drip), reduction of sense of smell (hyposmia) or headaches or facial pressure. Each of these symptoms are not necessarily specific to the nose—a sensation of a post-nasal drip may originate from stomach acid reflux (laryngo-pharyngeal reflux), headaches and facial pains may represent neurological conditions (e.g., migraines), ear or dental pathology (e.g., temporo-mandibular joint dysfunction), while hyposmia may present in patients with intracranial pathology (e.g., Alzheimer's disease or Parkinson's disease). Many sinonasal disorders also affect non-rhinologic domains of a patient's quality of life such as sleep disorders and mood changes.

Several validated questionnaires are routinely used in practice to comprehensively assess patient reported outcome measures. These questionnaires aid the clinician by helping establish the diagnosis, ascertain the quality of life impediment experienced by the patient and assess the response to treatment. Functional self-assessments include the Nasal Surgical Questionnaire, Nasal Obstruction Symptoms Evaluation (NOSE) Scale, and the Nasal Obstruction Septoplasty Effectiveness score. The Sinonasal Outcome Test (SNOT-22), consisting of 22 questions is the most commonly used questionnaire, particularly in patients suffering from chronic rhinosinusitis (CRS).

3.4.2 Assessment of the Nasal Cavity

The anterior nasal cavity can be assessed by direct visualization with a headlight and a Thudicum or Killian's nasal speculum Fig. 3.4. The remaining nasal cavity and sinuses can be assessed using endoscopy visualization. Different endoscopic scoring systems exist to evaluate pathologic states within the nose including polyps, discharge, edema, scarring, and crusting. The Modified Lund-Kennedy scoring system is often used for pre- and post-operative chronic rhinosinusitis (CRS) patients.

Evaluation of nasal patency and resistance can be performed using objective assessments such as rhinomanometry, acoustic rhinomanometry or peak nasal inspiratory flow (PNIF) meters. Although these instruments may be used in the selection of suitable patients for surgery, they are not routinely utilized in clinical practice as there is limited correlation between objective measures and subjective appreciation of nasal patency.

3.5 Common Disorders of the Nasal Cavity

3.5.1 Septal Deviation

Septal deviation describes a displacement of the nasal septum from the midline. The prevalence of nasal septal deviation has been reported to be as high as 80%, but many small deviations are of little symptomatic consequence. Nasal septal deviation can occur in any area of the nasal septum, which includes the cartilaginous anterior portion as well as the bony perpendicular plate of ethmoid and the vomer more posteriorly. Anterior deviations are more likely to cause nasal airway obstruction due to their proximity to the internal nasal valve which is the narrowest part of the nasal passage.

Septal deviation may be congenital, or occur during childhood as a result of asymmetric growth, or be secondary toa traumatic injury to the nose. Reference [14] described congenital septal deviations as usually smooth and shaped as either a "C" or an "S", while traumatic deviations were more often irregular, angulated or dislocated (Fig. 3.5). Deviations can be oriented anteroposteriorly or cranio-caudally. Occasionally, a more localized deviation of the bone can present unilaterally, and these are termed septal spurs spur.

Patients with symptomatic nasal obstruction secondary to septal deviation are often managed non-surgically in the first instance. Patients are assessed for allergy status and can be prescribed corticosteroid or antihistamine nasal sprays. Although these treatments do not correct the underlying anatomical deviation, they may reduce mucosal edema, reactive vascular changes and glandular hypertrophy enough to improve nasal airflow.

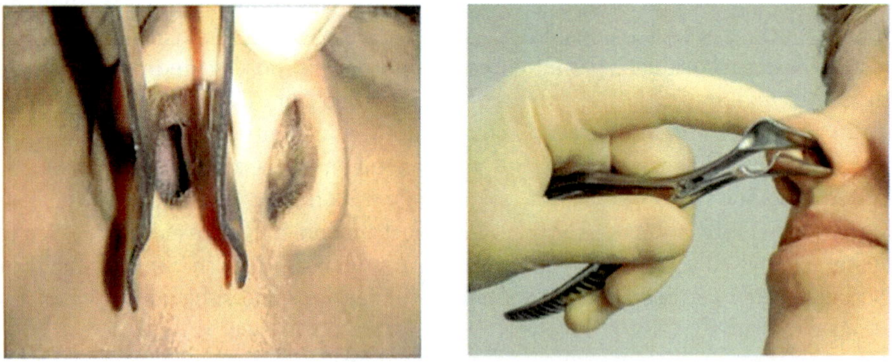

(a) Examination using Thudicum's speculum. Picture Courtesy: www.hkcfp.org.hk/

(b) Examination using Killian's speculum. Picture Courtesy: http://rhinitis.hawkelibrary.com

Fig. 3.4 Anterior nasal cavity assessed by direct visualization with a headlight and a Thudicum or Killian's nasal speculum

In patients who are recalcitrant to conservative measures, surgical intervention may be offered. A septoplasty involves correction of the deviated portion of the septal cartilage or bone while preserving the overlying septal mucosa. In some cases, when the deviated portion of cartilage lies very anteriorly, there may be concern that excision may cause substantial destabilization of nasal tip support. Therefore, an external rhinoplasty, which involves an external incision, may be required to ensure adequate anterior septal reconstruction.

3.5.2 Concha Bullosa

A concha bullosa occurs when the middle turbinate is pneumatized. This variant has a prevalence of up to 55% [3, 6]. Large unilateral conchae bullosae are often associated with nasal septal deviation to the contralateral side. A large concha bullosa may narrow the middle meatus on the same side, predisposing the patient to sinus outflow obstruction and rhinosinusitis. Several surgical techniques to address concha bullosa have been described, each with the primary aim of reducing the width of the middle turbinate. Techniques include crushing, stripping, marsupialization, medial and lateral laminectomy and complete middle turbinectomy.

3.5.3 Allergic Rhinitis

Allergic rhinitis is defined as an allergen-associated immunoglobulin E (IgE) mediated inflammation of the nasal mucosa. A patient with allergic rhinitis typically

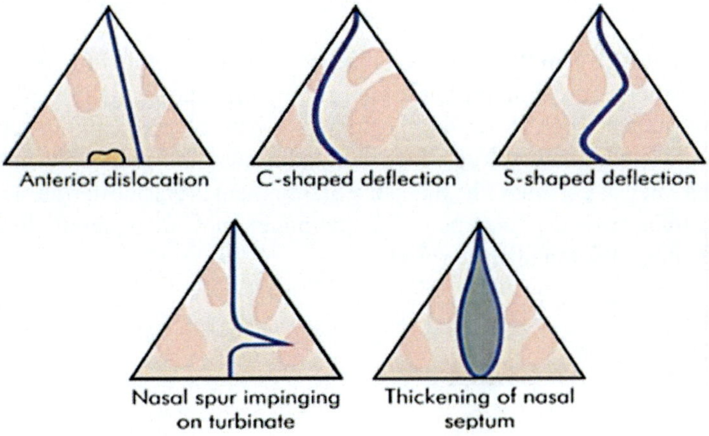

Fig. 3.5 Types of deviated nasal septum-Image from: Snapshots in Ear, Nose and Throat: Head and Neck Surgery by [22]

presents with nasal obstruction and congestion, nasal discharge, sneezing and nasal itching. Associated allergic conditions may also be present, including asthma, eczema and allergic conjunctivitis. Allergy is particularly common in modern urban populations and is believed to be associated with increased air pollution, indoor environments, a psychologically stressful lifestyle and reduced childhood exposure to bacteria resulting in an inappropriate activation of the immune system towards benign allergens. The diagnosis of allergic rhinitis is based on clinical features and confirmed with skin prick testing or RAST testing for common allergens. In a small proportion of patients, allergic rhinitis may be present without a positive skin test or positive results. This is a subset of disease known as local allergic rhinitis or entopy, which may require further specialized tests for diagnosis such as a nasal allergen provocation test. This test is confined to research institutions and not performed in routine clinical practice. The treatment of allergic rhinitis includes allergen avoidance, topical or systemic corticosteroids and antihistamines. Long term decongestant therapy is not advised due to a risk of rebound congestion (rhinitis medicamentosa). Some patients respond very well to allergen immunotherapy, which can be administered either subcutaneously or sublingually. When medical therapy for allergic rhinitis is not effective or not well tolerated, surgery may be considered. The main role of surgery in allergic rhinitis is to improve the nasal airway by reduction of enlarged inferior turbinates and the correction of any concurrent septal deviation. Various methods of turbinate reduction have been described and these are discussed in Chap. 4 (Surgery of the nasal cavity). Submucosal resection combined with lateral displacement has become a popular technique for nasal airway obstruction caused by inferior turbinate hypertrophy. Such surgery aims to preserve most of the turbinate mucosa and to date is believed to help maintain normal nasal air conditioning.

3.5.4 Non-allergic Rhinitis

A number of other pathologies may cause nasal mucosal inflammation without an identifiable allergic precipitant. These causes may vary substantially, and can include inhaled substances such as smoke, perfumes, and salts; medications such as non-steroidal anti-inflammatory drugs; foods, including alcohol; and, rarely, inhaled illicit substances, such as cocaine. The mainstay of treatment is corticosteroid sprays. Antihistamine treatments may also be trailled, although they are generally less effective in this condition than in allergic rhinitis.

3.5.5 Nasal Valve Collapse

The nasal valve is the narrowest part of the nasal airway, and is a cross-sectional area defined by the caudal end of the upper lateral cartilage, the head of the inferior turbinate and the septum. In a study of 1906 patients presenting with nasal airway obstruction, the prevalence of nasal valve collapse (NVC) was reported as 67% [5]. A reduction in the internal nasal valve cross-sectional area introduces substantial inspiratory negative pressures, as per Bernoulli's principle. Narrowing of cross-sectional area during inspiration may occur if the lateral nasal cartilage is thin, weak, detached or absent. This collapse is termed as dynamic or active nasal valve collapse. Static or passive nasal valve collapse refers to the condition where the nasal valve is permanently narrowed.

Clinical examination for dynamic valve collapse is performed by asking the subject whether nasal airflow is improved when applying gentle lateral traction to the ipsilateral cheek (Cottle's maneuver) or introducing a supporting instrument to the suspected site of valve collapse (modified Cottle's maneuver). In dynamic nasal valve collapse secondary to weakness of the lateral cartilage, various external and internal prostheses have been developed, which act like a permanent Cottle's maneuver. Surgical intervention can also be offered to strengthen the lateral cartilages and include alar batten grafts and lateral crural strut grafts. Treatment of static nasal valve collapse involves correcting the cause and may include septoplasty and reduction of the inferior turbinates.

3.5.6 Septal Perforation

A nasal septal perforation occurs when there is an internal communication between the left and right nasal cavities. Perforations most commonly occur as a result of septoplasty. Other causes include septal haematoma, long term vasoconstrictive drug application, cocaine use, vasculitis (granulomatosis with polyangiitis, formerly known as Wegener's granulomatosis) and medications such as bevacizumab. Septal perforations vary in size are most commonly found in the anterior cartilaginous region. Symptoms of septal perforation include obstruction, nasal crusting, whistling sounds while breathing and hyposmia secondary to the loss of normal laminar nasal airflow through each nasal cavity. Anterior perforations are usually more likely to cause symptoms while smaller perforations are more likely to cause a whistling sound. Nasal crusting likely occurs secondary to mucosal breakdown with exposure of cartilage or bone, particularly at the posterior margin of the perforation.

Symptomatic septal perforations may require treatment. Repair of smaller perforations have a higher success rate. Larger perforations are more likely to be managed conservatively. Closure of the perforation is typically performed in a layered fashion, with a graft substitute for the septal cartilage and a mucosal graft. Vascularized flaps have been demonstrated to have a good rate of closure, even in very large perforations. Pericranial flaps, using the external periosteum of the skull, have proven successful in very large perforations but involve a relatively extensive procedure. The alternative to repair is to use a prosthetic septal "button" made of silicone material.

3.5.7 Chronic Rhinosinusitis

Chronic rhinosinusitis (CRS) represents a spectrum of disorders that result from a variety of immunopathological mechanisms that lead to persistent inflammation of the paranasal sinus mucosa. The European Position Statement on Rhinosinusitis and Nasal Polyps (EPOS 2012) defines chronic rhinosinusitis as inflammation of the nose and sinuses with two or more symptoms lasting for more than 12 weeks [7]. One of these symptoms must include "nasal obstruction, blockage or congestion" or "anterior or posterior nasal discharge." Other symptoms that can be present include facial pains or pressure and hyposmia. Clinically CRS is classified in two broad groups or phenotypes: CRS with nasal polyps (CRSwNP) and CRS without NPs (CRSsNP). Both of these phenotypes are characterized by epithelial disruption, ciliary dysfunction, mucus gland hyperplasia, bacterial overgrowth and the formation of biofilms [11]. The role of microbes in the pathogenesis of CRS is largely unknown, but bacteria probably contribute to the persistence and severity of the disease [21].

Despite the widespread use of antibiotics in the treatment of CRS, there is some evidence that antibiotics may be paradoxically implicated in the pathogenesis of this condition. The repeated use of broad-spectrum antibiotics is associated with microbial dysbiosis which promotes the persistence of CRS and the emergence of more resistant bacterial strains [8, 13, 23]. The international consensus statement recommends the use of oral antibiotics from the macrolide group as an option in the treatment of CRS on the basis that some studies have demonstrated an improvement in symptoms and clinical scores assessed endoscopically [18]. Recommended best medical management for CRS also includes a prolonged period of nasal saline douching and topical or oral steroids.

FESS is indicated when patients are recalcitrant to medical therapy. FESS is a well-established minimally invasive surgical technique, the primary goal of which is to remove diseased tissue and facilitate ventilation, drainage and mucociliary clearance through widening the ostia [10–12]. However, some forms of CRS are driven by self-perpetuating immune-mediated processes, and in these cases, FESS alone may not eradicate the disease [15]. Surgery in these patients allows improved access of topical drugs and post-surgical surveillance [20].

In recent years, there has been a strong focus on the categorization of disease variants into endotypes based on their underlying pathophysiologic mechanisms. This highlights the inflammatory heterogeneity of the disease rather than the traditional phenotypic classification CRSwNP/Type 2 and CRSsNP/Type 1 dichotomy. It may provide more information on the risk for disease progression, recurrence and comorbid conditions, as well as identify suitable therapeutic targets. Endotypes can be defined in different ways according to circulating biomarkers, histopathological traits e.g., degree of inflammation, type of inflammatory cell predominance, level of remodeling and fibrosis, presence of polyps, and response to medical treatments [1].

References

1. S. Ahern, A. Cervin, Inflammation and endotyping in chronic rhinosinusitis – A paradigm shift (2019)
2. M. Bastir, P. Godoy, A. Rosas, Common features of sexual dimorphism in the cranial airways of different human populations. Am. J. Phys. Anthropol. **146**(3), 414–422 (2011)
3. W.E. Bolger, D.S. Parsons, C.A. Butzin, Paranasal sinus bony anatomic variations and mucosal abnormalities: Ct analysis for endoscopic sinus surgery. The Laryngoscope **101**(1), 56–64 (1991)
4. J.W. Carey, A.T. Steegmann Jr., Human nasal protrusion, latitude, and climate. Am. J. Phys. Anthropol. **56**(3), 313–319 (1981)
5. D.W. Clark, A.G. Del Signore, R. Raithatha, B.A. Senior, Nasal airway obstruction: prevalence and anatomic contributors. Ear, Nose Throat J. **97**(6), 173–176 (2018)
6. J. Earwaker, Anatomic variants in sinonasal ct. Radiographics **13**(2), 381–415 (1993)
7. W.J. Fokkens, V.J. Lund, J. Mullol, C. Bachert, I. Alobid, F. Baroody, N. Cohen, A. Cervin, R. Douglas, P. Gevaert, et al. Epos 2012: European position paper on rhinosinusitis and nasal polyps 2012. a summary for otorhinolaryngologists. Rhinology **50**(1), 1–12 (2012)
8. M. Hoggard, K. Biswas, M. Zoing, B. Wagner Mackenzie, M.W. Taylor, R.G. Douglas, Evidence of microbiota dysbiosis in chronic rhinosinusitis. Int. Forum Allergy Rhinol. (2017)
9. J.A. Keeler, A. Patki, C.R. Woodard, D.O. Frank-Ito, A computational study of nasal spray deposition pattern in four ethnic groups. Journal of aerosol medicine and pulmonary drug delivery **29**(2), 153–166 (2016)
10. D.W. Kennedy, S.J. Zinreich, A.E. Rosenbaum, M.E. Johns, Functional endoscopic sinus surgery: theory and diagnostic evaluation. Arch. Otolaryngol. **111**(9), 576–582 (1985)
11. J.L. Kennedy, M.A. Hubbard, P. Huyett, J.T. Patrie, L. Borish, S.C. Payne, Sino-nasal outcome test (snot-22): a predictor of postsurgical improvement in patients with chronic sinusitis. Ann. Allergy, Asthma Immunol. **111**(4), 246–251 (2013)
12. H. Khalil, D.A. Nunez, Functional endoscopic sinus surgery for chronic rhinosinusitis. Cochrane Database of Syst. Rev. **3** (2006)
13. J.T. Lee, D.N. Frank, V. Ramakrishnan, Microbiome of the paranasal sinuses: update and literature review. Am. J. Rhinol. Allergy (2016)
14. J. Lindahl, The deviated nasal septum. The Practitioner **173**(1035), 315–317 (1954)
15. Y. Matsuwaki, T. Ookushi, D. Asaka, E. Mori, T. Nakajima, T. Yoshida, J. Kojima, S. Chiba, N. Ootori, H. Moriyama, Chronic rhinosinusitis: risk factors for the recurrence of chronic rhinosinusitis based on 5-year follow-up after endoscopic sinus surgery. Int. Arch. Allergy Immunol. **146**(Suppl. 1), 77–81 (2008)
16. J. Medina, H. Hernandez, L.W. Tom, L. Bilaniuk, Development of the paranasal sinuses in children. Am. J. Rhinol. **11**(3), 203–210 (1997)
17. C.M.N.R.d. Miranda, C.P.d.M. Maranhão, F.M.N.R. Arraes, I.G. Padilha, L.d.P.G.d. Farias, M.S.d.A. Jatobá, A.C.M.d. Andrade, I.G. Padilha, L.d.P.G.d. Farias, M.S.d.A. Jatobá, A.C.M.d. Andrade, B.G. Padilha, Variações anatômicas das cavidades paranasais à tomografia computadorizada multislice: o que procurar? Radiologia Brasileira **44**(4), 256–262 (2011)
18. R.R. Orlandi, T.T. Kingdom, P.H. Hwang, T.L. Smith, J.A. Alt, F.M. Baroody, P.S. Batra, M. Bernal-Sprekelsen, N. Bhattacharyya, R.K. Chandra, A. Chiu, M.J. Citardi, N.A. Cohen, J. Delgaudio, M. Desrosiers, H.J. Dhong, R. Douglas, B. Ferguson, W.J. Fokkens, C. Georgalas, A. Goldberg, J. Gosepath, D.L. Hamilos, J.K. Han, R. Harvey, P. Hellings, C. Hopkins, R. Jankowski, A.R. Javer, R. Kern, S. Kountakis, M.L. Kowalski, A. Lane, D.C. Lanza, R. Lebowitz, H.M. Lee, S.Y. Lin, V. Lund, A. Luong, W. Mann, B.F. Marple, K.C. Mcmains, R. Metson, R. Naclerio, J.V. Nayak, N. Otori, J.N. Palmer, S.R. Parikh, D. Passali, A. Peters, J. Piccirillo, D.M. Poetker, A.J. Psaltis, H.H. Ramadan, V.R. Ramakrishnan, H. Riechelmann, H.J. Roh, L. Rudmik, R. Sacks, R.J. Schlosser, B.A. Senior, R. Sindwani, J.A. Stankiewicz, M. Stewart, B.K. Tan, E. Toskala, R. Voegels, D.Y. Wang, E.K. Weitzel, S. Wise, B.A. Woodworth, P.J. Wormald, E.D. Wright, B. Zhou, D.W. Kennedy, *International consensus statement on allergy and rhinology: rhinosinusitis* (Int, Forum Allergy Rhinol, 2016)

19. E. Seren, S. Seren, Morphological adaptation of the nasal valve area to climate. Med. Hypotheses **72**(4), 471–472 (2009)
20. K. Snidvongs, L. Kalish, R. Sacks, R. Sivasubramaniam, D. Cope, R.J. Harvey, Sinus surgery and delivery method influence the effectiveness of topical corticosteroids for chronic rhinosinusitis: systematic review and meta-analysis. Am. J. Rhinol. Allergy **27**(3), 221–233 (2013)
21. P.S. Stewart, J.W. Costerton, Antibiotic resistance of bacteria in biofilms (2001)
22. S. Swain, *Snapshots in Ear, Nose & Throat Head and Neck Surgery* (JP Medical Ltd, Bhubaneswar, India, 2016)
23. B. Wagner Mackenzie, D.W. Waite, M. Hoggard, R.G. Douglas, M.W. Taylor, K. Biswas, Bacterial community collapse: a meta-analysis of the sinonasal microbiota in chronic rhinosinusitis. Environ. Microbiol. (2017)
24. Y.F. Wen, H.M. Wong, R. Lin, G. Yin, C. McGrath, Inter-ethnic/racial facial variations: a systematic review and bayesian meta-analysis of photogrammetric studies. PloS one **10**, 8 (2015)
25. A. Zankl, L. Eberle, L. Molinari, A. Schinzel, Growth charts for nose length, nasal protrusion, and philtrum length from birth to 97 years. Am. J. Med. Genetics **111**(4), 388–391 (2002)

Chapter 4
Surgery of the Nasal Cavity

Kimberley Bradshaw and Narinder Singh

Abstract Early use of computational fluid dynamics (CFD) for studies in the nose primarily focussed on understanding airflow and physiology. With advanced techniques evolving CFD models have become more advanced, and the technology for the study of surgery has been improved through the analysis of pre-and post-operative models. This development has led to the use of "virtual" surgery in computer models to predict the results of existing procedures and to potentially create new procedures. This chapter covers pre-operative assessment and current methods of planning for nasal surgery, types of nasal surgery and an introduction to virtual surgery using CFD.

4.1 Introduction

Early use of computational fluid dynamics (CFD) in the nose was primarily to gain a greater understanding of nasal airflow and physiology. As experience has grown and techniques have improved, CFD of the nose and sinuses has become a valuable research tool, particularly as the existing "objective" testing methods, including rhinomanometry, acoustic rhinometry and peak nasal inspiratory flow (PNIF), are known to suffer significantly from several limitations. As CFD models have become more sophisticated, the technology has been developed to study the effects of surgery, through the analysis of pre- and post-operative models. Further evolution has led to the use of "virtual" surgery in computer models in order to predict the effects of known procedures and to potentially develop new procedures. Ultimately, this may lead to a new era of personalised precision medicine, where surgeons may use CFD to determine the ideal procedure for individual patients, through CFD-based virtual surgical planning.

K. Bradshaw (✉) · N. Singh
University of Sydney, Sydney, Australia
e-mail: kim.rose.bradshaw@gmail.com

K. Inthavong et al. (eds.), *Clinical and Biomedical Engineering in the Human Nose*, Biological and Medical Physics, Biomedical Engineering,
https://doi.org/10.1007/978-981-15-6716-2_4

4.2 Pre-operative Assessment of Nasal Function

Pre-operative assessment of nasal airway obstruction (NAO) begins with the clinician recording the patient's history and performing an examination. This includes anterior rhinoscopy, which involves using a headlight and speculum to examine the entrance of the nose, and nasendoscopy, which involves using a flexible or rigid endoscope to examine deep into the nasal cavity, often captured with a camera on a monitor (Fig. 4.1). Imaging studies are often used, particularly computed tomography (CT) scans and, less commonly, magnetic resonance imaging (MRI). Through these assessments, the surgeon will determine the most likely region of obstruction and the most appropriate procedure; based predominantly on their training and personal experience. There is limited objectivity in this process and the decision to operate is determined largely by the patient's description and severity of symptoms.

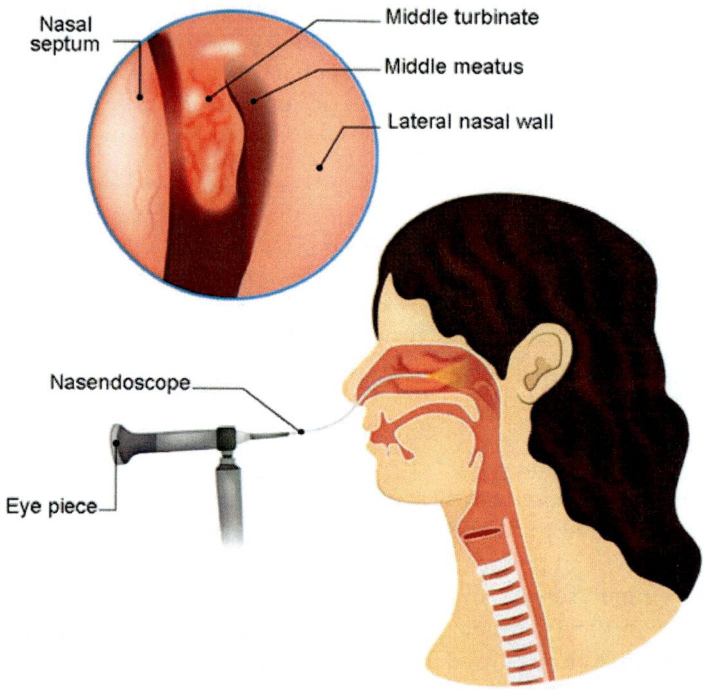

Fig. 4.1 Nasendoscopy technique and endoscopic view of the nasal cavity

4.2.1 Patient-Reported Outcome Measures (PROMs)

The Nasal obstruction symptom evaluation scale (NOSE) and Sinonasal outcome test (SNOT-22) scores have attempted to quantify patient reported outcomes and can be used as a comparative measure before and after intervention. However, they do not have much utility in guiding the choice of surgical intervention.

The SNOT-22 is a quality of life measurement tool for chronic rhinosinusitis (CRS), it has been used to assess the effectiveness of surgical intervention. The greatest effect post-surgery is seen in those with the highest SNOT-22 scores pre-operatively. "Runny nose" as a symptom predicts the greatest improvement in symptoms as does an absence of prior surgery. SNOT-22 is a helpful tool in quantifying changes in symptoms and can be used to predict the extent of post-operative improvement as well as being a useful auditing tool for surgeons, but has limited application for determining type or extent of intervention [24].

The NOSE score is used similarly for NAO and is the most commonly used score for reporting outcomes after nasal airway surgery [41]. The NOSE scale has five questions concerning the severity of airway obstruction, scoring 0 to 100, and has been validated for both septoplasty and functional septorhinoplasty. The NOSE score can be used to counsel patients regarding expected results following surgery [15]. Comparing scores at pre- and post-operative time intervals can show significant and durable improvement in NAO. NOSE scores have been shown to improve at the earliest time point of 1–3 months and remain improved beyond 10 months. The optimal duration for follow up of postoperative nasal obstruction severity using NOSE scores has not been established, however, it is suggested that early positive findings are associated with stable longer-term findings.

4.2.2 Objective Measures of NAO

There are currently three objective methods for analysing nasal airflow in use: rhinomanometry, acoustic rhinometry and peak nasal inspiratory flow (PNIF).

Rhinomanometry measures pressure and airflow during respiration to define resistance of the nasal airway. Anterior rhinomanometry measures unilateral airflow; both nasal passages are measured separately. A probe is placed in the nasal vestibule and the trans-nasal pressure difference is measured, producing a flow-pressure curve (Fig. 4.2). This is the most common form of rhinomanometry used in clinical practice. The alternative, posterior rhinomanometry is less commonly used, mostly in research settings. The main utility of rhinomanometry is in distinguishing between obstruction caused by mucosal hypertrophy versus structural deformity by comparing pre- and post-decongestion results. It does not provide detailed information on the exact location of the structural deformity or where the nasal airflow is most affected but can distinguish the side of greatest obstruction.

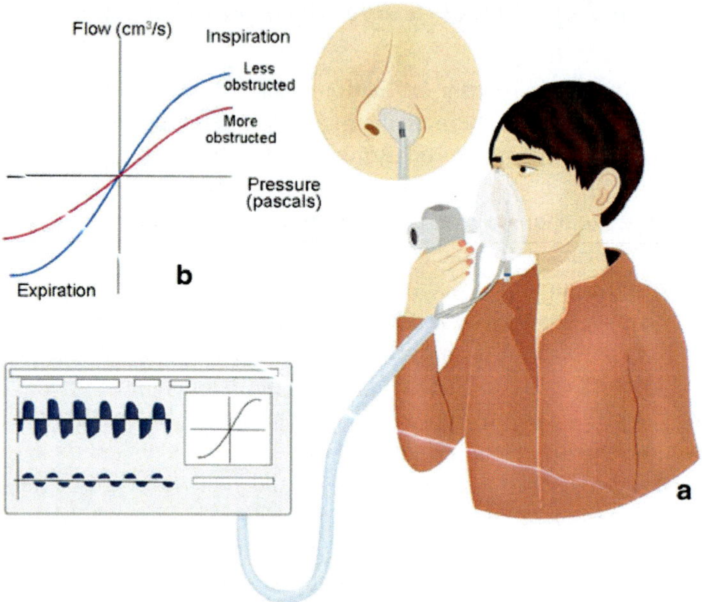

Fig. 4.2 **a** Diagram of Rhinomanometry **b** Flow-Pressure cure which can be used to diagnose nasal obstruction

Rhinomanometry works by considering how pressure and flow vary during a breathing cycle over time. The breathing cycle can be divided into four phases and is hence known as four-phase rhinomanometry (4PR). These phases include: the accelerating inspiratory phase, the decelerating inspiratory phase, the accelerating expiratory phase, and the decelerating expiratory phase. The diagnostic aim of 4PR is to measure the intranasal pressure, flow, and time variables necessary for maintaining an adequate oxygen supply through the nose. The technique for performing 4PR is called active rhinomanometry which uses the patient's own physiological airflow for assessing nasal ventilatory function, this technique is accepted as standard worldwide. As opposed to passive rhinomanometry which involves pumping a constant amount of air through the nasal passage and measuring the resultant pressure difference, this may be used in children when active rhinomanometry is not possible.

Active rhinomanometry involves closing one nostril with a measuring probe, as seen in Fig. 4.2a, while the contralateral nostril is left open, this is the side being tested. The patient holds a face mask, which is connected to the measuring device, tightly onto the face. The patient is asked to breathe through their nose keeping their mouth shut. The nasal passage acts as an extended tube and assumes the airway pressure of the nasopharynx equals the pressure at the naris of the non-tested side.

Laminar flow is based on a linear relationship between flow and pressure, as nasal obstruction occurs the amount of flow that can be generated plateaus sooner because of turbulence and airway resistance and the flow-pressure curve flattens (Fig. 4.2). If this curve can be corrected by nasal decongestion this suggests obstruction is due to mucosal disease. Decongestion causing less than a 35% decrease in resistance suggests a structural deformity. 4PR has been validated as a technically well-founded diagnostic method by CFD models.

Acoustic rhinometry uses sound waves reflected from the nasal walls to create two-dimensional imaging of the nasal cavity, quantifying the cross-sectional area at specific distances along the length of the nasal passage. The sound reflected from thes nasal airway is transformed into an area-distance plot. These dimensions have been shown to correlate well with radiologic measurements. The main use of acoustic rhinomanometry is to localize areas of constriction, which conventional rhinomanometry fails to do. However, technical challenges including malpositioning of the nasal tube can lead to distortion in measurement [16]. Acoustic rhinometry has greater accuracy at the nasal valve area, at the front of the nasal cavity.

Essentially, conventional rhinomanometry can determine nasal patency and reversibility and acoustic rhinometry may more accurately identify the site of obstruction, however its clinical utility to date is limited.

Peak nasal inspiratory flow (PNIF) is a gross measure of total nasal airflow, combining left and right sides. It is relatively easy to perform, and its main utility is for pre- and post-operative comparison. It is dependent on lung function, body position and patient cooperation. It does not correlate well with patient reported symptoms or NOSE score [13].

These tests do not represent the full scope of nasal function such as local flow, pressure changes and heat exchanges. Nor do they accurately predict the most effective surgical technique to improve nasal airflow and function (Table 4.1).

Table 4.1 General approach to assess patients

Patient history	Symptoms Previous medical treatment SNOT-22 symptoms questionnaire [23]
Clinical & Endoscopic assessment	Character of mucosa (e.g., mucosal hypertrophy, polyps, adhesions), appearance of sinus drainage, anatomical variations, structural abnormalities (e.g., septal deviation, nasal valve collapse)
Imaging: CT scan	Extent of disease with clinical correlation CT staging systems such as the Lund-Mackay system or 3D volumetric scoring provide an objective disease score [18, 31]. Frontal sinus drainage pathway

4.3 General Approach to Nasal Surgery

4.3.1 *Preoperative Approach*

Patient history is important in establishing an operative plan and should elicit information about subjective nasal airway obstruction, prior trauma, epistaxis, nasal decongestant and drug use. Endoscopic evaluation is a valuable adjunct.

4.3.2 *Peri-Operative Approach*

The patient is positioned supine (on the back) with arms by the side on the operating table. Nasal surgery is typically performed under a general anaesthetic but can be performed under local anaesthesia in an awake and sedated patient, in some parts of the world. Once the patient is anaesthetised the surgeon will prepare the nasal lining to reduce bleeding using medications such as adrenaline or Moffat's solution (adrenaline, sodium bicarbonate and cocaine), the choice being dependent on surgeon preference. The septum and turbinates will often also be injected with local anaesthetic to improve nasal decongestion and prevent intra-operative bleeding, which is essential to optimize visualisation during the procedure. A relatively low blood pressure and heart rate is achieved during anaesthesia for nasal surgery likewise to reduce bleeding and improve the visual field [4].

4.4 Septoplasty and Rhinoplasty

If a deviated nasal septum is present, causing obstruction to the nasal airway or limiting surgical access to the sinus cavities, a septoplasty is performed.

4.4.1 *Septoplasty*

Septal deformities are common and occur in 70-90% of the population. Even small deviations in key areas can adversely affect nasal airflow, mucociliary clearance and sometimes the external appearance of the nose. Choice of septoplasty technique is based on the nature and location of the deformity and surgeon skills and preference. In the traditional approach, the surgeon will use a headlight and speculum to visualise the nasal cavity. There are various surgical techniques to expose the cartilage of the septum. Firstly, an incision along the leading (caudal) edge of the septal cartilage, a "hemi-transfixion" incision, can be made to expose the entire cartilage, including the most anterior portion. Alternatively, a modified Killian incision can be made for

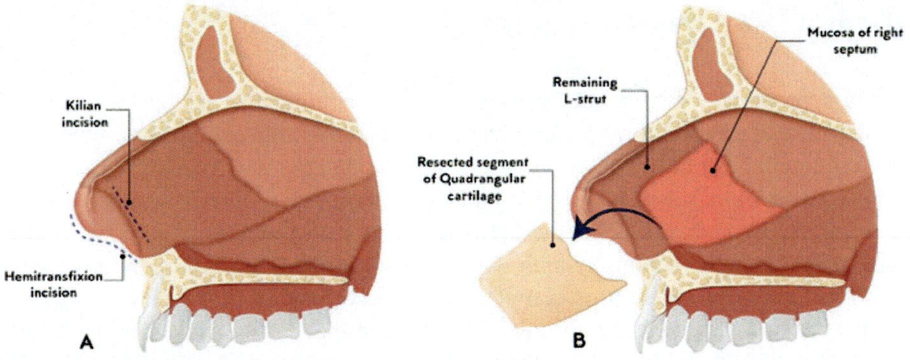

Fig. 4.3 **a** Standard hemi-transfixion incision and Killian incision. **b** Septal cartilage removal leaving behind L-strut

more posterior deflections or isolated spurs (Fig. 4.3). Mucosal flaps are raised, and the area of deviated cartilage and bone is removed, leaving behind a 15mm caudal and dorsal "L"-shaped strut for structural integrity. A portion of resected cartilage can be crushed or straightened and replaced to provide greater structural integrity, as required. The incisions and mucosal flaps are sutured together with absorbable suture material. Another technique is known as the "swinging door" or "doorstop" technique, in which a strip of cartilage that has been resting in the groove of the nasal floor is removed. The remaining cartilage, which is attached superiorly can then swing freely to the midline where it is secured to the nasal spine at the front of the septum, on the side opposite to the obstruction.

4.4.2 Endoscopic Septoplasty

A septoplasty can also be performed endoscopically with an approach similar to the standard hemi-transfixion incision or modified Killian's incision, but with greater visualisation of the septum. Endoscopic surgery involves the use of a fiberoptic light and camera to visualise and magnify the inside of the nasal cavity. Unlike the flexible nasendoscope used in the outpatient setting, the endoscopes used for nasal cavity surgery are rigid, but may have varying degrees of camera angle for a complete view inside the nasal cavity. The advantages of an endoscopic septoplasty are that the surgeon can perform a localized modified Killian's incision directly over the spur and remove it with an unobstructed view, minimizing the size of the flap elevated and the surgical trauma to the nasal mucosa. Other advantages include allowing for a better understanding of the lateral wall pathology associated with the septal deformity, reducing the need for an extensive flap dissection, and therefore reducing the risk of complications such as septal perforation. It can be useful in revision septoplasty when the flaps are frequently adherent and precise elevation is desirable [28].

4.4.3 Open Septorhinoplasty

Open septoplasty or open septorhinoplasty is a technique in which the cartilages of the nose are exposed by peeling back the nasal skin through a small incision in the columella, the fleshy external end of the nasal septum. It is required when the anterior septum is severely deviated and the septum is not otherwise amenable to a swinging door technique. Another situation where this technique may be required is a deviated dorsal septum, which deforms the nasal bridge. An open approach is also preferable in revision or complex cases. Several different techniques can be employed to address the deviation using cartilage that is removed from the septum or harvested from another area of the body, such as the ear or rib, to forms grafts. The cartilage grafts are used to reconstruct the inner architecture of the nose to widen the internal diameter (nasal valve), straighten the nasal bridge and tip, adjust tip rotation and projection, and strengthen the nasal side wall to prevent collapse and obstruction. The advantages of an open approach are that there is superior exposure of and access to the nasal tip cartilages and the septum therefore allowing more intensive remodeling of the nasal framework in difficult cases. The disadvantages of this approach are a longer and more complex procedure resulting in greater post-operative pain and swelling and a longer period of recovery.

4.5 Management of the Inferior Turbinate

The inferior turbinates play an important role in nasal physiology, containing parasympathetic nerve fibers, mucous glands, goblet cells and an abundant vasculature. The mucosal lining consists of ciliated pseudostratified columnar epithelium which captures larger inhaled particles and protects the lower respiratory tract, as well as allowing moisture exchange and playing a role in the sensation of airflow. Hence, aggressive reduction of turbinate tissue can lead to complications such as dryness, scarring, prolonged crusting, and the sensation of reduced airflow.

Inferior turbinate hypertrophy, causing nasal obstruction, can be treated by a range of techniques such as turbinoplasty (submucous resection, out-fracture), radiofrequency ablation (RFA), extra-mucosal or submucosal electrocautery, turbinectomy, laser-assisted resection or ablation, cryotherapy or ultrasound reduction [19].

Radiofrequency ablation of inferior turbinates can be performed in the outpatient setting. The anterior part of the turbinate is the most important given its proximity to the nasal valve (the narrowest region of the nose) and is easily visualised via anterior rhinoscopy with a headlight and speculum. Local anaesthetic is infiltrated into the turbinate and the radiofrequency probe is placed submucosally at three points along the turbinate length. Maximal reduction is seen at approximately 6 weeks post-operatively.

With larger turbinates a more aggressive approach is required, such as performing an inferior turbinoplasty using a microdebrider in the operating theatre setting. The

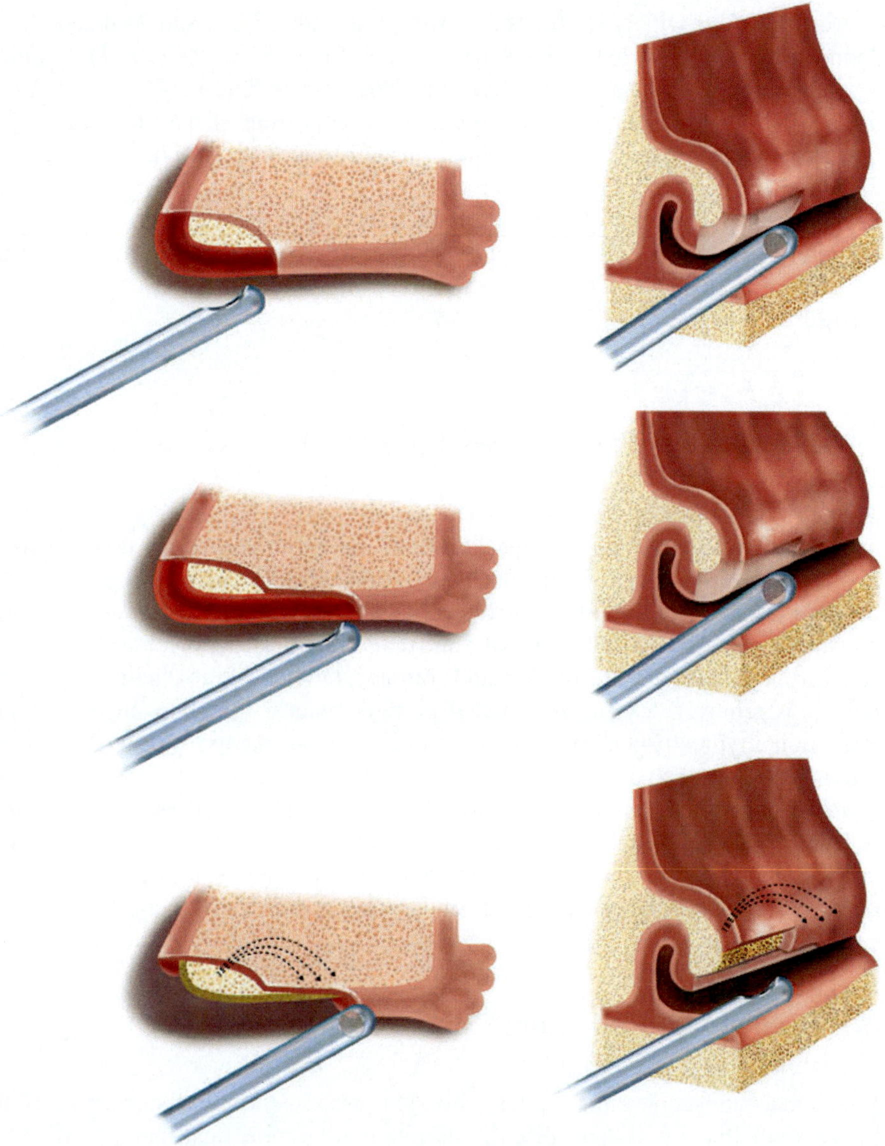

Fig. 4.4 Turbinate resection with a microdebrider

resection is performed submucosally with the elevation of a superiorly based mucosal flap (see Fig. 4.4). In this technique the bone of the turbinate is partially removed, the mucosal flap is lowered down, and the remnant turbinate attachment is outfractured towards the lateral wall. Care is taken to ensure there is no bleeding from the inferior turbinate artery by using electrocautery.

Despite being widely performed, surgeons do not have a standardized way to define turbinate hypertrophy or to select a patient for turbinate surgery. Additionally, a standardized method used to select the type of turbinate surgery needed does not exist. Both the osseous (bone) and soft-tissue portion of the turbinate may be hypertrophic. Patients with allergic rhinitis usually have hypertrophic mucosa, while unilateral enlargement of the bony structure is observed in patients with severe septal deviation [34]. The goal of turbinate reduction surgery is to eliminate as much of the inferior turbinate bulk while preserving most of the turbinate mucosa as this is believed to help maintain the normal physiology and function of the turbinate [2, 19, 45].

4.6 Management of the Middle Turbinate

As the middle turbinate has far less vasoerectile tissue compared with the inferior turbinate, it does not have as great a role in the regulation of air flow and is far less frequently a target for surgical intervention. Middle turbinate surgery mostly involves addressing anatomical variants, such as concha bullosa, which involves pneumatisation of the middle turbinate and may cause obstruction of airflow or of the sinus drainage pathway, the ostiomeatal complex. Often the lateral half of the concha bullosa is removed leaving the medial portion intact to function physiologically. Resection is typically performed using endoscopic instruments, often with a powered microdebrider.

Risks of middle turbinate resection include lateralisation leading to sinus obstruction and fracture at the skull base leading to leakage of cerebrospinal fluid from the cranial cavity into the nose. Lateralisation can be prevented using suture techniques or the placement of temporary splints in the early postoperative period.

4.7 Functional Endoscopic Sinus Surgery

Functional endoscopic sinus surgery (FESS) is used in chronic rhinosinusitis (CRS) patients who have persistent disease, despite maximal medical therapy. FESS is a well-established, minimally invasive surgical technique, in which sinus cells and ostia are opened using operating endoscopes [22]. The primary goal of FESS is to remove diseased tissue and facilitate ventilation, drainage and mucocillary clearance through restored patency of sinus drainage pathways (ostia). However, some forms of CRS are driven by self-perpetuating immune-mediated processes, and in these cases FESS alone may not eradicate the disease [32] The goal of surgery in these cases is to both improve sinus patency and to improve access for topical medications [39].

Historical surgical techniques used to treat CRS included antral lavage, external, intranasal and trans-antral ethmoidectomy, external frontal sinus surgery and the Caldwell-Luc procedure [21]. The surgical management of CRS was revolutionised with technical advancements in endoscopic systems, the advent of powered instrumentation and the recognition of the importance of restoring paranasal sinus function by re-establishing the physiologic pattern of mucociliary clearance through the anatomical ostia. Endoscopic sinus surgery had been researched and practiced in Europe particularly in Germany and Austria since the 1960s [20]. The term FESS was first introduced into the English-speaking world in 1985 [20, 22].

Major complications related to FESS include severe bleeding, damage to intra-orbital structures, loss of vision, damage to the nasolacrimal duct and cerebrospinal fluid leak. Very serious complications include carotid artery injury, cerebral deficit, meningitis or death [10, 27, 38]. In the modern endoscopic era, several large series quote a very low major complication rate demonstrating that this procedure can be performed safely [10, 27, 38]. In a series of 78,944 primary FESS cases, the major complication rate was 0.36% [27]. Factors that have been associated with an increased risk of postoperative complications include patient age, disease severity, extent of operation, and the surgeon's experience [10, 27, 38, 42, 43].

4.8 FESS in Detail

Uncinectomy

The first step in FESS involves removal of the uncinate process. The uncinate bone is a projection of the ethmoid bone which forms part of the medial wall of the maxillary sinus and covers the anterior part of the opening of the sinus. Once an Uncinectomy is performed, the base of infundibulum (drainage pathway of the frontal sinus), the anterior wall of ethmoid bulla and part of the maxillary natural ostium are exposed. This creates a larger opening for drainage of the maxillary sinus and for medications to penetrate. The superior attachment of the uncinate can be left intact, particularly if attached to the skull base or middle turbinate, to avoid forming adhesions around the frontal recess.

Maxillary Antrostomy

The natural opening of the maxillary sinus is identified within the middle meatus. It is usually at the same level as the inferior margin of the middle turbinate, anterior to the ethmoid bulla. The size of the enlarged ostial opening can be varied depending on the severity of disease and whether there has been previous surgery.

Ethmoidectomy

Unlike the maxillary sinus, which is a single cavity with a single opening, the ethmoid sinus consists of multiple interconnected cells. The basal lamella, or horizontal portion, of the middle turbinate, which attaches to the lateral nasal wall, separates the ethmoid labyrinth into two distinct anatomical and physiologic compartments. The anterior group of cells drains its secretions into the common drainage pathway together, the infundibulum, with the maxillary & frontal sinuses. The posterior group of cells drain into the superior meatus beneath the superior turbinate. The main anatomical landmark for ethmoidectomy is the ethmoid bulla, the most anterior ethmoid air cell, that can be seen as a projection from the lateral nasal wall. The ethmoid sinus is opened in an anterior to posterior direction parallel to the skull base and then dissection along the roof of the ethmoid bone is often performed in a posterior to anterior direction to avoid injuring the skull base.

Frontal Sinusotomy

Intranasal frontal sinusotomy is a potentially dangerous procedure as it risks injury to the orbit and brain. To visualise the frontal recess area, it is necessary to remove the anterior ethmoid cells, located anteroinferior to the frontal sinus drainage pathway. Then the frontal recess is enlarged using manual or powered instruments. The following table outlines the different types of frontal sinus procedure with progressively larger ostiotomies made for increased drainage, classifed according to Draf [46] and given in Table. 4.2.

Sphenoidotomy

The sphenoid sinus can be opened safely around 12 mm above the choana (the posterior opening of the nasal cavity into the nasopharynx) just lateral to the midline of the septum. After identification of the ostia the opening is enlarged in a lateral and inferior direction. The optic nerve and the internal carotid artery are located in the lateral and posterior wall of the sinus. Review of a CT paranasal sinuses should be undertaken pre-operatively and care taken to avoid these structures as damaging them can be catastrophic. The anterior face of the sella turcica containing the pituitary gland is situated posterior to the sinus. The cavernous sinus is located lateral to the sella. Tumours of the pituitary gland may be resected through the sphenoid sinus, in conjunction with a neurosurgeon.

Table 4.2 Draf classification of frontal sinus procedures

Type I	Anterior ethmoidectomy and drainage of fontal recess
	Frontal ostium left intact
Type IIA	Removal of all ethmoidal cells protruding into the frontal sinus
	Creating an opening between the middle turbinate medially and the lamina papyracea laterally
Type IIB	Removal of the frontal sinus floor between the nasal septum medially and the lamina papyracea laterally
Type III	Bilateral Draf IIB with removal of the upper part of the nasal septum and intersinus septum

4.9 Scope of CFD in Virtual Surgery

The first anatomically accurate 3D generated model of airflow in the nasal cavity was described in 1995 [25]. Prior to CFD techniques in otolaryngology it was difficult to assess variation in air flow in the nasal cavity and sinuses pre- and post-operatively. Cadaveric studies were easier to simulate however post-mortem tissue shrinkage distorts the nasal airway anatomy [11].

In recent years, researchers have begun to explore the use of CFD in the nose, firstly by modelling airflow patterns [25], then through the study of common deformities, such as nasal bone fractures [8], septal deviation [6], septal perforations [29] and inferior turbinate hypertrophy [7]. Early attempts to model surgical procedures have been performed, including the effects of turbinate surgery [9] and sinus surgery [50]. However, in these instances, researchers have largely only assessed the results of 'virtual surgery' on CFD models. Post-operative changes have been modelled in cadaveric specimens [48], in patients with unilateral sinus surgery [8] in single case reports of nasal obstruction surgery [36, 52] and in small-scale series of patients [44]. This body of research continues to grow.

CFD is also being used to look at sinus irrigation and medication delivery post-surgery. Previous studies of sinus irrigation are limited by their reliance on subjective human interpretation without quantifiable data and an inability to assess more than one sinus at a time. Methods using contrast-enhancing dyes or radioisotopes followed by computed tomography or scintigraphic imaging have been able to demonstrate sinus distribution of liquids, but only represent residual volumes and do not represent dynamic fluid movement or account for numerous variables simultaneously [35].

Virtual surgery and pre-operative planning are emerging applications of CFD technology. Part of routine surgical planning is determining which structures should be surgically altered to provide the most relief from symptoms, while still preserving normal function. Theoretically extensive surgery that involves all the anatomic structures would provide the widest nasal cavity possible and therefore the greatest relief of symptoms. However, such an approach may disrupt other nasal functions (air heating, humidification, filtration) and is not performed. The risk of always taking a minimally invasive approach is that there will not be complete resolution of symptoms. CFD is likely to have a significant role in the future in objectively weighing up the risks versus benefits of procedures and tailoring the approach to each patient individually.

4.10 Validity of CFD in Modelling the Nasal Cavity

CFD allows for simulation of fluid and air flow within a complex geometric shape. CFD requires reconstruction of the flow domain, or nasal cavity, with segmentation of finite volumes. In contrast to rhinomanometry, CFD can define more specific parameters of nasal airflow and particle flow with high resolution. A Digital Imaging and Communications of Medicine (DICOM) dataset is required, derived from a computed tomography scan (CT scan) of the paranasal sinuses, with minimum slice thickness of 0.6mm.

The shape of the nasal cavity has enormous heterogeneity due to the characteristics of structures such as the septum, turbinates and nasal side wall making comparison of individual nasal cavities challenging. It is therefore difficult to distinguish what is physiological versus pathological airflow.

Studies looking at the validity of CFD as a modelling tool have addressed this by creating an average geometric shape and airflow pattern from a collection of healthy nasal cavities of subjects who did not report any impaired breathing. This allows comparison with abnormal subjects who report for example nasal obstruction [17].

Nasal aerodynamics was initially tested with physical models. CFD has been subsequently validated by comparison with physical models. Garcia et al reported on the interplay between symptoms and airflow patterns in a patient with atrophic rhinitis. A physical replica in silicone of the patient's nasal geometry was made and particle image velocimetry (PIV) was used to visualise and measure the flow field [14]. Doorly et al performed similar experiments [11]. These physical models have demonstrated variation in airflow patterns in each nasal region, with slower flow in the olfactory region, rapid and unstable flow in the turbinate region and predominantly laminar flow through the middle meatus [26]. This confirms the relationship between nasal geometry and airflow and corresponds well with CFD simulations.

4.11 Nasal Airflow

CFD has been successfully used to simulate and analyse nasal airflow. Tan et al found that airflow accelerated in the nasal valve region due to increased resistance, being the narrowest part of the nose; there is also increased wall shear stress in this region. Airflow slows posteriorly and then again increases in the nasopharynx. Most of the airflow travels through the middle meatus and lower olfactory cleft. During inhalation the airflow forms vortices and turbulence at the anterior and inferior sections of the nasal cavity, but these do not occur during the expiratory phase [45]. These findings have been corroborated in other studies. Zhao and Jiang studied twenty-two healthy subjects and performed CFD simulation during resting respiration. The values of nasal resistance measured by rhinomanometry correlated well with CFD simulations. They also noted that there is significant variation of nasal airflow pattern among healthy subjects [51].

Within the nasal cavity airflow can take one of three paths, firstly from the outer edge of the nostril to the olfactory cleft, or from the central nostril to the middle meatus or finally from the base of the nostril to the inferior meatus. The predominant pathway as discussed is via the middle meatus. When there is septal deviation the usual flow pathways are disturbed. Velocity increases and the pathway traverses around the septal deviation. The normal turbulent flow in the nasal valve disappears [6]. Similarly in the case of inferior turbinate hypertrophy, nasal airflow is obstructed.

In moderate to severe blockage a higher velocity and shear stress was observed at the nasopharynx and dorsal nasal cavity. Nasal valve function is virtually non-existent in this disease state [30].

CFD is contributing significantly to the understanding of normal nasal airflow and airflow in diseased states. Authors are now using this knowledge to perform virtual surgery to determine the best approach to achieve optimum clinical outcomes for patients.

4.12 Virtual Septoplasty

Septoplasty, to correct septal deviation, is one of the most commonly performed surgical procedures by ear, nose and throat surgeons. A deviated nasal septum is the most common cause of nasal airway obstruction and affects patient quality of life. Surgical correction has variable success and virtual surgery planning using CFD has the potential to improve success rates.

Moghaddam et al selected a patient with nasal obstruction due to septal deviation to illustrate the concept of virtual surgery. They determined if septal geometry was abnormal pre-operatively, virtually corrected the deviation and then estimated the post-surgical airflow using CFD. The average dimension of the nasal septum from 47 healthy individuals was compared to the geometry of the nasal obstruction patient to identify the precise locations where the septum was likely to be causing obstruction. There was good correlation of cross-sectional airspace and nasal resistance between virtual surgery and actual surgery, giving objective and comparable outcomes [33].

Burgos et at. performed virtual surgery encompassing septoplasty, turbinoplasty, spreader grafts, lateral crural K-flap and various combinations to determine the best improvement in nasal airflow. Outcomes included airflow distribution, velocity profiles, pressure and wall shear stress. Interestingly, the combination of all the procedures did not provide the best nasal airflow. Septoplasty plus turbinoplasty obtained the best results and turbinoplasty alone gave almost similar results [5]. Ozlugedik found similar results, following the virtual operation there was a decrease in maximal intranasal air speed and increased flow volume through the middle meatus. Following septoplasty and partial lateral turbinectomy the total nasal resistance was reduced significantly [37].

CFD has the potential to significantly contribute to the pre-operative planning of patients with nasal obstruction by forseeing the aerodynamic effects of the operation.

4.13 Virtual Inferior Turbinoplasty

Inferior turbinoplasty is a commonly performed surgical procedure, from the least invasive technique of radiofrequency ablation to the more aggressive submucosal debridement using a micro-debrider. CFD has been used perform virtual surgery on inferior turbinates aiming to predict the best technique to increase airflow. Studies to date have shown variable results as to the best location and amount of turbinate to reduce. Lee et al. performed virtual surgery and found that a full inferior turbinectomy was the most effective to improve nasal airflow but found inconsistency in the optimal location for partial turbinectomy with some patients benefitting most from a posterior turbinectomy [30].

Virtual surgery has also been performed on the middle turbinate. Total airflow significantly increased and nasal resistance decreased following middle turbinectomy. There was also significant increased average flux to the olfactory cleft of phenylethyl alcohol (an odorant) [1]. A comparison of total middle versus inferior turbinectomy showed that total inferior turbinectomy greatly increases nasal airflow and decreases airway resistance but may also impair nasal air-conditioning capacity to a greater extent.

Virtual surgery may have a role in testing the efficacy of newer techniques. For example, CFD has been used to evaluation pyriform turbinoplasty and nasal wall lateralization. Pyriform turbinoplasty is an endoscopically performed procedure in which there is a submucosal resection of the bone of the frontal process of the maxilla and lacrimal bone, there is no mucosal resection of the turbinate. It demonstrated that nasal resistance was reduced and there was a more homogenous distribution of airflow along the inferior and middle turbinate post virtual surgery [40]. Thus, CFD is a good tool for testing novel techniques and predicting clinical outcomes.

4.14 Virtual Sinus Surgery

Virtual surgery with CFD has been used to demonstrate improvement in paranasal sinus ventilation post operatively. Middle meatal antrostomy of varying sizes can be performed on 3D models and has shown that the airflow to these sinuses has increased by up to 50% depending on the size of the antrostomy [47]. The average airflow into the maxillary sinus may increase by 18ml/s [12]. Similarly, virtual sphenoidotomy has been shown to increase airflow in the sphenoid sinus significantly from <1% preoperatively to 17% post operatively [3].

It has been demonstrated that after FESS flux in the common meatus decreases significantly and in the middle meatus and continuous ethmoid region there is an increase of flow by 10%. Airflow velocity in the sinuses also increased significantly after virtual FESS. Total nasal resistance was decreased and airflow exchange increase in the nasal sinuses most markedly in the maxillary sinus, corroborating previous evidence [49].

4.15 Virtual Surgery Software

As the technology of virtual surgery using CFD becomes more accessible commercial products are being developed to increase its usability. Flowgy (www.flowgy.com) is one such program that integrates 3D model building from CT images, 3D virtual surgery tools including first person and endoscopic views, automatic generation of surface and volume CFD mesh, CFD simulations of airflow producing variables including pressure, velocity, temperature, nasal resistance and flow rate and production of reports [5]. This technology eliminates the need for multiple programs to complete the 3D modelling from 2D CT data, creation of the 3D mesh model and perform the simulation as well as a program to visualise, interpret and present the data. This process takes a lot of time, several hours to days, and is not easily accessible to ENT surgeons who are not trained to use this software. This latest integrated software aims to overcome this issue and improve accessibility.

4.16 Limitations

CFD is still an emerging technology and still has limitations. One of the main limitations is the time required to perform the simulation. Although this is improving with experience and improved software capabilities, it still takes a few days to construct a 3D computer model from CT data, creating a mesh model of this and performing fluid or airflow simulations. The CT scan required for CFD modelling does expose the patient to a degree of radiation that needs to be taken into consideration. MRI is less accessible, more expensive and time consuming and, due to the nasal cycle and long acquisition time, may not produce sharp mucosal boundaries.

One difficulty as mentioned previously is the variable geometry of the nasal cavity even in healthy individuals. The nasal lining and nasal physiology is dynamic and regularly changes the geometry and functioning of the nasal cavity. Local blood flow, congestion and active mucous secretion cannot be easily simulated. CFD can only provide information from one static moment and assumes the nasal cavity as a single rigid structure.

In terms of performing virtual surgery, there are also limitations. As mentioned previously usability and accessibility for ENT surgeons has been difficult due to the need to use multiple different programs to obtain results, the need to learn to use these programs and even once this is mastered, the time it takes to produce a result has not been practical in the general ENT outpatient setting and to date has been used for research purposes only. With the advancement of integrated programs there is now software available that will streamline this process and make virtual surgery a real possibility in ENT practice in the near future.

4.17 Summary

CFD is still an emerging technology and has limitations in the application to virtual surgery of the nasal cavity. One of the main limitations is the time required to perform the simulation. Although this is improving with practice and improved software capabilities, it still takes a few days to construct a 3D computer model from CT data, creating a mesh model of this and running fluid or airflow simulations. The CT scan required for CFD modelling does expose the patient to a degree of radiation that needs to be taken into consideration. However, MRI is less accessible, more expensive and time consuming.

One difficulty as mentioned previously is the variable geometry of the nasal cavity even in healthy individuals. The nasal lining and nasal physiology is dynamic and regularly changes the geometry and functioning of the nasal cavity. Local blood flow, congestion and active mucous secretion cannot be easily simulated. CFD can only provide information from one static moment and assumes the nasal cavity as a single rigid structure.

References

1. S. Alam, C. Li, B. K. Z. K. L. T. Impact of middle turbinectomy on airflow to the olfactory cleft: a computational fluid dynamics study. Am. J. Rhinol. Allergy. **33**, 3, 263–268 (2019)
2. C. Aldren, N.S. Toley, Further studies on nasal sensation of airflow. Rhinology **29**, 59–55 (1991)
3. H. Bahmanzadeh, O. Abouali, M. Faramarzi, G. Ahmadi, Numerical simulation of airflow and micro-particle deposition in human nasal airway pre- and post-virtual sphenoidotomy surgery. Comput. Biol. Med. **61**, 8–18 (2015)
4. P. Boonmak, S. Boonmak, M. Laopaiboon, Deliberate hypotension with propofol under anaesthesia for functional endoscopic sinus surgery (fess). Cochrane Database Syst. Rev. **10** (2016)
5. M.A. Burgos, S.R.E.D.P.C.F.V.C.P.F.E.O.F. Sevilla GarcáA, Virtual surgery for patients with nasal obstruction: Use of computational fluid dynamics (mecomland®, digbody® & noseland®) to document objective flow parameters and optimise surgical results. Acta Otorrinolaringol Esp. **69**(3), 125–133 (2018)
6. X.B. Chen, H.P. Lee, V.F. Chong, Y. de Wang, Assessment of septal deviation effects on nasal air flow: a computational fluid dynamics model. The Laryngoscope **119**(9), 1730–1736 (2009)
7. X.B. Chen, H.P. Lee, V.F. Chong, Y. de Wang, Impact of inferior turbinate hypertrophy on the aerodynamic pattern and physiological functions of the turbulent airflow - a cfd simulation model. Rhinology **48**(2), 163–168 (2010)
8. X.B. Chen, H.P. Lee, V.F. Chong, Y. de Wang, Aerodynamic characteristics inside the rhino-sinonasal cavity after functional endoscopic sinus surgery. American journal of rhinology & allergy **25**(6), 388–392 (2011)
9. X.B. Chen, S.C. Leong, H.P. Lee, V.F. Chong, D.Y. Wang, Aerodynamic effects of inferior turbinate surgery on nasal airflow-a computational fluid dynamics model. Rhinology **48**(4), 394–400 (2010)
10. T.-W. Chou, P.-S. Chen, H.-C. Lin, K. Lee, H.-T. Tsai, J.-C. Lee, Y.-S. Leu, Y.-P. Wang, Multiple Analyses of Factors Related to Complications in Endoscopic Sinus Surgery **79** (2015)
11. D. Doorly, D.J. Taylor, P. Franke, R.C. Schroter, Experimental investigation of nasal airflow. J. Eng. Med. **222**(4), 439–453 (2008)

12. D.O. Frank, A.M. Zanation, V.H. Dhandha, K.A. McKinney, G.M. Fleischman, C.S. Ebert Jr., B.A. Senior, J.S. Kimbell, Quantification of airflow into the maxillary sinuses before and after functional endoscopic sinus surgery. Int. Forum Allergy Rhinol. **310**, 834–840 (2013)
13. J.C. Fuller, C.H. Bernstein, P.A. Levesque, R.W. Lindsay, Peak nasal inspiratory flow as an objective measure of nasal obstruction and functional septorhinoplasty outcomes. JAMA Facial Plast. Surg. **20**(2), 175–176 (2017)
14. G.J.M. Garcia, G. Mitchell, N. Bailie, D. Thornhill, J. Watterson, J.S. Kimbell, Visualization of nasal airflow patterns in a patient affected with atrophic rhinitis using particle image velocimetry. J. Phys.: Conf. Ser. **85** (2007)
15. D. Gerecci, F.J. Casanueva, J.C. Mace, A. Annen, D.M. Barrett, M.M. Kim, T.D. Wang, T.L. Smith, M. Loyo, Nasal obstruction symptom evaluation (NOSE) score outcomes after septorhinoplasty. Laryngoscope (2019)
16. L.F. Grymer, O. Hilberg, O.F. Pedersen, T.R. Rasmussen, Acoustic rhinometry: values from adults with subjective normal nasal patency. Rhinology **29**(1), 35–47 (1991)
17. T. Hildebrandt, J.J. Bruening, H. Lamecker, S. Zachow, W.J. Heppt, N. Schmidt, L. Goubergrits, Digital analysis of nasal airflow facilitating decision support in rhinosurgery. Facial Plast. Surg. **35**(1), 003–008 (2019)
18. C. Hopkins, J.P. Browne, R. Slack, V. Lund, P. Brown, The lund-mackay staging system for chronic rhinosinusitis: How is it used and what does it predict? Otolaryngol. Head Neck Surg. **137**(4), 555–561 (2007)
19. J. Jose, A.P. Coatesworth, Inferior turbinate surgery for nasal obstruction in allergic rhinitis after failed medical treatment. Cochrane Database of Systematic Reviews **12** (2010)
20. K.J. Kane, The early history and development of endoscopic sinonasal surgery in australia: 1985–2005. Australian Journal of Otolaryngology **1**, 1 (2018)
21. D. Kennedy, W. Bolger, S. Zinreich, *Diseases of the Sinuses: Diagnosis and Management* (B.C, Decker, 2001)
22. D.W. Kennedy, S.J. Zinreich, A.E. Rosenbaum, M.E. Johns, Functional endoscopic sinus surgery: Theory and diagnostic evaluation. JAMA Otolaryngology-Head & Neck Surgery **111**(9), 576–582 (1985)
23. J. Kennedy, M.A. Hubbard, P. Huyett, J. Patrie, L. Borish, S. Payne, *Sino-nasal Outcome Test (SNOT-22): A Predictor of Post-surgical Improvement in Patients with Chronic Sinusitis*, vol. 111 (2013)
24. J. Kennedy, M. Hubbard, P. Huyett, J. Patrie, L. Borish, S. Payne, Sino-nasal outcome test (snot-22): A predictor of post-surgical improvement in patients with chronic sinusitis. Ann. Allergy Asthma Immunol. **111**(4), 246–251 (2014)
25. K.P. Keyhani, P.W. Scherer, M.M. Mozell, Numerical simulation of airflow in the human nasal cavity. J. Biomech. Eng. **117**(4), 429 (1995)
26. J. Kook Kim, J.H. Yoon, C. Hoon Kim, T. Wook Nam, D. Bo Shim, H. Ae Shin, Particle image velocimetry measurements for the study of nasal airflow. Acta Otolaryngol. **126**(3), 282–287 (2006)
27. J.G. Krings, D. Kallogjeri, A. Wineland, K.G. Nepple, J.F. Piccirillo, A.E. Getz, Complications of primary and revision functional endoscopic sinus surgery for chronic rhinosinusitis. The Laryngoscope **124**(4), 838–845 (2014)
28. S. Kulkarni, V. Kulkarni, b. k. b. m. b. c. s. v. Endoscopic septoplasty: A retrospective analysis of 415 cases. Ind. J. Otolaryngol. Head Neck Surg. **67**(3), 248–254 (2015)
29. H.P. Lee, R.R. Garlapati, V.F. Chong, D.Y. Wang, Effects of septal perforation on nasal airflow: computer simulation study. The Journal of laryngology and otology **124**(1), 48–54 (2010)
30. H.P. Lee, H.J. Poh, C. F. W. D. Changes of airflow pattern in inferior turbinate hypertrophy: a computational fluid dynamics model. Am. J. Rhinol. Allergy **23**(2), 153–158 (2009)
31. M.M. Likness, J.F. Pallanch, D.A. Sherris, H. Kita, T.L. Mashtare, J.U. Ponikau, Computed tomography scans as an objective measure of disease severity in chronic rhinosinusitis. Otolaryngology-Head and Neck Surgery **150**(2), 305–311 (2013)
32. Y. Matsuwaki, T. Ookushi, D. Asaka, E. Mori, T. Nakajima, T. Yoshida et al., Chronic rhinosinusitis: risk factors for the recurrence of chronic rhinosinusitis based on 5-year follow-up after endoscopic sinus surgery. Int. Arch. Allergy Immunol. **146**(Suppl 1), 77–81 (2008)

33. M. Moghaddam, G. Garcia, F.-I. D. K. J. . R. J. Virtual septoplasty: a method to predict surgical outcomes for patients with nasal airway obstruction. Int. J. Comput. Assist. Radiol. Surg. (2020)
34. S. Mucci, A. Sismanis, Inferior Partial Turbinectomy: An Effective Procedure for Chronic Rhinitis **73** (1994)
35. D.E. Olson, R. B. RL, H. J. Radiographic comparison of three methods for nasal saline irrigation. The Laryngoscope **112**, 1394–1398 (2002)
36. J.S. Rhee, C. D. e. a. Role of virtual surgery in preoperative planning: assessing the individual components of functional nasal airway surgery. Arch. Facial Plast. Surg. **16**, 354–359 (2012)
37. S. Ozlugedik, G. Nakiboglu, C. S. A. E. E. T. S. A. I. T. Numerical study of the aerodynamic effects of septoplasty and partial lateral turbinectomy. The Laryngoscope **118**(2) (2009)
38. V. Siedek, E. Pilzweger, C. Betz, A. Berghaus, A. Leunig, Complications in endonasal sinus surgery: a 5-year retrospective study of 2,596 patients. Eur. Arch. Otorhinolaryngol. **270**(1), 141–148 (2013)
39. K. Snidvongs, L. Kalish, R. Sacks, R. Sivasubramaniam, D. Cope, R.J. Harvey, Sinus surgery and delivery method influence the effectiveness of topical corticosteroids for chronic rhinosinusitis: systematic review and meta-analysis. Am. J. Rhinol. Allergy **27**(3), 221–233 (2013)
40. F. Sommer, D. Simmen, B. H. J. N. H. T. S. M. L. J. Effects of nasal wall lateralization and pyriform turbinoplasty on nasal air conditioning. Otorhinolaryngol. Head Neck Surg. **2**(2), 1–5 (2016)
41. C. Spiekermann, E. Savvas, R. C. S. M. Adaptation and validation of the nasal obstruction symptom evaluation scale in german language. Health Q. Life Outcomes **172** (2018)
42. J.A. Stankiewicz, Complications in endoscopic intranasal ethmoidectomy: an update. Laryngoscope **99**, 686–690 (1989)
43. J.A. Stankiewicz, Cerebrospinal fluid fistula and endoscopic sinus surgery. Laryngoscope **101**, 250–256 (1991)
44. C.D. Sullivan, G.J. Garcia, D.O. Frank-Ito, J.S. Kimbell, J.S. Rhee, Perception of better nasal patency correlates with increased mucosal cooling after surgery for nasal obstruction. Otolaryngol. Head Neck Surg. **150**(1), 139–147 (2014)
45. J. Tan, D. Han, J. Wang, T. Liu, T. Wang, H. Zang, Y. Li, X. Wang, Numerical simulation of normal nasal cavity airflow in chinese adult: a computational flow dynamics model. Eur. Arch. OtoRhinoLaryngol. **269**, 881–889 (2012)
46. R. Weber, W. Draf, B. Kratzsch, W. Hosemann, S.D. Schaefer, Modern concepts of frontal sinus surgery. Laryngoscope (2001)
47. Wofford MR, Kimbell JS, F.-I. D. D. V. M. K. F. G. E. C. J. Z. A. S. B. A computational study of functional endoscopic sinus surgery and maxillary sinus drug delivery. Rhinology **53**(1), 41–48 (2015)
48. G.X. Xiong, J.M. Zhan, K.J. Zuo, L.W. Rong, J.F. Li, G. Xu, Use of computational fluid dynamics to study the influence of the uncinate process on nasal airflow. The Journal of laryngology and otology **125**(1), 30–37 (2011)
49. G.X. Xiong, J.F. Li, G.L. Jiang, J.M. Zhan, L.W. Rong, G. Xu, Effect of endoscopic sinus surgery on airflow of the nasal cavity and paranasal sinuses: a computational fluid dynamics study. Chin. J. Otolaryngol. Head Neck Surg. **44**(11), 911–917 (2009)
50. K. Zhao, E.A. Pribitkin, B.J. Cowart, D. Rosen, P.W. Scherer, P. Dalton, Numerical modeling of nasal obstruction and endoscopic surgical intervention: outcome to airflow and olfaction. Am. J. Rhinol. **20**(3), 308–316 (2006)
51. K. Zhao, J. Jiang, What is normal nasal airflow? a computational study of 22 healthy adults. Int. Forum Allergy Rhinol. **4**, 6435–6446 (2014)
52. J.H. Zhu, K.M. Lim, K.T. Thong, H.P. Lee, Assessment of airflow ventilation in human nasal cavity and maxillary sinus before and after targeted sinonasal surgery: a numerical case study. Respir Physiol. Neurobiol. **192**, 29–36 (2014)

Chapter 5
Computational Reconstruction of the Human Nasal Airway

Jose Luis Cercos-Pita

Abstract Reconstructing the human airways begins with image acquisition which involves the procurement of medical images typically from computed tomography (CT) or magnetic resonance (MR) imaging. This produces a 3D matrix of a series of 2D cross-sections, that contains information about tissues and anatomic structures. Reconstructing the airway from the medical scans involves segmentation of the region of interest. The segmentation step extracts a contiguous airway region which is ready for 3D model processing. This chapter discusses CT and MR scan imaging, followed by some common segmentation algorithms based on thresholding, edge detection, and region characteristics.

5.1 Introduction

Reconstructing the human airways can be divided into three stages: image acquisition, segmentation and surface/volume reconstruction. A typical flow chart depicting some common steps is given in Fig. 5.1. Image acquisition involves the procurement of medical images that can be obtained from various sources such as computed tomography (CT) or magnetic resonance (MR) imaging . Although non-invasive, CT scans deliver hundreds of X-ray beams through the body to formulate a 3-dimensional image for interpretation. Due to the nature of this modality, radiation exposure may be a concern. MR imaging also produces a 3 dimensional image, however instead of radiation, incorporates the use of magnetic fields and radio frequency pulses and therefore does not expose the patient to ionizing radiation. MRI also distinguishes mucosal structures more accurately than CT images, however assessment of the bony nasal turbinates often needs to be inferred from its surrounding mucosa, since actual bone structures appear dark (low-signal) on MRI scans and may not be visible. A 3D matrix of a series of 2D cross-sections, separated by an interval distance, contains

J. L. Cercos-Pita (✉)
Postdoctoral researcher, Department of Surgical Sciences, Hedenstierna laboratory, Uppsala Universitet, Uppsala, Sweden
e-mail: joseluis.cercospita@surgsci.uu.se

K. Inthavong et al. (eds.), *Clinical and Biomedical Engineering in the Human Nose*, Biological and Medical Physics, Biomedical Engineering,
https://doi.org/10.1007/978-981-15-6716-2_5

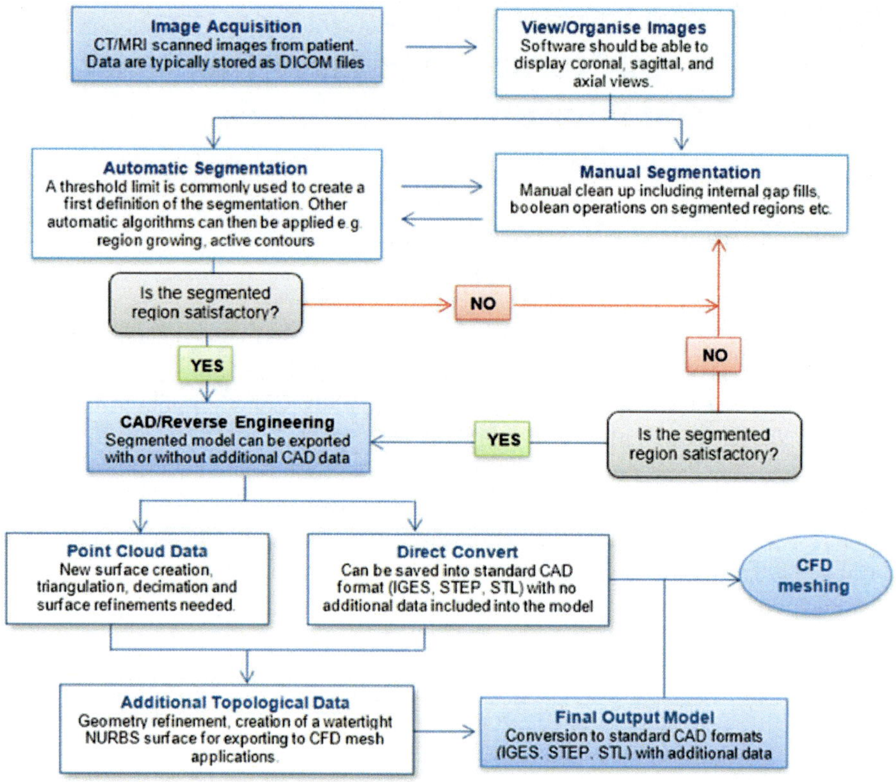

Fig. 5.1 A generic work flow is shown as a flowchart. After segmentation the output file can be directly read into a CFD meshing software or additional topological data can be included into the model. An advantage of including the topological data is that the file becomes more compatible with CFD software

information about tissues and structures which are distinguished from one another by differences in brightness or greyscale. Reconstructing the morphological structures from the medical image series involves segmentation of the region of interest. Segmentation involves the partitioning of these images into a number of homogeneous segments, such that any two neighboring segments yields a heterogeneous segment. Some common algorithms based on thresholding, edge detection, and region characteristics are explored in this chapter.

5.2 Acquisition of Imaging Data

Unless only a simplified model is required, which are typically restricted to research applications, the representative surface generation of the upper respiratory tract is invariably seeded by 3D volumetric data. Concepts behind 3D volumetric data are complex and exceeds the scope of this book. Fortunately, in medical applications it is commonly accepted to use a simplified case of voxels-based data, i.e. 3D Lattice discretized data. This approach has been widely adopted in computer graphics, and has extended to several medical applications [13]. A voxel is the natural 3D extension of an image pixel (2D), and indeed voxel-based data can be interpreted from a set of aligned scanned images, so-called slices, which is the base unit of a common Medical Imaging formatting standard called DICOM [40]. Since each cross-sectional slice is separated by a known thickness, described in the pixel data in the 2D-slice, changes into a volume element known as the voxel.

Hence, the first step of surface reconstruction is the acquisition of the images set, which also contains information of the type of material at each voxel, like bones, metal implants, or even air. The materials are represented by its density, which are captured by the imaging modality. For the nasal cavity, we consider two main acquisition methodologies: Computed Tomography (CT) and Magnetic Resonance (MR).

5.2.1 Computed Tomography

Roentgen [53] first described in 1895 x-ray radiation as a disruptive technology that would completely transform the noninvasive diagnosis techniques [16]. Even though the 2D radiographies were a great step forward, 3D applications were also investigated from the very beginning, leading to the theory published by Radon [51, 52], in 1917 which allows reconstruction of 3D images on top of projection radiographies taken from different angles [12]. Computed Tomography (CT) was later developed in 1972 by Sir Godfrey Hounsfield, which involved using multiple x-rays to create a series of planar cross-sectional images along an axis. The simplest CT scanner is a punctual x-ray beam emitter, with an associated punctual radiation sensor. The emitter is calibrated to emit at a known intensity $\mathcal{I}_0$, such that the measuring intensity reaching the sensor, $\mathcal{I}$, is an indirect measurement of the radiodensity. The radiodensity is therefore the material attenuation factor, and various relative intensities correspond to various tissue densities. For example, bone absorbs more X-rays than soft tissue and therefore has the most contrast relative to the air space.

During a typical CT procedure, the patient is positioned supine on a table. The table then moves the patient through a rotating gantry (a toroid-shaped device), which houses the x-ray tube and electronic x-ray detectors. Multiple x-rays are taken in thin cross-sections in the region of interest along the person's body forming slices (like slicing a loaf of bread). The detectors collect the x-ray information from each cross-section, transferring them onto computer that combines them into a single image. Most modern incarnations of CT scanners are based on Helical CT technology [26], where a linear array of collimated emitters, producing parallel X-Ray beams, is disposed in opposition of the sensors array (Fig. 5.2). Both, emitters and sensors rotate to produce the different α projections, while the patient table moves to produce the slices. In helical (spiral) CT continuous rotation occurs which averts the need to stop and start, reducing the radiation dose. These helical CT scanners provide quick

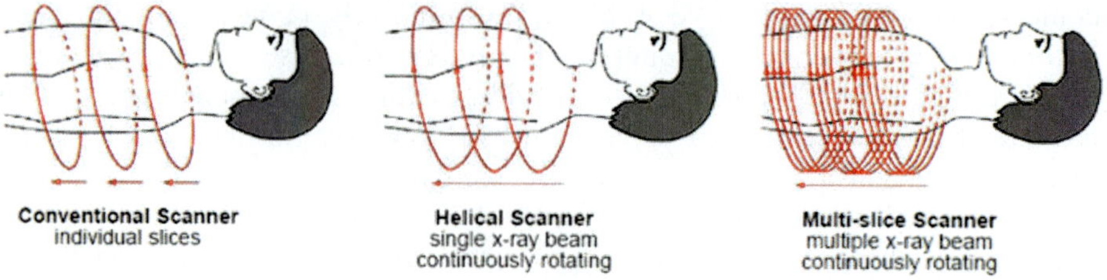

Fig. 5.2 Different types of computed tomography scanners

Fig. 5.3 Greyscale values interpreted as Hounsfield (Hu) units

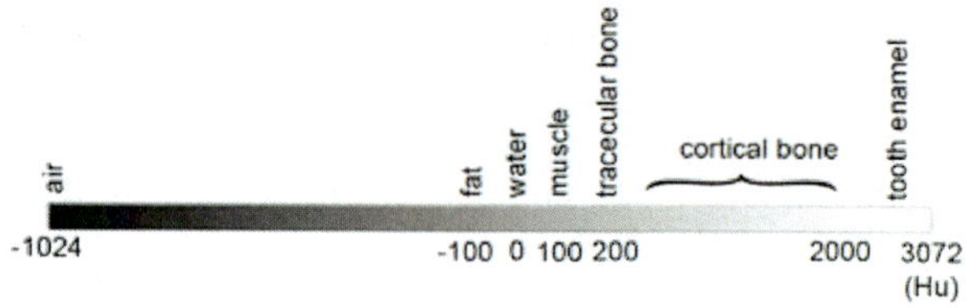

scans of entire anatomic regions such as the lungs, where scans must be taken with the patient in a single position.

As the x-ray beam passes through the body, it is absorbed or redirected (i.e. scattered) by different anatomical structures, which reduces the strength of the x-ray beam received. The degree to which an x-ray beam is reduced is called attenuation and, in each image, pixels are assigned a numerical value (gray scale) given as Hounsfield units (Fig. 5.3).

CT has several strengths that make it a popular choice amongst clinicians. CT machines are relatively inexpensive, and the associated operational costs are lower than other 3 dimensional modalities like MR [57]. CT also does not suffer from significant technician variability in the way that ultrasound does. The imaging can be completed in five minutes, and provides images with excellent spatial resolution between 0.5 and 1.0 mm [32]. The resulting contrast is quite homogeneous and well calibrated by the Hounsfield units [35].

On the other hand, there are several disadvantages to CT imaging: x-ray produces ionizing radiation as previously described; the contrast resolution is low which means that slight differences in density of soft tissues may not be detected; and it is prone to metal artifacts. While the low contrast resolution can be neglected in the context of nasal cavity reconstructions, the effects of metal artifact needs to be considered in the reconstruction pipeline and is discussed in Sect. 5.3.1.

5.2.2 Magnetic Resonance

An MR scanner uses magnetic fields and radio waves to obtain cross-sectional images of a section of the body in a similar fashion to CT imaging. The main difference is that

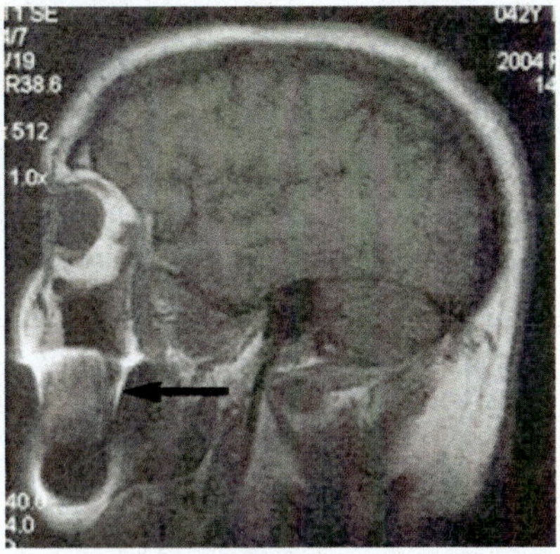

Fig. 5.4 A coronal slice with metal artifacts caused by metallic dental implants. Extracted from Erasmus et al. [19], licensed under CC-BY 4.0 terms

the emitting energy is from induced nuclear magnetic resonance, and therefore does not expose the patient to ionizing radiation [2, 50]. An MR scanner also consists of a table that slides into a large cylinder. Inside the cylinder is a large superconducting magnet, which is usually made up of many coils of wire (typically 1.5 T, which is approximately 50,000 times greater than the Earth's magnetic field). When the MR is in operation an electric current passes through the coils creating a weak oscillating magnetic field as well as a strong static one.

The main benefit of MR is the lack of ionizing radiation. Additionally, MR offers a high contrast resolution, however this is a negligible benefit when nasal cavities are to be reconstructed as there already is a large difference between air and tissues. However, MR imaging is slower and a significantly more expensive methodology. By nature, MR is not based on photon counting, and therefore there is no reference values available for use as threshold settings, (e.g. the Hounsfield units in CT). This makes it harder to automate the reconstruction task. MR is perniciously affected by the presence of ferromagnetic materials, which critically alter the magnetic field.

In Fig. 5.4 a metal artifact caused by dental implants is presented. Metal artifact reduction in MR is a complex task, which usually requires a great a priori knowledge of the implant [24], with a very little room for automation.

5.3 Image Enhancement

Before addressing image segmentation, it is common to perform an image enhancement. This is one of the most computationally intensive parts of the reconstruction process. Unsurprisingly, it is not uncommon to find applications where image enhancement has been neglected, especially in those provided by CT machine manufacturers, which are designed to visualize certain structures right after the image acquisition. When a 3D surface is required for Computational Fluid Dynamics simulations, even small differences in the 3D surface can potentially lead to large differences in the resulting flow. Hence, some image pre-processing is almost certainly required in this application.

We focus on the correction of image artifacts and noise reduction, an image enhancement process that is generally applied to all nasal cavity reconstruction applications [21]. Contrast enhancement is another image enhancement technique that is occasionally used, depending on the selected segmentation technique. These are discussed in Sect. 5.4. In the following, the image enhancement is discussed through i) Metal Artifact Reduction, and ii) smoothing (noise reduction). It is advisable to keep these techniques separated, since both the governing physical phenomena and the resulting image artifacts have completely different natures.

5.3.1 Metal Artifact Reduction (MAR)

Metal artifacts in CT are produced by a physical phenomenon known as beam hardening, first described bv Brooks et al. [4] in 1976. This occurs when dielectric materials like metals are featured by a non-uniform attenuation factor, $\mu(\mathcal{E})$, which depends on the photon energy, $\mathcal{E}$. An example of a metal artifact affected image is demonstrated in Fig. 5.5.

Common metallic artifacts in the context of nasal cavity reconstruction are dental implants. Both CT and MRI scans can be significantly affected by metal implants. However, the artifacts obtained significantly differs depending on whether CT or MR has been performed, mainly because they are associated to different physical phenomena. Herein we focus on metal artifact reduction techniques for CT, presenting two approaches: alternative reconstruction methods, and reconstructed images correction. Currently, there is no single technique which can remove the artefacts produced by metal implants, without producing new artifacts. The methods manipulate the raw sinogram data measured by the CT machine [31, 60].

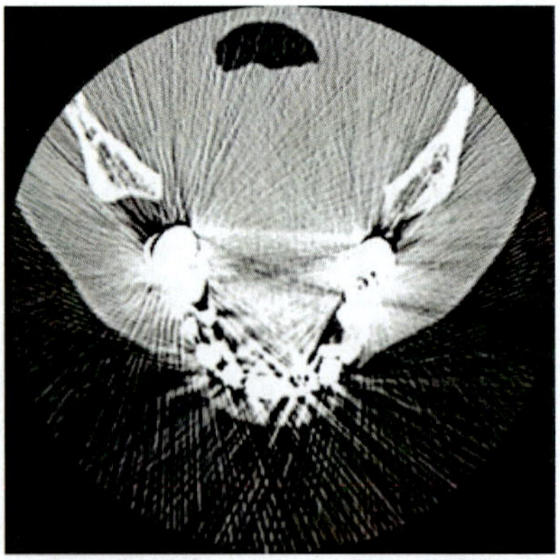

Fig. 5.5 Arbitrary axial slice with metal artifacts caused by metallic dental fillings

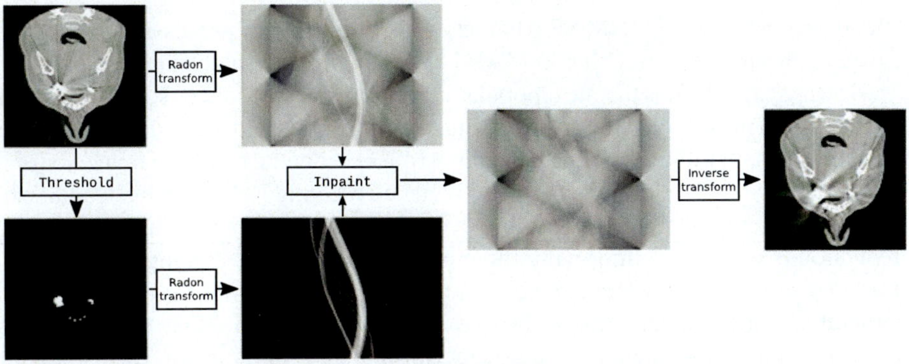

Fig. 5.6 Schematic view of traditional MAR algorithm. Extracted from [9]

An example of the 'reconstructed images correction' method is schematically depicted in Fig. 5.6. This method is based on the sinogram inpaint, described in Glover et al. [23] and formalized later in Kalender et al. [27].

The process begins with the original image affected by metal artifacts, $\mu_0(x, y)$, from which the metal implants imprint is extracted by a threshold operation. Effectively, a binary image is forward projected, producing a non-binary sinogram, i.e. the resulting sinogram is no longer a mask. One would be tempted to consider as the metal mask the non-null set of pixels, which is in fact a consistent approach. However, that is not necessarily the most accurate approach, and depends both in the resolution of the image and the implementation of the Radon transform. Indeed, the vast majority of Radon transform implementations are based on rotating and resampling the image, subsequently integrating by columns. Considering all the non-null pixels of the result can artificially enlarge the area affected.

The next step of the MAR algorithm is inpainting the original sinogram, where all the pixels marked by the metal mask sinogram, to produce a corrected sinogram. In the most traditional MAR algorithms the inpaint process consists of row by row linear interpolations. Finally, the corrected sinogram is projected back to produce the corrected image.

Metal Artifact Reduction therefore relies on a key concept; beam rays passing through a metal body are affected by beam hardening, so such information must be considered invalid. Unfortunately, as can be appreciated in Fig. 5.6, the resulting image is far from perfect, effectively removing former artifacts, exposing at the same time new artifacts. This led some researchers to implement different inpaint methodologies which may eventually avoid inducing new artifacts [15, 29, 54, 62, 64]. Bal and Spies [1] drafted the possibility of pre-processing the image to produce an uncorrected sinogram, better suited to become inpainted without producing new artifacts. More specifically, an adaptive filter was applied to the original image to smooth the white and dark streaks induced by the metal implants. This idea was extended later by Meyer et al. [38], introducing for the first time the Normalized-

Metal Artifact Removal (NMAR). In fact, NMAR consists of processing the original image with such adaptive-filtered MAR [1], producing the so-called prior-image.

Nowadays NMAR is the most popular algorithm, but still carries some limitations as it does still introduce some artifacts. Indeed, Meyer himself suggested 2 years later a postprocessing filter [39]. This idea of post-processing the resulting image to try to correct the remaining artifacts has been revisited recently [22], taking advantage on the rapid development that of learning over the last years. Traditional metal artifact reduction is effectively improving the quality of the images, making them suitable for the reconstruction, when the metal implants are relatively far from the important structures. Unfortunately, this technique is inadequate when either metal implants are close to the nasal cavity, e.g. upper incisor implants, or there are dense materials which are not causing significant metal artifacts. In this case the application of metal artifact reduction will introduce artifacts without any overall improvement. Finally, it has been demonstrated in the past that NMAR can be successfully applied for the correction of head and neck CT slices [30], becoming probably a common feature of nasal cavity surface reconstruction applications in a close future.

5.3.2 *Edge-Preserving Smoothing*

In many cases of imaging, the signal-to-noise ratio is low enough to inexorably produce imagery affected by noise, which perniciously affects the segmentation process, making noise reduction a common step within nasal cavity surface reconstruction pipelines. Unfortunately, reducing image noise while preserving sharp edges is still nowadays an open question in computer science, something in fact very surprising given that there is literature about this topic reported by Weiner as early as 1949 [63]. This gives an idea of the complexity of the problem. Herein we introduce some popular algorithms, discussing their potential strengths and caveats. This is not intended to be an extensive review of the vast variety of image denoising methodologies already discussed in the literature, but an overview to introduce the reader to get involved in the topic.

5.3.2.1 Gauss Filter

The most popular smoothing technique is the Gauss filter which involves a convolution of each pixel. The Gaussian filter has demonstrated to be a powerful and efficient way to remove noise. Unfortunately, it removes both edges and image features with the same efficiency, often producing blurred images, as demonstrated in Fig. 5.7b. The Gauss filtering provides noise-free blurry images, where both sharp edges and thin structures are heavily affected, and sometimes even destroyed.

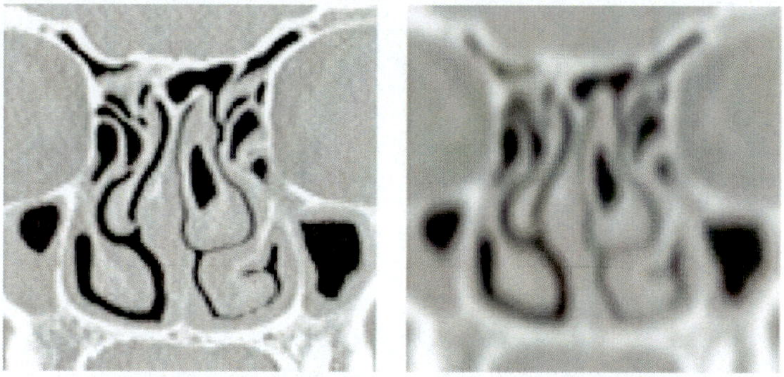

(a) original image (b) Gauss-filter

Fig. 5.7 **a** An arbitrary coronal slice of a nasal cavity, affected by noise and fuzzy thin features. **b** The same coronal slice of after applying a Gauss filter

5.3.2.2 Alternative Convolutions

The remarkable noise-removal performance of the Gauss filter motivated the development of some techniques to exploit the strengths of convolutions, while adding edge-preserving features. One of the most popular approaches is the Median filter [25]. In the Gaussian filter, the convolution is a weighted average of each pixel on top of its closest neighbours, whereas the Median filter considers the median of the closest neighbours. The Median filter has significantly improved edge-preserving features, at the cost of diminished noise removal effectiveness. In Fig. 5.8b, the result of applying a Median filter is depicted.

Another widely used convolution-based approach is the Bilateral filter [59] which is based on modifying the convolution to include radiosity differences, and is demonstrated in Fig. 5.8c. The Bilateral filter retains the excellent noise-removal performance from the Gauss filter, while preserving sharp edges. However, tuning the bilateral filter σ_μ parameter is not a trivial task. Furthermore, the convolution kernel, W_{σ,σ_μ}, is not a geometrical function and instead depends on the image radiosity itself. This makes it impossible to use a pre-computed template to integrate at each pixel, significantly increasing the computational cost.

5.3.2.3 Adaptive Filtering

Similar to convolutions, the benefits of Gauss filter noise removal performance were applied with subsequent filters to restore perniciously affected sharp edges, with a number of adaptive filters. Perhaps the most traditional approach is the Unsharp Mask, widely used in photography [55]. This filter takes a blurred image and produces a sharpened version with dependence on a filter intensity factor. The filter measures the unsharping effect of the Gauss filter blurring the image again, subsequently

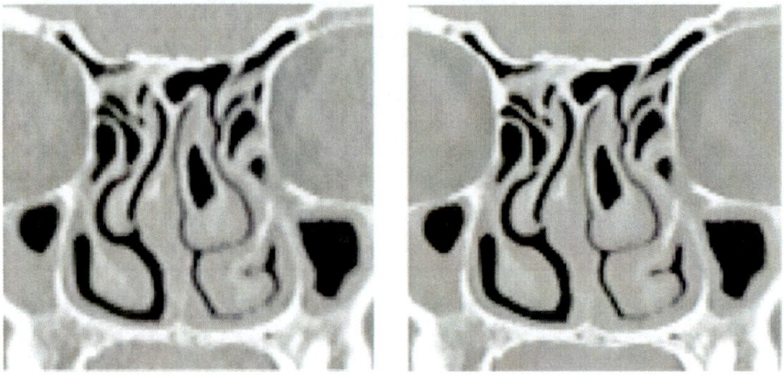

(a) Median Filter (b) Bilateral Filter

Fig. 5.8 Resulting images after applying different convolutions. **a** Median filter with 3 pixel diameter. **b** Bilateral filter with $\sigma = 2$ and $\sigma_\mu = 1/10$ normalized pixel radiosity, $0 \leq \mu \leq 1$, is considered

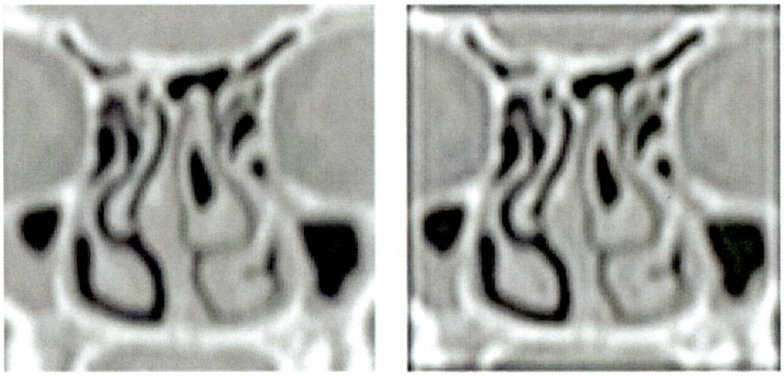

(a) (b)

Fig. 5.9 Resulting images after applying adaptive filters. **a** Unsharp mask **b** Wiener filter

subtracting the image itself. Such information can be therefore be used with an aim to invert the effect of the previous Gauss filter, provided that the noise has been already removed. Since this method is based only on Gauss convolution, it is computationally efficient. Unfortunately, as it increases the number of parameters to be tuned, its efficiency drops and is outperformed by other methods. Another popular adaptive filter is the image deconvolution, usually known as Wiener filter [63] which approximates the original image, inverting the convolution in frequency space. Figure 5.9a demonstrates the resulting image for Unsharp mask, while Fig. 5.9b demonstrates the Weiner filter. The latter has been able to partially restore the sharpen edges. It is important to note that the Wiener filter produces some artifacts. For this reason, most modern implementations offer a stabilization hyperparameter, which can be even stochastically computed [43].

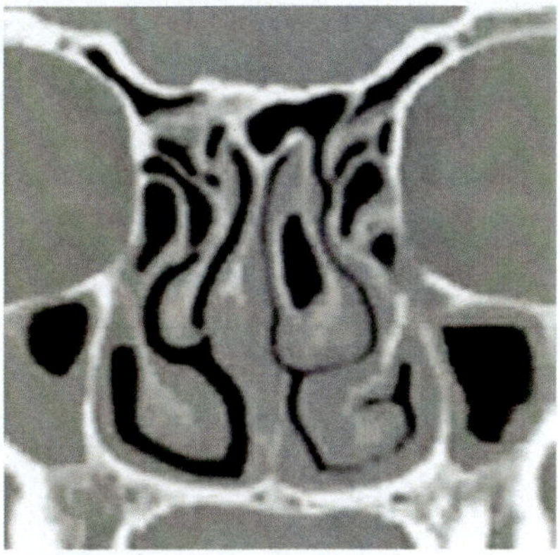

Fig. 5.10 Image filtered by Wavelets (DB1 function)

5.3.2.4 Wavelets

Wavelets are a generalization of Fourier Transform, where functions other than $\exp(i\omega t)$ can be considered. The underlying idea however, is the same; mitigating modes that are widely related with white noise. Wavelets have shown good edge-preserving smoothing properties for many years [36]. Figure 5.10 demonstrates an example of a smoothed image applying a *DB1 function* within VisuShrink Wavelets model. This filter is certainly prone to produce staircasing artifacts. There is no doubt that the resulting image has been significantly de-noised, while the sharp edges and thin structures remained almost unaltered.

The resulting image depends on the wavelet function, on the modes mitigation model, and even on the hyperparameters selected for each model. Indeed, tuning Wavelets models for image denoising can be a difficult task. For this reason, some researchers are developing machine learning aided Wavelets filters [61].

5.3.2.5 Anisotropic Diffusion

A different yet very promising approach is considering the problem as a differential equation, and more specifically as a diffusion process [47]. It is well known that pure diffusion equations tends to send energy from high frequency modes to low frequency ones.

A pure diffusion model would produce blurred images, in a similar manner to that demonstrated in Fig. 5.7b, and to avoid this, a non-constant diffusion value may be considered to provide anisotropy to the filter. In Fig. 5.11 the resulting image using a Perona–Malik filter is used. As demonstrated the filter significantly removes noise, while preserving sharp edges and thin features. Several parameters are used to tune the filter, which makes it a difficult task as it becomes an iterative process. The filter is also the most computationally intensive of all discussed here.

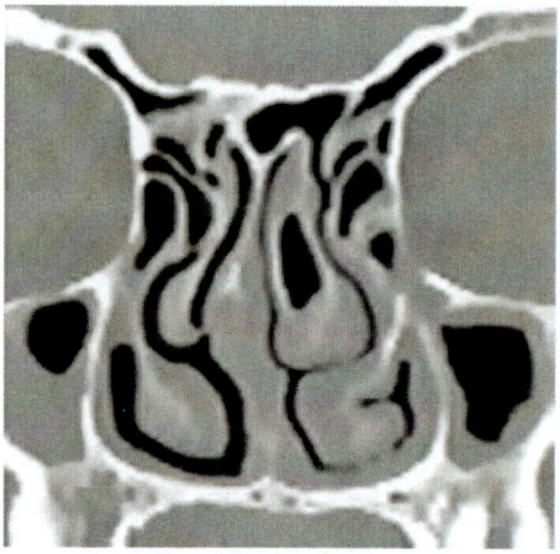

Fig. 5.11 Denoised image applying a Perona–Malik anisotropic diffusion

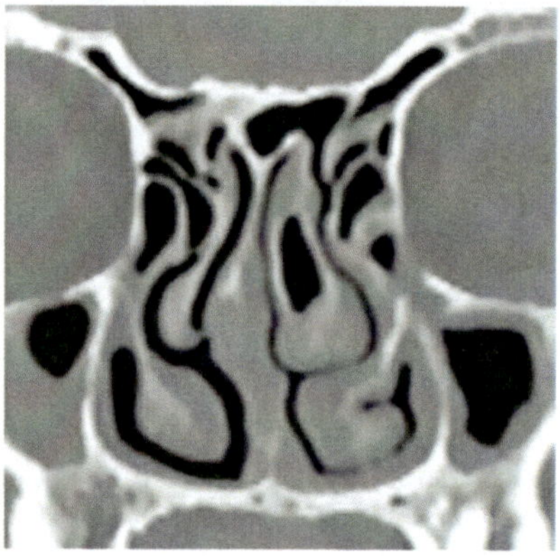

Fig. 5.12 Total variation minimized image

5.3.2.6 Total Variation Minimization

Total Variation minimization is a relatively recent methodology [10]. In Fig. 5.12 the de-noised image applying a Chambolle Total Variation minimization filter is depicted. This filter also de-noises the image, preserving the sharp edges, and even thin structures at some point. Unfortunately, Total Variation minimization algorithms are computationally expensive by nature, and cannot be easily extended to 3D images.

5.4 Image Segmentation

Image segmentation is the classification of each 3D image voxel within a limited number of categories. For the nasal cavity reconstruction, it is only required to assign a label to the air voxels. This would invite us to think that, in this context, we are always assigning just 2 labels to the image. However, the reality is this depends on whether the air surrounding the head shall be neglected or not. For simplicity, the air volume split will not be discussed here, although an algorithm to this end has previously been described [9].

A question arises regarding the different tissue regions. However in the nasal cavity reconstruction, only air and soft tissue labels are necessary, so the process can be simplified by cropping the radiodensity values to an acceptable range for the two regions, prior to segmentation. This is effectively replacing bones by soft tissues. There are several segmentation algorithms described in the literature [18, 20, 34]. Typically different segmentation methodologies are grouped based on their underlying algorithms.

Segmentation methodologies can be also classified by their automation grade [42], since there are many practical benefits of fully automatic approaches. However, it is appropriate to issue certain caveats regarding bias errors when human intervention is introduced [17]. In the context of the nasal cavity reconstruction, a whole range of degrees of automation have been trialled: Computer-Aided manual segmentation [48, 49], Semi-Automatic segmentation [6, 14], and fully automatic segmentation [5, 9]. We discuss the most commonly applied segmentation algorithms below.

5.4.1 Threshold-Based Segmentation

The simplest approach to the image segmentation is to consider air voxels as all those which have a radiodensity value lower than an arbitrary threshold value, t. This leads to different results obtained for different threshold values. Figure 5.13 demonstrates two different Threshold-based segmentations. As can be appreciated, neither segmentation was able to capture the thin structures, which is one of the greatest challenges in segmenting the nasal cavity.

If a larger threshold value t was used, then more structures at the interface between air and surfaces (blurred pixels) are captured. However, enlarging the threshold value has also a detrimental effect, since structures separated by thin tissues may be artificially connected. Therefore, defining an optimal threshold value becomes critical. Variational methods can be considered [45], leading to the result in Fig. 5.13b. Additionally, more complex clustering Machine Learning methodologies can be considered [1].

5.4.2 Edge-Based Segmentation

A popular segmentation approach is to use an edge detection algorithm to define the air volume boundary [56]. The voxels completely enclosed within the edge boundaries are subsequently labelled as air. Figure 5.14 demonstrates an image after the Canny edge detector [7] has been applied. This technique is well suited for detecting thin structures compared with threshold-based algorithms. However, a lack of sta-

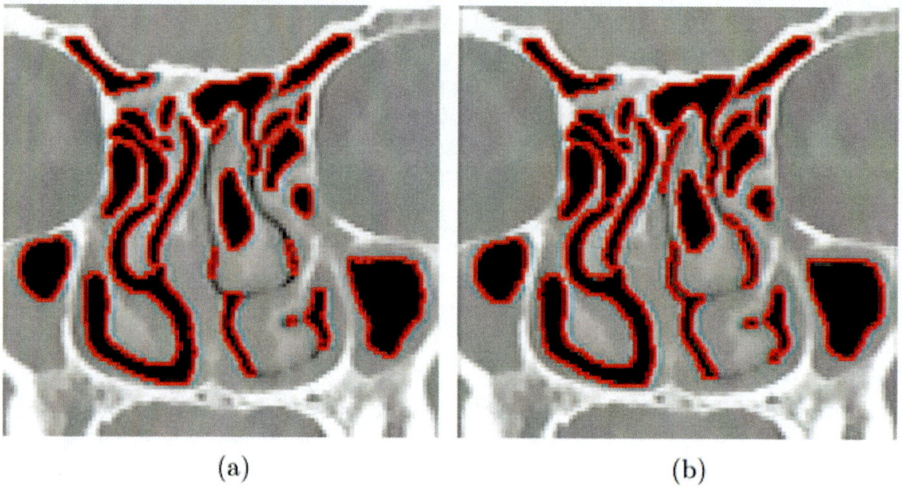

Fig. 5.13 Considered air volume using threshold-based segmentation, normalized pixel radiosity, $0 \leq \mu \leq 1$, is considered. The border of the considered air volumes, computed using Laplace border detector, is highlighted in red. **a** $t = 0.2$. **b** $t \sim 0.27$, computed using a variational method [45]

Fig. 5.14 Determining air volume using edge-based segmentation where the border highlighted in red is computed using Laplace border detector

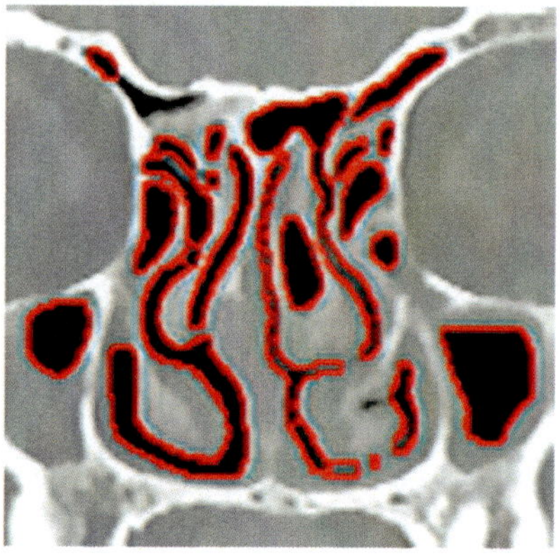

bility should be considered as a main drawback, i.e. tiny errors in the edge detection may potentially lead to false air volumes.

5.4.3 Hybrid Techniques

A natural progression is to combine the main strengths of threshold and edge-based methodologies. There is no doubt that the most popular approach along this line is Region-based segmentation, involving computation of image gradients [58] followed by a watershed segmentation. In Fig. 5.15a (Left) a Region-based segmentation is depicted.

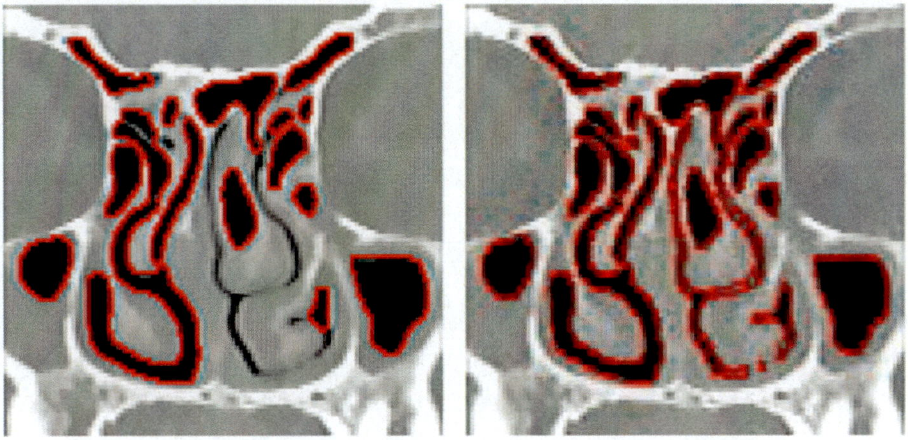

Fig. 5.15 Considered air volume using hybrid techniques where the border highlighted in red is computed using Laplace border detector. **a** Region-based segmentation. **b** Edge-based contrast enhancement [9]

It should be commented that Region-based segmentation techniques are not intended to achieve subpixel accuracy. Indeed, it is unable to detect thin structures. Conversely, a Canny edge detection [7] could be applied to carry out an image contrast enhancement prior to the Threshold-based segmentation. This is demonstrated in Fig. 5.15b. This allows some of the thin structure detection features of Edge-based segmentation methodologies to be preserved, while being based as a Thresholding technique, and therefore unaffected by stability issues. The main drawback of this method is that thin tissues can be missed due to the more relaxed Threshold-based segmentation.

5.4.4 Energy Minimization

A completely different segmentation methodology is Energy minimization. Such energy is composed of 2 main terms, a proper energy contribution, or cost, and a term that penalizes the assignment of significantly different labels than the expected based on radiodensity. Among the Energy minimization approaches the most popular is the max-flow/min-cut [3], which has successfully been applied for the reconstruction of nasal cavities [33]. Figure 5.16 demonstrates the result of applying max-flow/min-cut. In this specific application the results are similar to those obtained with Otsu's threshold method [45].

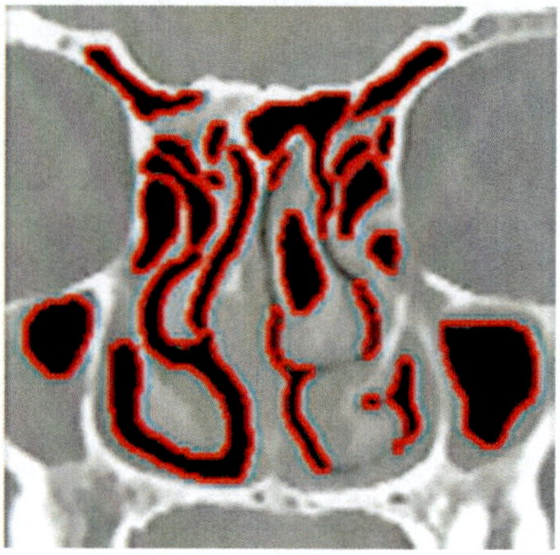

Fig. 5.16 Considered air volume applying max-flow/min-cut segmentation where the border was computed using Laplace border detector, highlighted in red

5.4.5 *Deep Learning*

Some researchers are taking advantage on the recent rapid computational power growth to address segmentations applying Deep learning [34]. Deep learning is a completely different paradigm where a neural network is fed with already segmented images, and the algorithm itself creates an optimized strategy to predict further segmentation. Training a neural network requires large datasets of segmented images.

5.5 Nasal Cavity Reconstructions, NASAL-Geom

NASAL-Geom (developed by the author, Cercos [9]) is a fully automatic surface generator, initially developed as a core component of NASAL-Flow system (https://nasalsystems.es/en/nasalflow-en). NASAL-Flow is a web service where medical professionals can upload CTs, containing nasal cavities, to automatically carry out stationary flow CFD simulations. The web service offers static reports, interactive visualization tools and even virtual surgery utilities.

NASAL-Geom has been released as an independent free software (under Affero-GPL v3 license terms), and can be downloaded from its software repository (https://gitlab.com/sanguinariojoe/nasal-geom). It features capabilities of automatically loading nasal cavity CTs and reconstructing surfaces of the internal nasal cavity, the external face, the pharynx and nostrils patches.

Currently, MAR, noise removal and contrast enhancement filters have been implemented. Both NASAL-Geom and NASAL-Flowhave been previously validated with a number of CFD studies utilizing CT scans from different machines, methodologies and resolutions [9]. In Fig. 5.17 we compare geometries the same nasal cavity surface, one semiautomatically reconstructed by Osman et al. [44] and the one reconstructed with NASAL-Geom [9]. For instance, sinus air cells have been truncated in the semiautomatic reconstruction, but retained in the NASAL GEOM model. Furthermore,

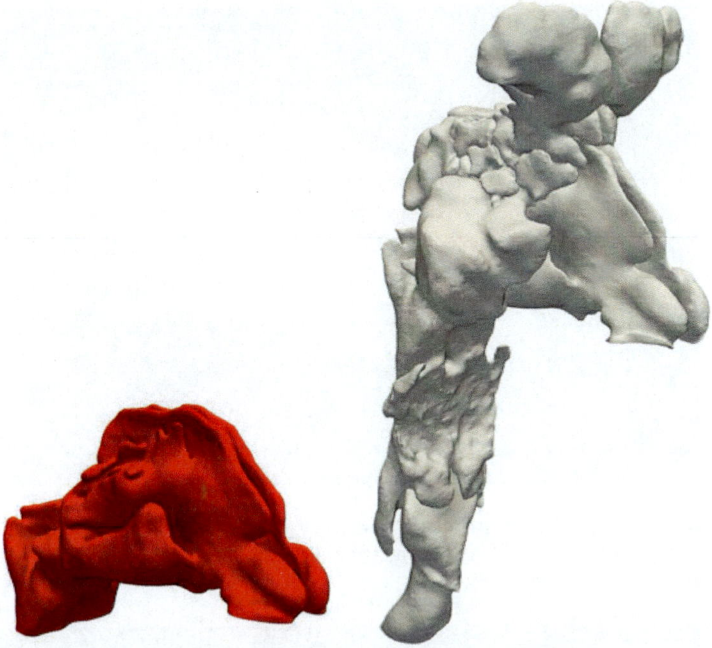

Fig. 5.17 The generated 3D surfaces. Left: The 3D geometry generated by [44]. Right: The 3D geometry generated by NASAL-Geom. Figure adapted from the one already published in [9]

the semiautomatic surface has been clipped to an axial plane at the nasopharynx, right above the metal implants, while the entire pharynx was modeled successfully using NASAL-Geom. More subtle differences can also beappreciated along the nostrils edges. In Fig. 5.18, detail on the surfaces close to the right nostril are provided. The surface on the left has been extended and subsequently sliced by an axial plane, while the one generated with NASAL-Geom remains unaltered. NASAL-Geom also provides patches for the nostrils, to eventually produce watertight nasal cavity surfaces.

Finally, in Fig. 5.19 an axial slice of both surfaces, overlapped with the original CT, is shown. Incidentally it should be remarked that slices of the actual produced surfaces are presented, in contrast to the Figs. depicted along Sect. 5.4, where the border of the segmented voxelizations were presented instead.As can be appreciated NASAL-Geom outperforms the semiautomatic surface generation, capturing the features of the air cavity more accurately. While Nasal-Geom is one tool available, we list other software in the Appendix, that can be used as alternatives.

Fig. 5.18 Detail of the differences at the right nostril. Red: The nasal cavity generated in [44]. Gray: The nasal cavity generated by NASAL-Geom. Blue: The facial surface generated by NASAL-Geom. Figure taken from [9]

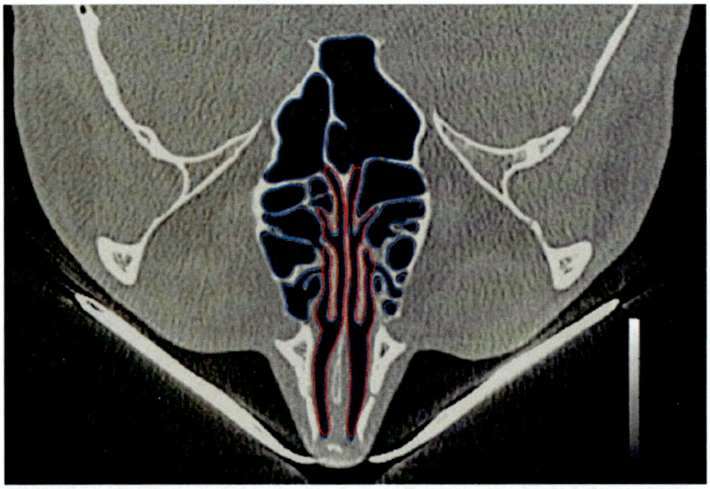

Fig. 5.19 Arbitrary axial slice of the surfaces and the CT. Red: Slice of the geometry generated in [44]. Blue: Slice of the surface generated by NASAL-Geom. Figure taken from [9]

5.6 Controversies, Challenges and Future Directions in Reconstruction

There is no doubt that the nasal cavity reconstruction will largely be fully automated in the near future. Greater automation levels will dramatically reduce costs and may reduce human biases in the reconstruction process.

While fully automatic tools [28] have been described in the past, they still remain relatively scarce. Machine Learning is a field that has been exponentially developing in recent years and promises to be a powerful tool to automatize complex tasks that previously only humans could perform. The most promising Machine Learning technique for image processing is Deep Learning. Unfortunately, when a Deep Learning neural network is trained, it is not possible to afterwards analyze it, introducing a level of uncertainty which may become difficult to accept, particularly in the context of medical applications. Nevertheless, novel research of deep learning in this field is already in the published literature [11, 37] although, the robustness of these kinds of neural networks still needs to be determined [8, 46]. Further evolution of the methodologies and computational power in next years will probably make room to extensively apply Machine Learning in a wide variety of complex tasks, such as medical imaging segmentation [34, 41].

References

1. M. Bal, L. Spies, Metal artifact reduction in CT using tissue-class modeling and adaptive prefiltering. Medical Phys. **33**(8), 2852–2859 (2006)
2. F. Bloch, Nuclear induction. Phys. Rev. **70**(7–8), 460 (1946)
3. Y. Boyko, V. Kolmogorov, Computing geodesics and minimal surfaces via graph cuts, in *null* (IEEE, 2003), p. 26
4. R. Brooks, G. Di Chiro, Beam hardening in x-ray reconstructive tomography. Phys. Med. Biol. **21**(3), 390 (1976)
5. N.L. Bui, S.H. Ong, K.W.C. Foong, Automatic segmentation of the nasal cavity and paranasal sinuses from cone-beam CT images. Int. J. Comput. Assist. Radiol. Surg. **10**(8), 1269–1277 (2015)
6. M. Burgos, E. Sanmiguel-Rojas, C. Del Pino, M. Sevilla-García, F. Esteban-Ortega, New CFD tools to evaluate nasal airflow. European Archives of Oto-Rhino-Laryngology **274**(8), 3121–3128 (2017)
7. J. Canny, A computational approach to edge detection. IEEE Trans. Pattern Anal. Mach. Intell. **6**, 679–698 (1986)
8. N. Carlini, D. Wagner, Towards evaluating the robustness of neural networks, in *2017 IEEE Symposium on Security and Privacy (SP)* (IEEE, 2017), pp. 39–57
9. J. Cercos-Pita, I. Cal, D. Duque, G. de Moreta, NASAL-Geom, a free upper respiratory tract 3D model reconstruction software. Comput. Phys. Commun. **223**, 55–68 (2018)
10. A. Chambolle, An algorithm for total variation minimization and applications. J. Math. Imaging Vis. **20**(1–2), 89–97 (2004)
11. X. Chen, Y. Duan, R. Houthooft, J. Schulman, I. Sutskever, P. Abbeel, Infogan: Interpretable representation learning by information maximizing generative adversarial nets, in *Advances in neural information processing systems* (2016), pp. 2172–2180

12. A. Chmielewski, M. Haji-Saeid, Radiation technologies: past, present and future. Radiat. Phys. Chem. **71**(1–2), 17–21 (2004)
13. T.J. Cullip, U. Neumann, Accelerating volume reconstruction with 3D texture hardware
14. P. Dastidar, T. Heinonen, J. Numminen, M. Rautiainen, E. Laasonen, Semi-automatic segmentation of computed tomographic images in volumetric estimation of nasal airway. Eur. Archi. Oto-Rhino-Laryngology **256**(4), 192–198 (1999)
15. X. Duan, L. Zhang, Y. Xiao, J. Cheng, Z. Chen, Y. Xing, Metal artifact reduction in ct images by sinogram tv inpainting, in *Nuclear Science Symposium Conference Record, 2008. NSS'08. IEEE* (IEEE, 2008), pp. 4175–4177
16. R. Eisenberg, *Radiology: an illustrated history* (Mosby Inc., St. Louis, 1992)
17. A. Eklund, T. Nichols, H. Knutsson, Cluster failure: Why fMRI inferences for spatial extent have inflated false-positive rates, in *Proceedings of the National Academy of Sciences* (2016), p. 201602413
18. A. Elnakib, G. Gimel'farb, J.S. Suri, A. El-Baz, Medical image segmentation: a brief survey, in *Multi Modality State-of-the-Art Medical Image Segmentation and Registration Methodologies* (Springer, 2011), pp. 1–39
19. L. Erasmus, D. Hurter, M. Naudé, H. Kritzinger, S. Acho, A short overview of MRI artefacts. SA J. Radiol. **8**, 2 (2004)
20. K.-S. Fu, J. Mui, A survey on image segmentation. Pattern Recognit. **13**(1), 3–16 (1981)
21. G. Gerig, O. Kubler, R. Kikinis, F.A. Jolesz, Nonlinear anisotropic filtering of MRI data. IEEE Trans. Med. Imaging **11**(2), 221–232 (1992)
22. L. Gjesteby, Q. Yang, Y. Xi, B. Claus, Y. Jin, B. De Man, G. Wang, Reducing metal streak artifacts in CT images via deep learning: Pilot results, in *The 14th international meeting on fully three-dimensional image reconstruction in radiology and nuclear medicine* (2017), vol. 14, pp. 611–614
23. G. Glover, N. Pelc, An algorithm for the reduction of metal clip artifacts in CT reconstructions. Medical Phys. **8**(6), 799–807 (1981)
24. C.A. Harris, L.M. White, Metal artifact reduction in musculoskeletal magnetic resonance imaging. Orthopedic Clinics **37**(3), 349–359 (2006)
25. T. Huang, G. Yang, G. Tang, A fast two-dimensional median filtering algorithm. IEEE Trans. Acoust. Speech Signal Process. **27**(1), 13–18 (1979)
26. W. Kalender, Technical foundations of spiral CT. J. Belge De Radiologie **78**(2), 68–74 (1995)
27. W. Kalender, R. Hebel, J. Ebersberger, Reduction of CT artifacts caused by metallic implants. Radiology **164**(2), 576–577 (1987)
28. A.K. Kono, K. Ishii, K. Sofue, N. Miyamoto, S. Sakamoto, E. Mori, Fully automatic differential diagnosis system for dementia with Lewy bodies and Alzheimer's disease using FDG-PET and 3D-SSP. Eur. J. Nucl. Med. Mol. Imaging **34**(9), 1490–1497 (2007)
29. B. Kratz, T. Knopp, J. Müller, M. Oehler, T.M. Buzug, Non-equispaced Fourier Transform vs. polynomial-based metal artifact reduction in computed tomography, in *Bildverarbeitung für die Medizin 2008* (Springer, 2008), pp. 21–25
30. M. Lell, E. Meyer, M. Kuefner, M. May, R. Raupach, M. Uder, M. Kachelrieß, Normalized metal artifact reduction in head and neck computed tomography. Investig. Radiol. **47**(7), 415–421 (2012)
31. C. Lemmens, D. Faul, J. Nuyts, Suppression of metal artifacts in CT using a reconstruction procedure that combines MAP and projection completion. IEEE Trans. Med. Imaging **28**(2), 250–260 (2009)
32. E. Lin, A. Alessio, What are the basic concepts of temporal, contrast, and spatial resolution in cardiac CT? J. Cardiovasc. Comput. Tomography **3**(6), 403–408 (2009)
33. B.N. Linh, Modeling of the human upper airway from multimodal 3D dentofacial images. Ph.D. thesis (2014)
34. G. Litjens, T. Kooi, B.E. Bejnordi, A.A.A. Setio, F. Ciompi, M. Ghafoorian, J.A. Van Der Laak, B. Van Ginneken, C.I. Sánchez, A survey on deep learning in medical image analysis. Med. Image Anal. **42**, 60–88 (2017)

35. P. Mah, T. Reeves, W. McDavid, Deriving Hounsfield units using grey levels in cone beam computed tomography. Dentomaxillofacial Radiol. **39**(6), 323–335 (2010)
36. S.G. Mallat, Multifrequency channel decompositions of images and wavelet models. IEEE Trans. Acoust. Speech Signal Process. **37**(12), 2091–2110 (1989)
37. D.A. Melis, T. Jaakkola, Towards robust interpretability with self-explaining neural networks, in *Advances in Neural Information Processing Systems* (2018), pp. 7775–7784
38. E. Meyer, R. Raupach, M. Lell, B. Schmidt, M. Kachelrieß, Normalized metal artifact reduction (NMAR) in computed tomography. Med. Phys. **37**(10), 5482–5493 (2010)
39. E. Meyer, R. Raupach, M. Lell, B. Schmidt, M. Kachelrieß, Frequency split metal artifact reduction (FSMAR) in computed tomography. Med. Phys. **39**(4), 1904–1916 (2012)
40. P. Mildenberger, M. Eichelberg, E. Martin, Introduction to the DICOM standard. Eur. Radiol. **12**(4), 920–927 (2002)
41. F. Milletari, N. Navab, S.-A. Ahmadi, V-net: fully convolutional neural networks for volumetric medical image segmentation, in *2016 Fourth International Conference on 3D Vision (3DV)* (IEEE, 2016), pp. 565–571
42. S.D. Olabarriaga, A.W. Smeulders, Interaction in the segmentation of medical images: a survey. Med. Image Anal. **5**(2), 127–142 (2001)
43. F. Orieux, J.-F. Giovannelli, T. Rodet, Bayesian estimation of regularization and point spread function parameters for Wiener-Hunt deconvolution. JOSA A **27**(7), 1593–1607 (2010)
44. J. Osman, F. Großmann, K. Brosien, U. Kertzscher, L. Goubergrits, T. Hildebrandt, Assessment of nasal resistance using computational fluid dynamics. Curr. Dir. Biomed. Eng. **2**(1), 617–621 (2016)
45. N. Otsu, A threshold selection method from gray-level histograms. IEEE Trans. Syst. Man Cybern. **9**(1), 62–66 (1979)
46. K. Patan, M. Witczak, J. Korbicz, Towards robustness in neural network based fault diagnosis. Int. J. Appl. Math. Comput. Sci. **18**(4), 443–454 (2008)
47. P. Perona, J. Malik, Scale-space and edge detection using anisotropic diffusion. IEEE Trans. Pattern Anal. Mach. Intell. **12**(7), 629–639 (1990)
48. S. Pirner, K. Tingelhoff, I. Wagner, R. Westphal, M. Rilk, F. Wahl, F. Bootz, K.W. Eichhorn, CT-based manual segmentation and evaluation of paranasal sinuses. Eur. Archives of Oto-Rhino-Laryngology **266**(4), 507–518 (2009)
49. D. Prasun, N. Jura, H. Tomi, R. Pertti, R. Markus, L. Erkki, Nasal airway volumetric measurement using segmented HRCT images and acoustic rhinometry. Amer. J. Rhinol. **13**(2), 97–104 (1999)
50. E. Purcell, R. Pound, N. Bloembergen, Nuclear magnetic resonance absorption in hydrogen gas. Phys. Rev. **70**(11–12), 986 (1946)
51. J. Radon, Über die bestimmung von funktionen durch ihre integralwerte längs gewisser mannigfaltigkeiten. Ber. Saechsische Akad. Wiss **29**, 262 (1917)
52. J. Radon, On the determination of functions from their integral values along certain manifolds. IEEE Trans. Med. Imaging **5**(4), 170–176 (1986)
53. W. Roentgen, Ueber eine Art von Strahlen. Sittzungsber Phys.-Med. Ges. Wuerzburg, Dez. 132 (1895)
54. M. Safdari, A. Karimian, M. Yazdchi, A new method for metal artifact reduction in CT scan images. Iran. J. Med. Phys. **10**(2), 139–146 (2013)
55. W.F. Schreiber, Wirephoto quality improvement by unsharp masking. Pattern Recognit. **2**(2), 117–121 (1970)
56. N. Senthilkumaran, R. Rajesh, Edge detection techniques for image segmentation-a survey of soft computing approaches. Int. J. Recent Trends Eng. **1**(2), 250 (2009)
57. C.L. Sistrom, N.L. McKay, Costs, charges, and revenues for hospital diagnostic imaging procedures: differences by modality and hospital characteristics. J. Amer. Coll. Radiol. **2**(6), 511–519 (2005)
58. I. Sobel, An Isotropic 3×3 Image Gradient Operator. Presentation at Stanford A.I. Project 1968 (2014)

59. C. Tomasi, R. Manduchi, Bilateral filtering for gray and color images, in *Sixth International Conference on Computer Vision, 1998* (IEEE, 1998), pp. 839–846
60. M. Vannier, Iterative deblurring for CT metal artifact reduction. IEEE Trans. Med. Imaging **15**(5), 651 (1996)
61. X.-Y. Wang, H.-Y. Yang, Z.-K. Fu, A new wavelet-based image denoising using undecimated discrete wavelet transform and least squares support vector machine. Expert Syst. Appl. **37**(10), 7040–7049 (2010)
62. J. Wei, L. Chen, G. Sandison, Y. Liang, L. Xu, X-ray CT high-density artefact suppression in the presence of bones. Phys. Med. Biol. **49**(24), 5407 (2004)
63. N. Wiener, *Extrapolation, Interpolation and Smoothing of Stationary Time Series-with Engineering Applications* (MIT Press, New York, 1949)
64. S. Zhao, D. Robeltson, G. Wang, B. Whiting, K. Bae, X-ray CT metal artifact reduction using wavelets: an application for imaging total hip prostheses. IEEE Trans. Med. Imaging **19**(12), 1238–1247 (2000)

Chapter 6
Computational Meshing for CFD Simulations

Andreas Lintermann

Abstract In CFD modelling, small cells or elements are created to fill this volume. They constitute a mesh where each cell represents a discrete space that represents the flow locally. Mathematical equations that represent the flow physics are then applied to each cell of the mesh. Generating a high quality mesh is extremely important to obtain reliable solutions and to guarantee numerical stability. This chapter begins with a basic introduction to a typical workflow and guidelines for generating high quality meshes, and concludes with some more advanced topics, i.e., how to generate meshes in parallel, a discussion on mesh quality, and examples on the application of lattice-Boltzmann methods to simulate flow without any turbulence modelling on highly-resolved meshes.

6.1 Introduction

The segmented airway region from CT/MRI scans provides surface boundary information (see Chap. 5). The internal space, where inhaled air or particles pass through, is an enclosed volume formed by the surface boundaries. In CFD modelling, small cells or elements are created to fill this volume. They constitute a mesh and thereby discretise the space. The mathematical equations that represent the flow physics (This chapter and Chap. 7) are then applied to each cell of the mesh. The end result is the ability to store time-dependent flow properties such as pressure, velocity, or temperature in each individual cell.

Generating a high quality mesh is extremely important to obtain reliable solutions and to guarantee numerical stability. In the CFD community, mesh generation is also referred to as grid generation, which has become a separate discipline in itself and remains a very active area of research and development. This is evident with

A. Lintermann (✉)
Forschungszentrum Jülich GmbH, Jülich, Germany
e-mail: Lintermann@fz-juelich.de

K. Inthavong et al. (eds.), *Clinical and Biomedical Engineering in the Human Nose*, Biological and Medical Physics, Biomedical Engineering,
https://doi.org/10.1007/978-981-15-6716-2_6

many existing commercial codes on the market having their own powerful, built-in mesh generators, as well as with a number of independent grid generation packages available.

Generating a mesh for the first time is a daunting process, where decisions have to be made on the arrangement of discrete points (nodes) throughout the computational domain, and the type of connections for each point. The quality of the mesh leads to either success or failure of the numerical simulation.

6.2 Mesh Types

The nasal airway volume is treated as a 3D geometry and therefore the mesh elements/cells are generally hexahedral, tetrahedral, square pyramids (pyramids, extruded triangles, wedges or triangular prisms) or polyhedral, see Fig. 6.1a. It is useful to envision the mesh topology as a hierarchical system, where higher topology assumes the existence of topologies beneath it. Figure 6.1b shows a 3D cell inherit the lower topologies, i.e., a volume is made up from faces, a face is made from lines, and a line is made from vertices. A common meshing approach is to use a bottom-up approach, where firstly vertices, lines, and faces (from the segmentation step) are created to form a surface mesh and to subsequently fill the enclosed volume with 3D cells, i.e., the mesh elements. This means that for boundary conforming meshes care should be taken in generating good quality surfaces from the segmented airway models.

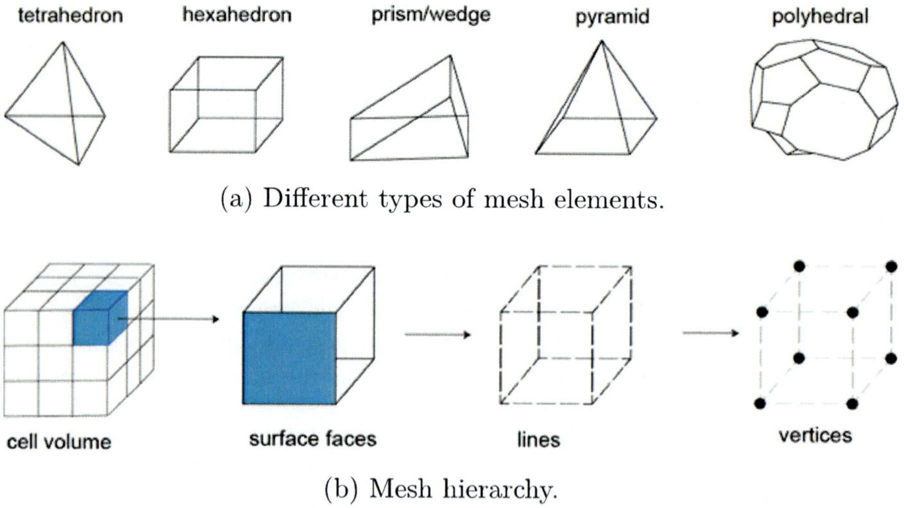

Fig. 6.1 **a** Types of mesh elements including tetrahedron, hexahedron, prism/wedge, pyramid, and polyhedral. **b** Mesh topology demonstrating the hierarchy for each level of meshing

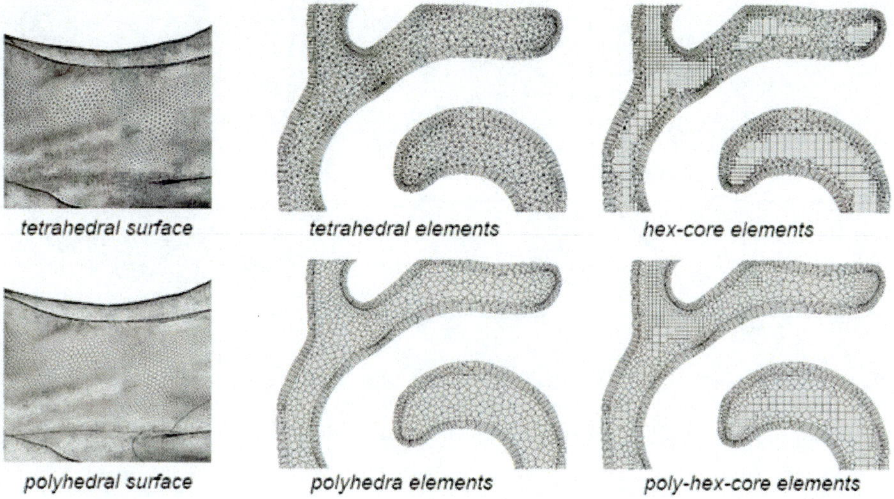

(a) Surface elements.

(b) Different kinds of internal meshes.

Fig. 6.2 Unstructured meshes are typically used for complex geometries where **a** irregular shaped 2D surface elements can wrap around the surface. Thereafter, **b** the internal mesh may contain 3D elements such as tetrahedral, polyhedral, hex-core, and a combination of polyhedral-hex-core elements

6.2.1 Unstructured Meshes

The nasal airway geometry is highly irregular and does not fit exactly to orthogonal lines or Cartesian coordinates. As a result, an irregular unstructured surface consisting of 2D elements can be created to wrap around the contoured geometry. Figure 6.2a exemplary shows surfaces consisting of tetrahedral and polyhedral elements. The surface can internally be filled with the 3D unstructured elements of the same kind or can be a hybrid mix of different elements including structured and prism layers, see Fig. 6.2b. An unstructured grid is typically identified by irregular connectivity. To store the data for the computation, new data structures, e.g., edge based, face based, or cell based structures are needed to hold the connectivity scheme. A node of a cell can be connected to any number of nodes from its neighbours, which may lead to complex connectivity dependencies. The descriptions of some of these connectivity algorithms include Delauney, advancing front, and quadtree/octree (see Sect. 6.3).

Unstructured CFD meshing technologies have matured and their use has become more prevalent with commercial CFD vendors. They provide the user with simpler automated methods in achieving a meshed model and a user-friendly interface.[1] The typical workflow of generating an unstructured mesh involves:

[1]A large number of grid generation software and open source codes exist with a listing of some available software given in the appendix.

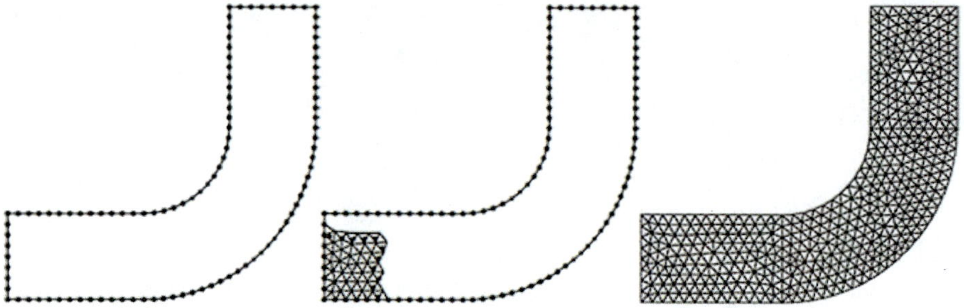

Fig. 6.3 Mesh construction for a 2D 90° bend pipe based on the advancing front method. The initial front is established on the edge mesh setting. It advances by recursively adding new points and to subsequently creating a triangular element for each added point

- detection of the edges of a surface and establishment of nodes to form the edges of each surface
- creation of new nodes on the surface and connecting them to fill the surface with 2D elements such as triangles or polygonal 2D shapes (advancing front method—an example is shown in Fig. 6.3)
- for 3D geometries the interior volume (and in 2D the interior area) is filled and all the elements are reconnected (Delauney method).

Recently, there has been considerable development towards the creation of meshes containing polyhedral cells in the interior domain. Such meshes have the flexibility of an unstructured mesh and do not suffer from the overhead associated with large tetrahedral meshes. The application of such cells is still in its infancy. Nevertheless, polyhedral meshing has shown to have tremendous advantages over tetrahedral meshing with regard to the attained accuracy and efficiency of corresponding numerical computations.

6.2.2 Structured Meshes

By definition, a structured mesh is one that has eight nodal corner points in three dimensions and is characterised by regular connectivity. It uses a Cartesian system with an (i, j, k) indexing to locate neighbouring cells. On a regular orthogonal geometry, creating a structured mesh is the simplest technique to undertake, since a Cartesian mesh can be conveniently generated. Figure 6.4 shows an example of a structured mesh with the corresponding indexing.

When the geometry exhibits curvatures, as for example shown in Fig. 6.5, an orthogonal structured mesh applied to the geometry requires adjustments resulting in staircase-like steps in the region of the curved sections. These kinds of meshes require

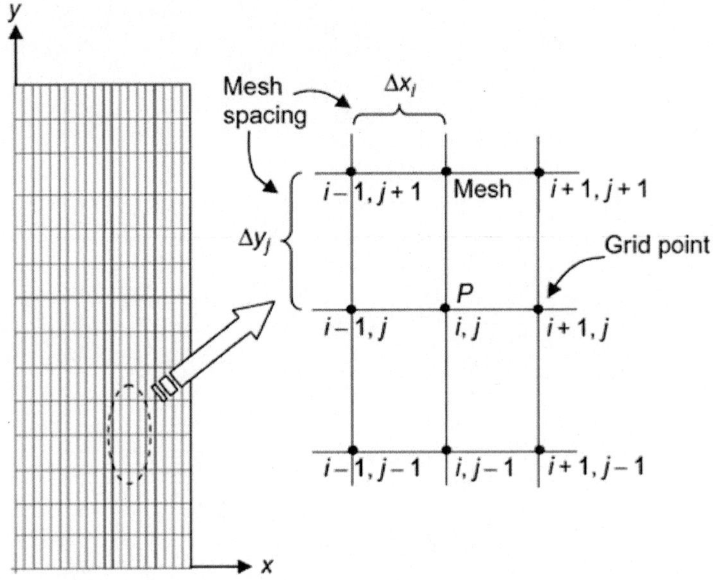

Fig. 6.4 Ordered indices for a structured mesh of a 2D orthogonal domain

Fig. 6.5 An example of a mesh using staircase-like steps for a 90° bend geometry

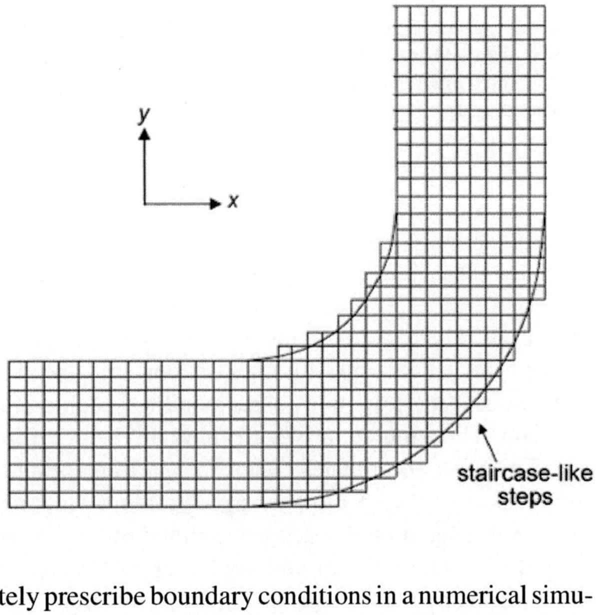

additional treatment to accurately prescribe boundary conditions in a numerical simulation and to accurately compute wall-shear stresses, heat fluxes, and boundary-layer effects.

An approach to overcome these issues, is the generation of a body-fitted mesh. Therefore, an orthogonal mesh is applied to the geometry and a mapping of the distorted region in physical space onto a rectangular region in the curvilinear coordinate space is performed. This is in general done through coordinate transformation

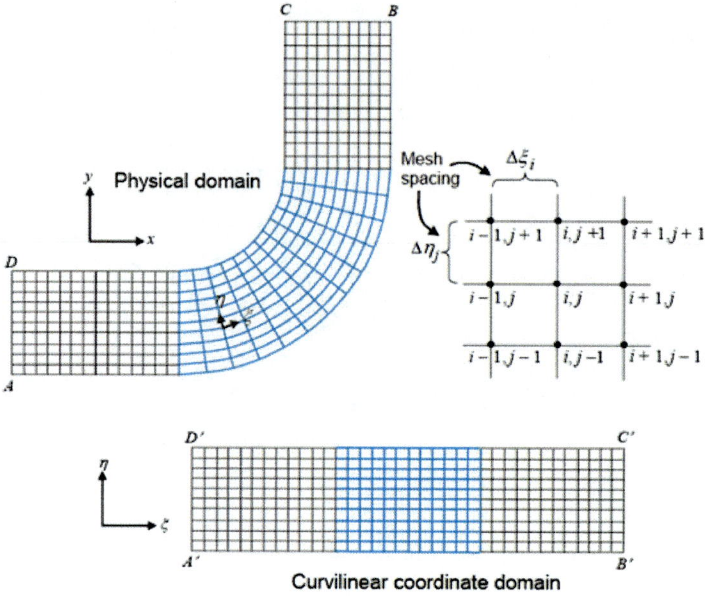

Fig. 6.6 Body-fitted mesh applied to a curved geometry. A transformation of coordinates is requires

Fig. 6.7 A body-fitted mesh which creates skewed cells, and a block-structured mesh (using O-grid)

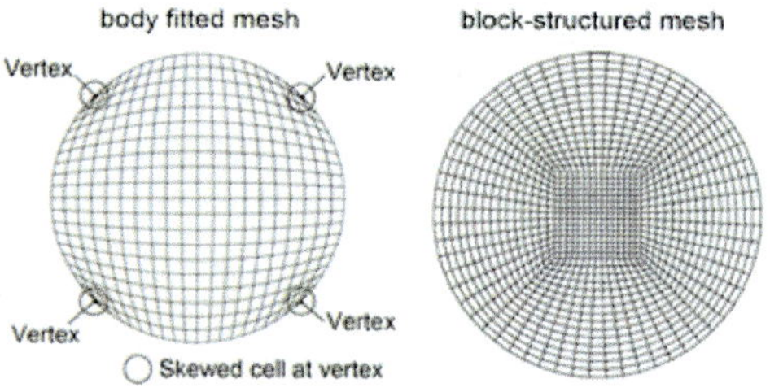

functions. For example, creating a body-fitted mesh for a 90° bend geometry, the walls need to coincide with lines of constant η, see Fig. 6.6, while in the streamwise direction of the flow lines correspond to specific values of ξ in the computational domain. The simulation is performed on the curvilinear coordinates rather than on the Cartesian coordinates. Therefore, a transformation function is used to switch domains from $f(x, y)$ to $f(\eta, \xi)$ to yield new independent variables.

Another technique to adapt structured meshes to non-orthogonal geometries is to use block-structured or multiblock meshes, where a number of structured blocks is assembled and connected. This method is effective for complicated shapes, where it is difficult to apply a single block. Take for example the circle shown on the left side in Fig. 6.7, where a body-fitted mesh leads to highly skewed and deformed cells at the perimeter of the geometry. These highly skewed cells generally lead to numerical instabilities and erroneous results in a simulation.

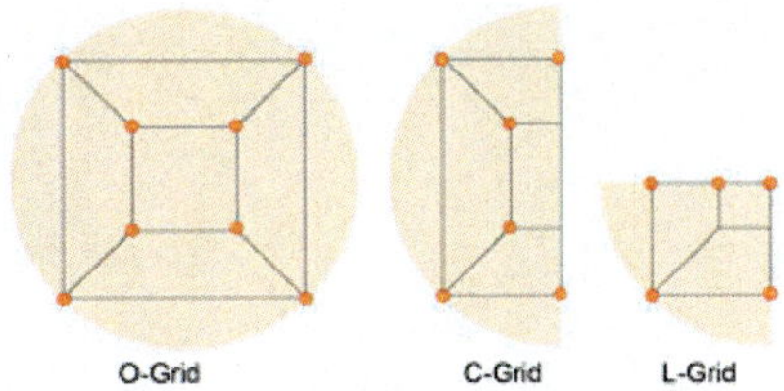

Fig. 6.8 Block structures used to map onto curved sections of a circle (O-grid), semi-circle (C-grid), and quarter-circle (L-grid)

Using a block-structured mesh, as shown on the right side of Fig. 6.7, the outer perimeter takes the shape of an annular grid. Hence the term O-grid has been established for these kind of meshes. Additionally, a square block fills the centre of this mesh. Other block-structure topologies include O-, C-, or L-grids. An O-grid has lines of points where the last point wraps around and meets the first point, hence creating a circular 'O' shape. Similarly, a C-grid has lines that bend in a semicircle shaped like the letter 'C' (essentially half an O-grid), while an L-grid forms an L-shape derived from a quarter of an O-grid. Examples are depicted in Fig. 6.8.

Further examples to overcome the issue of staircase-like approximations of curved walls are applying a cut-cell method, i.e., reshaping the staircase elements to fit the geometry, or to include a treatment in the computation, cf. [5, 31, 49]. This approach is also followed by the meshing method presented in Sect. 6.3, where a technique to generate hierarchical Cartesian meshes for nasal cavity geometries in parallel is described.

6.2.3 Structured Versus Unstructured Meshes

A structured mesh has the advantage that its layout is simple due to the direct and straightforward connections between adjacent cells. For example, a three-dimensional cell is connected by six neighbouring cells with points easily accessible by triple indices (i, j, k), incrementing or decrementing with each adjacent cell. This indexing makes CFD solvers simple to program and leads to a small memory footprint. However, the creation of structured meshes for complex geometries such as the nasal airway is highly time consuming as it involves a lot of manual work. Therefore, in most literature, unstructured tetrahedral, hexahedral, or polyhedral meshes are used (see, e.g., the meshes depicted in Fig. 6.9).

An unstructured mesh has the advantage that it is much more suited to fit within complex shapes. Furthermore, the volume filling can, e.g., be performed automatically using a Delauney triangulation or a hierarchical approach, see Sect. 6.3. In regions of high curvatures, however, tetrahedrals tend to become over-stretched leading to high aspect ratios that affect the skewness of the cells—this is of course not the case for hexahedral cells. Additionally, tetrahedral cells can be difficult to align with the flow direction. These two problems can impede convergence and may lead to artificial errors in the solution, known as numerical diffusion. This common source

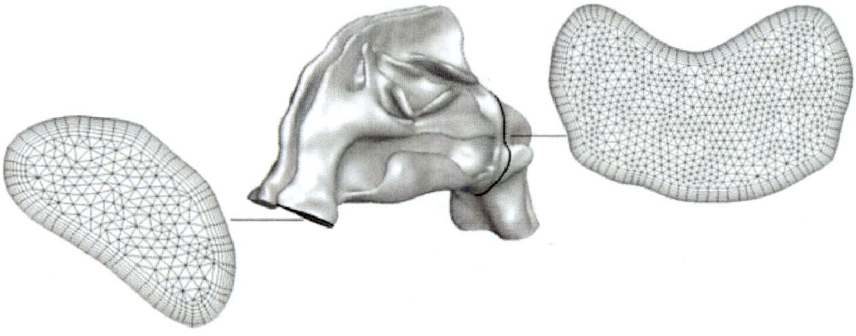

(a) Tetrahedral mesh for the inner core region and prism layers lining the wall region.

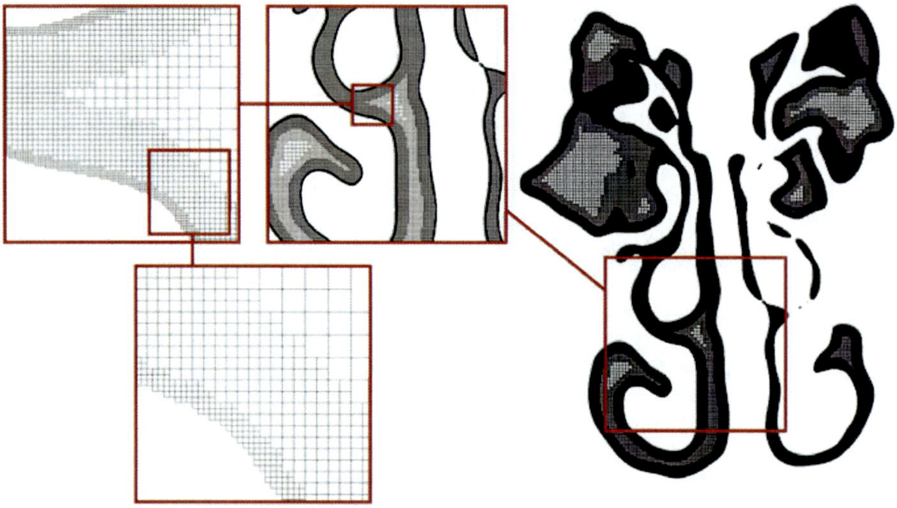

(b) Boundary-refined computational mesh consisting in total of $1.8 \cdot 10^9$ elements [33]. Only every second grid line is shown.

Fig. 6.9 Two types of boundary-refined meshes

of error is also called false diffusion as it is a product of numerical error and does not represent a physically occurring phenomenon. The other major drawback of unstructured meshes is the treatment of the nodal points of any cell. Unlike in structured meshes, they cannot be accessed by triple indices (i, j, k) in three dimensions. Instead additional data calculations are needed to connect an arbitrary number of neighbouring nodes. This requires additional computational memory and further complicates the solution algorithms that are used to solve the flow field variables. This typically results in increased computational times compared to structured-mesh-based codes to obtain a solution.

In the -wall region, both types of meshing algorithms need to achieve a resolution that has reasonable accuracy in resolving boundary layer flows. This is achieved by either adding prism layers along walls or by refining hexahedral cells in hierarchical manner. An example for mixed hybrid meshes consisting of tetrahedra and prisms in the vicinity of the walls is shown in Fig. 6.9a. Another example, where a hierarchical Cartesian mesh is boundary-refined, is shown in Fig. 6.9b.

6.2.4 Polyhedral Meshing

An alternative to tetrahedral meshes is the use of polyhedral elements, which have become increasingly popular [40]. The major advantage of using polyhedra compared to tetrahedra is their larger number of adjacent neighbouring cells (typically ten, compared to four for tetrahedra). This allows greater resolution in gradient approximations. Polyhedra can be viewed as cells created from combining tetrahedral cells. Thereby, the overall cell count is decreased [10] and highly skewed cells are reduced. This makes polyhedral cells less sensitive to stretching due to their irregular shape. Due to the increased number of nodes and faces per cell, and increased number of neighbours, polyhedral meshes require more computing operations per cell. This is, however, compensated by the rapid time to reach convergence.

6.3 Parallel Mesh Generation of Hierarchical Cartesian Meshes

The intense increase of computational power over the last decades has enabled simulation software to simulate physical phenomena at continuously growing resolution and accuracy. The pace of algorithm development lacks behind the pace of hardware development and hence the full potential of todays computer architectures is in many cases left unexploited, especially when it comes to CFD simulations.

At small scale, the generation of computational meshes in serial for simple geometric shapes is not a big challenge and efficient algorithms already exist for quite some time [52, 53]. The considered geometries become, however, continuously complex and in general the complexity of the topology of computational meshes follows this behaviour. This more and more necessitates to write efficient parallel algorithms that can create such computational meshes, e.g., for intricate anatomical shapes such as the nasal cavity, consisting of hundreds of millions up to billions of elements [6, 22, 33, 36]. Since memory limitations prevent the generation of such meshes in serial, algorithm developers need to think about multi-level communication-reducing parallelisation strategies, decomposition methods, load-balancing, and parallel I/O. This renders corresponding software development an issue of computer science, and to be more precise, of high-performance computing (HPC).

Hierarchical Cartesian meshes are an exceptional tool to discretise the space for CFD computations. They allow for a straightforward mesh generation, their rectangular grid lines enable to easily apply high-order schemes, and refinement and coarsening can be performed on the fly [6, 33]. Latter feature is in combination with a dynamic load-balancing scheme a powerful tool to run large-scale balanced simulations on HPC systems [21, 48]. Their drawbacks result from their infeasibility for simulations of flow around bodies at high REYNOLDS numbers with small boundary layer thicknesses. That is, the near-wall region is populated with a massive amount of mesh elements in streamwise direction although the gradient, e.g., of the fluid velocity, in this direction is rather small. Boundary-fitted meshes or meshes combining near-wall prism layers and tetrahedral elements are in such cases more advantageous [38, 39], require, however, in many cases either intense manual work or their generation is quite complex and not easy to parallelism.

Since the REYNOLDS number is rather low in respiratory flow, i.e., it is usually on the order of $Re \in \mathcal{O}(10^4)$ [17, 18, 30], hierarchical Cartesian meshes are good candidates to simulate the corresponding flow. In the following, a parallel mesh generation technique is described that enabled the generation of such meshes and scales up to $\mathcal{O}(10^5)$ cores to generate $\mathcal{O}(10^{11})$ mesh elements. For more details, the interested reader is referred to the original publication [33], where the algorithm is first described. Application to biomedical flows can be found in [27, 30, 34, 35] and Sect. 6.6 briefly presents some examples. More non-biomedical applications can be found in [7, 41, 42, 49, 50], i.e., for the simulation of jets [7], direct particle simulations [49, 50], or rotating fan setups [41, 42].

6.3.1 Description of the Parallel Meshing Algorithm

The algorithm to generate meshes consists of a serial and a parallel part. The serial part executes the same algorithm on each message passing interface (MPI) process $n \in \{1, \ldots, N_m\}$, where N_m is the total number of spawned MPI processes on some cluster system. In contrast, the parallel part performs individual work on separate sub-domains of the output of the serial part. In the following, first the serial part is explained before the algorithmic details of the parallel part are discussed.

In the initial stage, each process reads the geometry from disk, i.e., the nasal cavity surface, which is stored in standard tessellation language (STL) format and usually stems from computer tomography (CT) images [34], cf. Chap. 5. From the triangles an alternating digital tree (ADT) [4] is generated to efficiently retrieve triangle `ids` for triangle / mesh element intersections and inside / outside detections. Around the geometry an initial cube element on a hierarchy level l_0 with an edge length $\delta x(l_0)$ of the maximum extent of the geometry is generated. A subdivision algorithm continuously splits this cube element into its eight children and creates parent-child relationships and element-neighbourhood information. This constitutes a mesh hierarchy representing an octree in three dimensions, see Fig. 6.10.

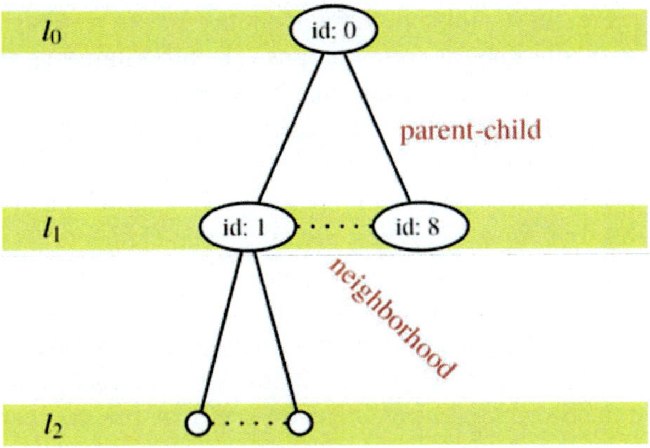

Fig. 6.10 Hierarchical octree representation of the Cartesian mesh with parent-child relationships and neighbourhood information. Levels l_0, the initial cubic element around the geometry, and levels l_1 and l_2 are shown

This refinement is continuously performed until a mesh with levels l_0 up to a user-defined level l_a exists. In each iteration, elements that live outside the geometry, i.e., cells that do not need to be considered in the computation, are removed from the octree. This is done by identifying the elements that intersect the geometry and use those as a boundary for a recursive flagging algorithm that separates inside from outside cells. The outside elements are then deleted from memory. Subsequently levels $l < l_a$ are removed from the octree, a space filling Hilbert curve [46] is placed into the remaining elements, and the Hilbert `ids` are used to subdivide the mesh for further parallel processing. That is, each MPI process removes those elements it is not responsible for, but keeps those elements that are neighbours to the remaining elements. These elements, so called halo elements, are in a later stage responsible to create a consistent neighbourhood information across all MPI processes before the mesh is written to disk.

In the subsequent parallel phase, each MPI process continues to uniformly refine its local mesh hierarchy to a level $l_b > l_a$ under removal of the corresponding outside elements. Since it is desirable to have a high resolution in the vicinity of the nasal cavity tissue surface or in regions with high velocity gradients, e.g., for an accurate computation of wall-shear stresses, the mesh is furthermore region-refined. Two distinct methods can be used separately or in combination. Patch refinement allows the user to define volumetric regions in space that are to be further refined. Elements residing in these regions are flagged for further subdivision until a level l_c^α is reached. In contrast, boundary-refinement starts by measuring the distances in units of elements of level l_b from the wall and identifies those elements with a distance $d \leq d_u$, where d_u is a user-defined constant. These cells are then further refined. Based on a distance rule, the measurement and refinement is repeated on the subsequent levels until a fine resolution on level l_c^β is reached. It should be noted that since the distance measurement is a recursive process it needs to take distance information from neighbouring MPI processes into account. While for the boundary refinement the restriction to have only level distances of maximum 1 between neighbouring elements is implicitly solved by applying the distance rule and providing a decent value for d_u, in patch refinement the user is responsible to guarantee a mesh output that is usable for computations. The algorithm features a load-balancing strategy to avoid imbalances due to excessive local refinement and out-of memory failures.

In the final stage, the global neighbourhood and a unique identification for all elements is generated across MPI boundaries via the halo elements on all levels. The number of elements in the subtrees of the elements on all levels are counted and used as for a weighting measure. The elements on refinement level l_a are used to generate a list $\mathcal{L} = \{e_1, \ldots, e_k\}$. The simplest approach is placing all elements on level l_a into $\mathcal{L}$, leading to a Hilbert-sorted list. In a subsequent simulation, the entries of the list $\mathcal{L}$ and their corresponding subtrees are then evenly distributed among the MPI processes. However, undesirable weightings on level l_a that exceed a certain threshold, e.g., an element $e^+ \in \mathcal{L}$ that has compared to other elements on level l_a many more descendants in the octree, might cause an imbalance at simulation start. Such cases are treated by replacing e^+ by its child elements $c(e^+)$. That is, $\mathcal{L} = \{\ldots, e^+, \ldots\}$ is replaced by $\mathcal{L}^\sim = \{\ldots, \{c(e^+)\}, \ldots\}$. The mesh and the list $\mathcal{L}^\sim$ are finally written to disk using parallel I/O with parallel NetCDF [26] or HDF5 [11]. The option to write a parallel geometry file to also distribute the triangles of the STL among the MPI processes in the simulation exists at this stage as well [27].

6.3.2 *Meshing Example and Parallel Performance*

The algorithm can now be employed to generate large-scale meshes with billions of elements on HPC systems. The availability of list $\mathcal{L}$ at the end of the meshing process allows to use a quasi arbitrary number of MPI processes N_s for the simulation with $o \leq N_s \leq |\mathcal{L}|$. The number N_s is independent from the number of MPI ranks used for the mesh generation N_m and has a lower bound of o, which is determined by the amount of available memory per process, i.e., at process number o it is assumed that the memory is completely exhausted. Figure 6.9b shows an example of a mesh of a nasal cavity with $1.8 \cdot 10^9$ elements with boundary-refined regions. The base level is at $l_a = 8$ and the uniform refinement level at $l_b = 10$. On the finest level $l_c^\beta = 13$ the mesh resolution is at $12.651 \cdot 10^{-3}$ mm. The parallel performance can be analyzed by considering the strong scalability of the meshing process and the absolute run time for several cases. Such analyses were performed on the HERMIT[2] and JUQUEEN[3] HPC systems at the High-Performance Computing Center Stuttgart (HLRS) and the Jülich Supercomputing Centre (JSC) for cubic domains. The meshes to be generated consisted of $C_1 = 9.82 \cdot 10^9$, $C_2 = 78.54 \cdot 10^9$, and $C_3 = 0.64 \cdot 10^{12}$. On HERMIT, cases C_1 and C_3 were tested, where C_1 was scaled from 4, 096 up to the full machine with 112, 768 cores and C_3 was generated on the full machine. A scaling for the latter case was not possible due to memory limitations. On JUQUEEN, cases C_1 and C_2 were tested, i.e., case C_1 was scaled from 4, 096 up to 131, 072 cores (1/4 of the

[2]HERMIT is the predecessor of the currently installed HAZEL HEN system at HLRS Stuttgart.

[3]JUQUEEN is the predecessor of the currently installed JUWELS system at JSC.

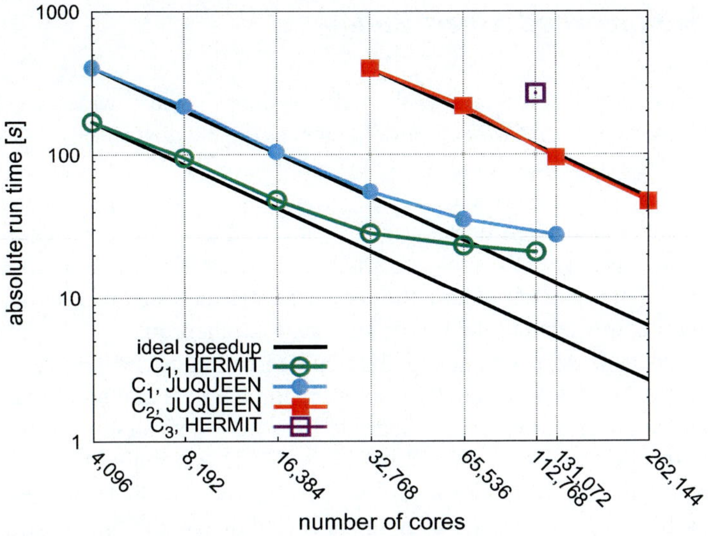

Fig. 6.11 Results of strong scaling experiments on the HERMIT and JUQUEEN HPC systems at HLRS and JSC. Three different cases with $\mathcal{C}_1 = 9.82 \cdot 10^9$, $\mathcal{C}_2 = 78.54 \cdot 10^9$, and $\mathcal{C}_3 = 0.64 \cdot 10^{12}$ elements were considered to mesh cubic domains [33]

system) and $\mathcal{C}_2$ from 32, 768 up to 262, 144 cores (1/2 of the system). Figure 6.11 shows the results of the scaling experiment. Obviously, case $\mathcal{C}_1$ scales well across both systems up to 32, 768 cores and experiences a slight drop in efficiency up to the maximum tested core counts. The JUQUEEN system performs slightly better than the HERMIT from a scalability point of view, the wall clock time is, however, better on HERMIT, i.e., it is 20.95 s on 112, 768 HERMIT cores compared to 27.63 s on 131, 072 JUQUEEN cores. Considering case $\mathcal{C}_2$, a perfect scalability is visible up to 262, 144 cores and even a slight superlinear behavior is visible for both core counts 131, 072 and 262, 144. The fastest execution is measured for 262, 144 JUQUEEN cores with 47.21 s. The large mesh $\mathcal{C}_3$ is generated in 267.99 s on the full HERMIT.

6.4 Quality of Meshes

Generating a quality mesh is a challenging step, which can be considered as an art on its own. Frequently, users' experiences in mesh design dictate the final mesh quality. There are, however, some general guidelines that can help to form a good basis to work from.

6.4.1 Skewness and Aspect Ratio

One primary source determining mesh quality is the cell shape, which can be characterised by its skewness and aspect ratio. Skewness or grid distortion, relates the angle θ between grid lines, where angles $\theta \approx 90°$ (orthogonal) are desired, see Fig. 6.12a. This is obviously the case for Cartesian meshes. For other types of elements, the mesh is considered skewed if it is $\theta \leq 45°$ or $\theta \geq 135°$. This classification can be applied to all types of cells and face shapes, and can always be applied to prism layers. For triangles and tetrahedra, the skewness is determined by relating the actual shape to their equivalent equilateral triangle and tetrahedron.

Large aspect ratios should always be avoided in important flow regions, where rapid changes, e.g., due to jets, flow separation, attachment and recirculation, occur. Distorted cells in these regions can degrade the solution accuracy and may result in poor convergence. An aspect ratio in the range of 0.2 to 5 within the interior region is desirable, while for boundary layers in the vicinity of a wall, the condition may be relaxed. In Cartesian meshes with uniform grid distance δx in all directions, the aspect ratio is obviously always 1.0. Figure 6.12b shows an examples of different aspect ratios for different kinds of cell types.

To summarise, a quality mesh should ensure that the following conditions hold:

- minimal cell skewness
- minimal changes in cell volumes between adjacent cells

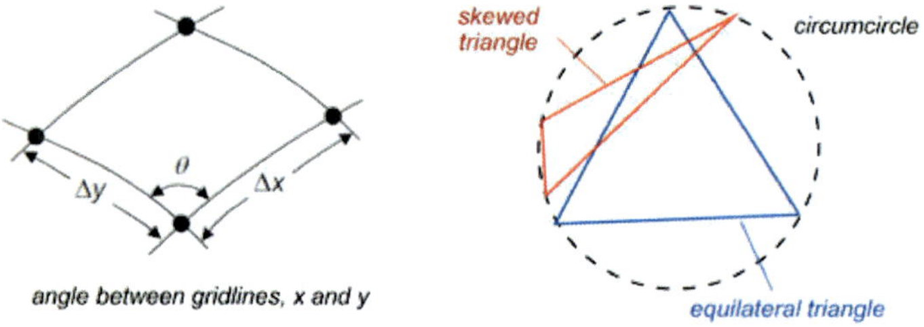

(a) Mesh skewness determined by the angle made between gridlines or based on the equivalent area (volume in 3D) of the actual cell relative to its equilateral cell.

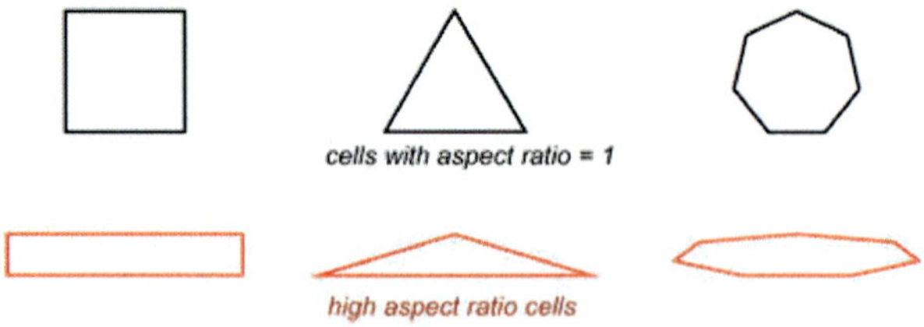

(b) Differently shaped cells with aspect ratios of 1.0 and their respective high aspect ratio cells.

Fig. 6.12 Skewness and aspect ratio of mesh elements

- suitable aspect ratios (depending on the cell location)
- sufficient resolution in regions of rapid flow changes, e.g., in regions with sharp gradients or boundary layers

6.4.2 *Mesh Independence*

Mesh independence refers to a mesh being optimised to have the minimum number of cells to obtain a solution that will not change if any further mesh refinement is made, i.e., the solution becomes independent of the resolution for decreasing element sizes and increasing numbers of cells. This typically involves monitoring a fluid flow parameter of interest, e.g., the velocity or the pressure, under successive grid refinement. Ideally, at least three significant, different grid resolutions should be evaluated, where each subsequent mesh is refined or coarsened. Upon refinement, the grid cells become smaller and the number of cells in the flow domain increases. If a global mesh refinement is not feasible, selective local refinement of the mesh in critical flow regions should be applied, cf. Sect. 6.2.3.

In the CFD community, mesh independence is evaluated by estimates of the discretisation error using Richardson extrapolation [44]. More recently, the grid convergence index (GCI) was introduced by Roache [45] as a method to uniformly evaluate the quality of CFD results. Inthavong et al. [20] propose an alternative method that uses a 2D cross-sectional plane subtraction method, where scalar values, e.g., the velocity magnitude, from a cross-sectional plane are interpolated onto a regularly spaced grid. The interpolated grid values from any two meshed models can then be compared by subtracting the values from each cell. In [20], six planes across the nasal cavity are used to represent the entire domain. The method is also extended to 3D by introducing a volume subtraction.

Performing mesh independence studies on a small number of patient models is feasible. However, where large numbers of patients, e.g., over 20, need to be treated, the task of performing these studies becomes extremely time consuming. In Chap. 1, Fig. 1.2b, the number of mesh elements used for the CFD simulations increased from order of tens of thousands of elements to millions of elements. The large variations in mesh numbers among different studies has led to investigations of what constitutes a sufficiently resolved mesh. Vinchurkar and Longest [54], Frank-Ito et al. [12], and Bass and Worth Longest [1] recommended prism cells meshes to resolve the near wall boundary layers, with Frank-Ito et al. [12] suggesting the use of 4 million tetrahedral cells with three prism layers for near-wall regions.

It should, however, be noted that a general estimation of the number of cells required to accurately run a simulation cannot be given. Flow phenomena such as jets, flow separation, and even turbulence strongly depend on the considered REYNOLDS number, which might locally vary and hence require an additional increase of the resolution. Furthermore, the method to solve the governing equations of fluid mechanics also plays an important role. While RANS (Reynolds Averaged Navier Stokes) computations might already reach grid independence at a relatively low numbers of cells, large eddy simulations (LES) and direct numerical simulations (DNS) require a massive amount of cells. This is exemplarily shown for a highly-resolved nasal cavity simulation in Sect. 6.6, where a sub-millimeter resolution and a number of cells $n > 100 \cdot 10^6$ is required [30, 32].

6.5 Advanced Mesh Applications: Lattice-Boltzmann Methods to Simulate Complex Flows in Complex Geometries

Many different numerical methods exist to simulate complex compressible or incompressible flow in intricate geometries. Fortunately, the flow in the nasal cavity is incompressible and it is not necessary to consider MACH numbers M that exceed the incompressibility regime limit of roughly $M \approx 0.3$ [17, 18, 30]. A straightforward approach is to numerically solve the governing equations of fluid mechanics, i.e., the Navier-Stokes equations, directly by means of, e.g., finite volume, finite difference, finite element, or discontinuous Galerkin methods. In these methods, in the incompressible limit, it is required to equate the pressure by solving the pressure Poisson equation.

In the last few decades, the lattice-Boltzmann (LB) method gained a lot of popularity. Instead of describing the fluid mechanics on a macroscopic level as done by the above mentioned computational methods, the LB method considers the statistics of particle interactions and distributions in a finite volume and derives the macroscopic flow variables from the statistics. The standard LB methods are quasi-incompressible, do, however, not require to solve a Poisson equation for the pressure as the pressure is an implicit solution of the numerically solved LB equation. Both DNS, which resolves all scales, and with some extensions, LES, which models non-resolved scales, can be performed with LB methods. Furthermore, the compute kernels can easily be parallelised leading to high parallel efficiency. In the LB method, complex boundaries can be treated in a straightforward manner with a high order of accuracy. This excels the LB method for the simulation of intricate flows in complex geometries such as the human respiratory tract.

In the following, a brief introduction is given to the physics of fluids from a statistical point of view and to different LB methods that are suited for the simulation of respiratory flows including LES modelling approaches. For more details on the fundamentals of fluid mechanics, the interested reader is referred to Chap. 7.

6.5.1 The Boltzmann Equation

Ludwig Boltzmann (1844–1906) developed a general equation to describe the evolution of the particle probability distribution functions (PPDF) $f(\mathbf{x}, ¸, t)$ following the laws of gas kinetics, i.e., the conservation of the number of particles and the statistical description of particle movement and collisions. The so called Boltzmann equation is given by

$$\overbrace{\underbrace{\frac{\partial f}{\partial t}}_{\text{temporal change}} + \underbrace{\boldsymbol{\xi} \cdot \nabla_{\mathbf{x}} f}_{\text{convective term}} + \underbrace{\frac{\mathbf{F}}{m} \cdot \nabla_{\boldsymbol{\xi}} f}_{\text{external forces}}}^{\text{transport}} = \overbrace{\int_{\boldsymbol{\xi}_\beta} \int_{\Omega} \boldsymbol{\xi}_r C(\boldsymbol{\xi}_r, \Omega)(f'_\alpha f'_\beta - f_\alpha f_\beta)\, d\Omega\, d\boldsymbol{\xi}_\beta}^{\text{collision}} . \tag{6.1}$$

The left-hand side (LHS) describes the collision step with the temporal change of the PPDFs, the convective term, and the influence of external forces. The right-hand side (RHS) describes the collision of particles in a statistical sense in an infinitesimal fluid volume. In (6.1) the quantity t represents the time, $\boldsymbol{\xi} = (\xi_1, \xi_2, \xi_3)^T$ is the vector of the molecular velocity, $\boldsymbol{\xi}_r = \boldsymbol{\xi}_\alpha - \boldsymbol{\xi}_\beta$ is the relative velocity vector between two particles with velocity vectors $\boldsymbol{\xi}_\alpha$ and $\boldsymbol{\xi}_\beta$, and $\mathbf{F}$ is the external forcing vector. Furthermore, $C(\boldsymbol{\xi}_r, \Omega)$ is the differential cross section of the collision, in which the relative momenta of the colliding particles turns through an angle θ into the element of the solid angle $d\Omega$, and PPDFs with $<'>$ are PPDFs after a collision. The nabla operators ∇ with subscripts $\mathbf{x}$ and $\boldsymbol{\xi}$ represent differentiation operators with respect to the spatial directions $\mathbf{x} = (x_1, x_2, x_3)$ and the velocity directions of $\boldsymbol{\xi}$. Via a Chapman–Enskog development the Navier-Stokes equations can be recovered from the Boltzmann equation, see e.g. [2].

Equation 6.1 is an integro-differential equation and hence hard to solve. Bhatnagar, Gross, and Krook (BGK) therefore developed the BGK equation [3], which replaces the RHS integral formulation of (6.1) by a relaxation towards a thermodynamical equilibrium, i.e., the lattice-BGK equation is given by

$$\frac{\partial f}{\partial t} + \boldsymbol{\xi} \cdot \nabla_{\mathbf{x}} f + \frac{\mathbf{F}}{m} \cdot \nabla_{\boldsymbol{\xi}} f = \omega_c(F - f), \tag{6.2}$$

where F is the PPDF at equilibrium and ω_c is the collision frequency.

6.5.2 Computational Lattice-Boltzmann Methods

Different numerical LB methods exist. The single relaxation (SRT) and multiple relaxation time (MRT) methods, are however, the most frequently used methods. Both are, depending on the considered REYNOLDS number, well suited for the simulation of respiratory flows and are discussed in the following before some details on boundary conditions and extensions for LES and refined meshes are presented. Finally, results of some performance analyses are shown.

6.5.2.1 Single Relaxation Time Method

To numerically solve (6.2), the equation is discretised in i-many space directions and in time, leading to the lattice-BGK equation [43]

$$f\left(\mathbf{x} + {}_{¸i}\delta t, t + \delta t\right) = f_i\left(\mathbf{x}, t\right) + \omega_x \delta t \cdot \left(f_i^{eq}\left(\mathbf{x}, t\right) - f_i\left(\mathbf{x}, t\right)\right) \quad (6.3)$$

with discretised equilibria PPDFs f_i^{eq}, and time increments δt. The equilibrium distribution function is given by

$$f_i^{eq}(\mathbf{x}, t) = \rho t_p \underbrace{\left[1 + \frac{v_a \xi_{i,a}}{c_s^2} + \frac{v_a v_b}{2c_s^2} \cdot \left(\frac{\xi_{i,a}\xi_{i,b}}{c_s^2} - \delta_{ab}\right)\right]}_{\chi}, \quad a, b \in \{1, 2, 3\} \quad (6.4)$$

where ρ is the non-dimensional density, t_p is a discretisation-method-direction-dependent weighting factor, $\xi_{i,a}$ and $\xi_{i,b}$ are components of ¸, v_a and v_b are components of the non-dimensional velocity vector $\mathbf{v} = (v_1, v_2, v_3)^T$, $c_s = 1/\sqrt{3}$ is the non-dimensional speed of sound, and δ_{ab} is the Kronecker delta. The discrete collision frequency is given by

$$\omega_x = \frac{c_s^2}{\nu + \delta t c_s^2/2} \quad (6.5)$$

with viscosity ν.

The number of directions i is determined by choosing a discretisation model. Qian et al. [43] developed the $DxQi$ model, which discretises the space in x dimensions with i-many directions. Examples in three dimensions are the $D3Q15$, $D3Q19$, and the $D3Q27$ models, where the former two models have been shown to not preserve rotational invariance [23, 56]. The last PPDF, i.e., $i \in \{15, 19, 27\}$ is always reserved for the rest-particle distribution. Figure 6.13 shows the directions for the different models in a Cartesian mesh setup, cf Sect. 6.3.

The derivation of f^{eq} employs a small MACH number approximation, which leads to the limitation of the LB method to quasi-incompressible flow and a decoupling of the energy equation from the equations of momentum conservation. To simulate thermal flow, the thermal LB (TLB) method is often used, which solves an additional scalar transport equation for the temperature T [16]

$$g\left(\mathbf{x} + {}_{¸i}\delta t, t + \delta t\right) = g_i\left(\mathbf{x}, t\right) + \omega_{x,T} \delta t \cdot \left(g_i^{eq}\left(\mathbf{x}, t\right) - g_i\left(\mathbf{x}, t\right)\right), \quad (6.6)$$

where the equilibrium is given by $g_i^{eq} = T t_p \chi$ and the collision frequency $\omega_{x,T}$ is a function of the heat conduction coefficient given by the PRANDTL number.

All macroscopic variables can be derived from the moments of the PPDFs, i.e.,

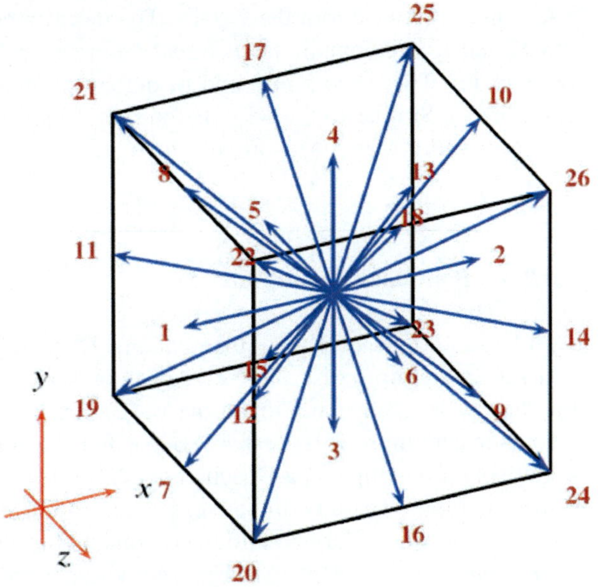

Fig. 6.13 $D3Qi$ discretisation scheme in three dimensions with directions $i \in \{1, \ldots, 27\}$. The remaining particle-distribution has subscript $i \in \{15, 19, 27\}$ for the models $D3Q15$, $D3Q19$, and $D3Q27$

$$\text{density:} \quad \rho = \sum_i f_i(\mathbf{x}, t), \tag{6.7}$$

$$\text{temperature:} \quad T = \sum_i g_i(\mathbf{x}, t), \tag{6.8}$$

$$\text{momentum:} \quad \rho v_a = \sum_i \xi_i^a f_i^a(\mathbf{x}, t), \tag{6.9}$$

where ξ_i^a are directions having a component $\xi_{i,a} \neq 0$ and f_i^a are the corresponding PPDFs. The pressure can be obtained from the density via $p = (1/3) \cdot \rho$.

From a algorithmic point of view, (6.3) and (6.6) are split into a collision step by solving their RHSs and storing the results in a temporary array, and into a propagation step (LHS) in which this information is forwarded to all neighbouring elements.

6.5.2.2 Multiple Relaxation Time Method

For higher REYNOLDS number flows, the SRT method might suffer from instabilities [24]. Therefore, MRT methods [8] have been developed, which relax, in contrast to the SRT methods, in moment space. In vector notation, the MRT equation is given by

$$\mathbf{f}(\mathbf{x} + \xi_i \delta t, t + \delta t) = \mathbf{f}(\mathbf{x}, t) - \underline{\mathbf{M}}^{-1} \underline{\mathbf{K}}_{MRT} \cdot \left[\mathbf{m}(\mathbf{x}, t) - \mathbf{m}^{eq}(\mathbf{x}, t) \right], \tag{6.10}$$

with **f** being the vector of the PPDFs. The moment coefficient and relaxation matrices $\underline{\mathbf{M}}$ and $\underline{\mathbf{K}}_{MRT}$ contain the different moments and the moment-specific collision frequencies. The vectors $\mathbf{m}^{eq}$ and $\mathbf{m}$ define the moments at equilibrium and non-equilibrium. Similar to the SRT method, the algorithm can be split into a collision and propagation step. For more information, especially for more details on the setup of the various matrices and vectors, the reader is referred to [8].

6.5.2.3 Boundary Conditions

At inlets and outlets, von Neumann and/or Dirichlet conditions are frequently prescribed. In a simple case, the velocity is prescribed via a Dirichlet condition and the density is extrapolated from the next inner computational elements via a von Neumann condition. This creates a source and is commonly used for inlets. A simple outlet condition employs a Dirichlet condition for the density and a von Neumann condition for the velocity. In both cases, the equilibrium can be computed via (6.4). More complex boundary condition combinations [30] iteratively lower the pressure in the sense that the density is reduced, and extrapolate the velocity at the outlet. At the same time the momentum is extrapolated at the inlet via solving the equation of Saint-Vernant/Wanzel [47] in LB context. This yields a realistic inspiration behaviour if the former boundary condition is prescribed at the pharynx and the latter at the nostrils.

To realize no-slip conditions at tissue walls, interpolated bounce-back conditions [5], which perform a reflective collision of the wall-pointing PPDFs, are frequently used. They are of second-order accuracy and reflect the PPDFs weighted by the distance of the element center to the wall in direction i. There exist further popular schemes such as those from Ginzburg and D'Humières [14] and Yu et al. [57], which also deliver highly accurate results.

6.5.2.4 Large-Eddy Simulations

To perform LES computations, the governing equations are filtered in space and time and a modelling assumption for non-resolved scales is introduced. For simplicity, the following describes modelling via the Smagorinsky approach [19, 51]. In LB context, the viscosity ν in ω_x is replaced by $\nu_{LES} = \nu + \nu_t$, where

$$\nu_t = (C_s \delta x)^2 \sqrt{\underbrace{2\left(\frac{\omega_x}{2\rho c_s^2}\sum_i \left[f_i(\mathbf{x},t) - f_i^{eq}(\mathbf{x},t)\right]\xi_{i,a}\xi_{i,b}\right)^2}_{\text{squared filtered strain rate}}} \tag{6.11}$$

is the turbulent viscosity with mesh resolution δx and Smagorinsky constant C_s.

6.5.2.5 Mesh Refinement

If computations are to be employed on refined meshes, some modifications to the algorithm are necessary [9]. Missing incoming PPDFs need to be reconstructed from neighbours on a different level. To obey the laws of physics, it is required to keep the viscosity constant across mesh level interfaces. This goes a long with an adaption of the time step δt and yields transformations

$$f_{i,\lambda+1}(\mathbf{x}, t) = \tilde{f}_i^{eq}(\mathbf{x}, t) + \Theta_{\lambda+1} \cdot (\tilde{f}_{i,\lambda}(\mathbf{x}, t) - \tilde{f}_i^{eq}(\mathbf{x}, t)) \tag{6.12}$$

$$f_{i,\lambda}(\mathbf{x}, t) = f_i^{eq}(\mathbf{x}, t) + \Theta_{\lambda} \cdot \left(f_{i,\lambda+1}(\mathbf{x}, t) - f_i^{eq}(\mathbf{x}, t)\right) \tag{6.13}$$

for levels λ and $\lambda + 1$. The PPDFs with a tilde <˜> are interpolated from the other mesh hierarchy. The transformations are functions of Θ_λ and $\Theta_{\lambda+1}$ given by

$$\Theta_\lambda = \frac{\delta x_\lambda}{\delta x_{\lambda+1}} \cdot \frac{\omega_{x,\lambda+1}}{\omega_{x,\lambda}} \tag{6.14}$$

$$\Theta_{\lambda+1} = \frac{\delta x_{\lambda+1}}{\delta x_\lambda} \cdot \frac{\omega_{x,\lambda}}{\omega_{x,\lambda+1}} \tag{6.15}$$

6.5.2.6 Performance of LB Methods

To analyze the parallel performance of LB methods, strong scaling experiments are performed for the SRT method on a cubic periodic domain with uniform mesh refinement, consisting of $1.1225 \cdot 10^9$ mesh elements [31, 32].

The experiments are run on the systems JURECA, JUQUEEN, and HAZEL HEN HPC systems located at JSC and HLRS. From Fig. 6.14 it is obvious that the LB method shows a very good strong scaling behaviour on all three systems, i.e., on JURECA an almost linear behaviour is visible up to 128 nodes, on HAZEL HEN the code scales well up to 512 nodes, and on the massively parallel system JUQUEEN a good scalability up to 8, 192 nodes with a slight decrease in parallel efficiency up to 16, 384 nodes is visible.

6.5.3 Summary and Conclusions on Using Lattice-Bolztmann Methods

Lattice-Boltzmann methods are a powerful alternative to conventional CFD methods such as finite volume, finite difference, finite element, and discontinuous Galerkin methods. Derived from the Boltzmann equation the single relaxation time and the multiple relaxation time methods are great options to simulate intricate flow in complex geometries. Especially their ability to implicitly compute the pressure and not to solve a Poisson equation in the incompressibility limit excels them. Furthermore,

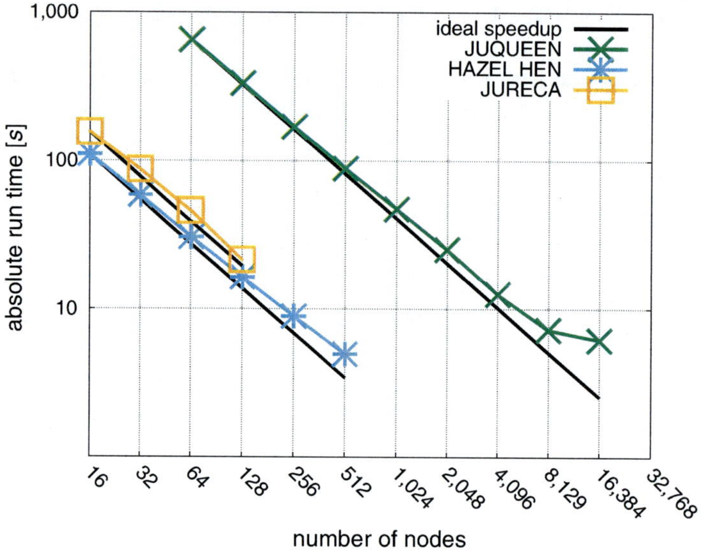

Fig. 6.14 Results of strong scaling experiments on the three different HPC systems JURECA, JUQUEEN, and HAZEL HEN. The mesh is uniformly refined and consists of $1.1225 \cdot 10^9$ mesh elements. [32]

their high parallel efficiency and their straightforward implementation on hierarchical Cartesian meshes suit them well for large-scale simulations on HPC systems. Complex geometries can be represented by accurate boundary conditions and there exist modelling approaches for LES computations and refinement strategies. All these features, together with their great applicability for low MACH number and REYNOLDS number flows, make them an extraordinary method for the simulation of respiratory flows, especially when applied in combination with the meshing tools presented in Sect. 6.3.

6.6 Advanced Mesh Applications: High Resolution Simulations without Any Turbulence Modelling

The computational simulation method in conjunction with the available computational resources determine a maximum resolution that can be achieved. Depending on this resolution, different approaches with increasing levels of modelling can be used to simulate flow.

A DNS resolves all scales of the flow and introduces no additional modelling. In contrast, an LES solves the spatially and temporally filtered governing equations and uses sub-grid scale (SGS) models such as the Smagorinsky [51] or dynamic Smagorinsky [13] models to reconstruct scales that are not resolved by the mesh, see

Sect. 6.5.2. A computationally less expensive but also less accurate method than DNS and LES is to solve the RANS equations and to employ additional turbulence models such as the k-ϵ [25], k-ω [37], or the k-ω SST [15] models. These models introduce additional equations to solve for the turbulent kinetic energy k, the rate of dissipation ϵ, and the specific dissipation of the turbulence kinetic energy into internal thermal energy ω. The flow in the nasal cavity is at respiration at rest laminar and in some regions transitional but not necessarily turbulent [17, 18, 30]. RANS was developed for fully turbulent flow and the Reynolds averaging assumes decent fluctuations around a mean flow. Furthermore, model constants need to be individually tuned per case delivering inconsistencies in the solutions. The results obtained by this method hence allow for only a tentative evaluation of the flow in the nasal cavity.

It is obvious that DNS and even LES computations are out of reach for researchers without any access to HPC systems and that hence RANS simulations, e.g., with commercial software, remain their only option to simulate flow. Considering, however, the trend in hardware and parallel software development it is assumed that LES and DNS will become available sooner or later as desktop applications.

In the following, results for highly resolved simulations of the flow in the nasal cavity are presented that do not employ any kind of turbulence modelling, i.e., the equations of fluid mechanics are solved as is by means of an LB method on fine meshes that capture all important flow scales, see Sects. 6.3 and 6.5. The results are based on the findings in [28–30, 32, 55] and make use of the framework Zonal Flow Solver (ZFS) [31]. The interested reader is referred to these publications for more details.

6.6.1 Mesh Resolution Required for Fully-Resolved Simulations

The mesh resolution is an important parameter for the accuracy of a simulation (see Sect. 6.4). It is of crucial interest to accurately determine various parameters that determine the quality of a nasal cavity, e.g., the pressure loss and the temperature increase as measures of the respiratory and heating capabilities. Furthermore, the wall-shear stress, which helps to identify regions of irritation, is of interest. Especially regions with high velocity gradients such as in the vicinity of the wall require a high resolution to accurately predict the aforementioned parameters.

To analyse the dependency of the mesh resolution on the simulation result, it is in most cases sufficient to perform grid convergence studies. This is done by solving the same case on successively refined meshes and to determine the change of the solution with respect to the essential flow parameters. Figure 6.15 shows a comparison of the temporally averaged velocity profiles at the pharynx for two resolutions, i.e., for a case G_{st} with $92.6 \cdot 10^6$ elements and $\delta x_{st} = 93.569 \cdot 10^{-3}$ mm, and a case G_f with $724 \cdot 10^6$ elements and $\delta x_f = \delta x_{st}/2$. Although some smaller differences are visible in the pharynx center region, especially in the vicinity of the walls, the curves of

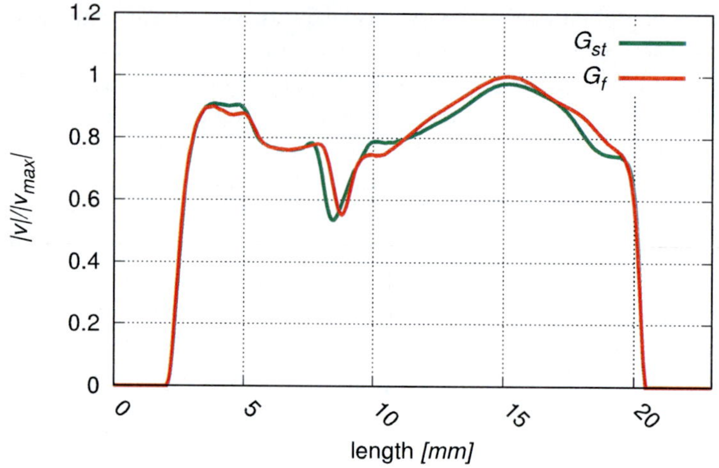

Fig. 6.15 Grid convergence analysis. Two profiles at the pharynx for mesh resolutions of $\delta x_{st} = 93.569 \cdot 10^{-3}$ mm (G_{st} with $92.6 \cdot 10^6$ elements) and $\delta x_f = \delta x_{st}/2$ (G_f with $724 \cdot 10^6$ elements) are juxtaposed. The velocity magnitude $|v|$ is normalised by the overall velocity magnitude $|v_{max}|$ along the profile line [30]

both profiles coincide. Considering furthermore the total pressure loss $\delta p_t = (p_s + p_d)_{pharynx} - (p_s + p_d)_{nostril}$, i.e., the difference of the sums of the static pressure p_s and the dynamic pressure $p_d = (\rho/2)|\mathbf{v}|^2$ at the nostrils and pharynx, the difference between G_f and G_{st} is only 2.1‰ and 0.8‰ for the left and right nasal cavity. The temperature difference $\delta T = T_{pharynx} - T_{nostril}$ can be evaluated similarly to the pressure loss as the difference in temperature between the nostrils and pharynx cross-sections. This yields a small difference of only $0.1985K$ between G_f and G_{st}. To summarise, the grid convergence analysis renders the resolution of G_{st} to be sufficient to simulate flow at respiration at rest.

Another method to evaluate a resolution is considering the power spectral density of the velocity components for various resolutions. Details on this method can be found in Lintermann et al. [30].

6.6.2 *Examples of Simulation Results*

In the following, several example results are presented that were generated by the highly scalable LB flow solver that operates on hierarchical Cartesian meshes (see Sect. 6.3 and 6.5). All simulations were run on supercomputers and the resolutions are in the sub-millimeter range to cover all essential flow phenomena. Three different cases N_1, N_2, and N_3 are considered that underwent an evaluation by medical experts beforehand. Figure 6.16 shows frontal CT-cross-sections of the different cases. Case N_1 was considered a healthy nasal cavity with only slightly swollen turbinates on

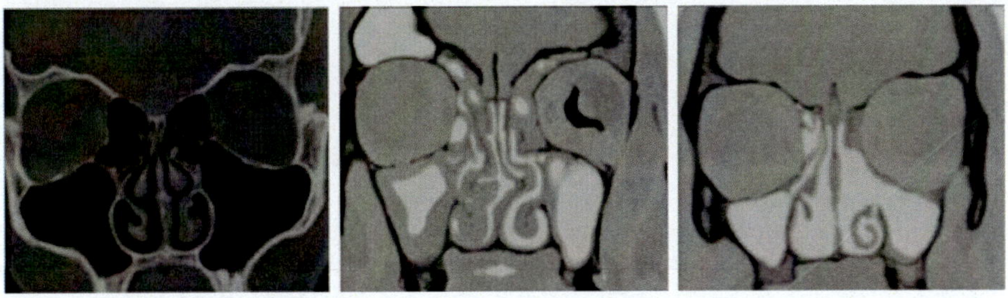

Fig. 6.16 Frontal CT-cross-sections of the nasal cavities N_1, N_2, and N_3 from left to right [30]

the left side. Case N_2 suffered from a bent septum, and swollen lower and center turbinates on the right side. Nasal cavity N_3 previously underwent a surgery in which the lower turbinate on the right side and the center turbinate on the left side were removed. Furthermore, a large orifice to the paranasal sinus on the left side existed. For further details, the interested reader is referred to the original publications [30, 32].

There are various results of a respiratory flow simulation that are of interest. As mentioned before, the difficulty to inhale can be quantified by considering the total pressure loss in the nasal cavity.

This is also a good parameter to juxtapose nasal cavities, i.e., to classify nasal cavities by their respiratory capabilities. Such a juxtaposition is shown in Fig. 6.17a for three nasal cavities N_1, N_2, and N_3. Obviously, case N_3 has the lowest pressure loss on both sides as the respiratory tract misses several turbinates. However, compared to N_1 the advantage is not excessive. In contrast, the swollen turbinates in N_2 lead to a high pressure loss and hence to impaired respiration. To furthermore find local phenomena responsible for the pressure loss, the total pressure along a stream line can be considered. The distribution of the mass flux in the nasal cavity can be analysed by measuring it in different cross sections or by visualising stream lines (see Fig. 6.18a). Another interesting parameter is the capability of a nasal cavity to heat the inhaled air. Figure 6.17b shows an evaluation for the same nasal cavities as in Fig. 6.17a. Undisturbed respiration in case N_3 comes with a diminished heating capability compared to cases N_1 and N_2, which almost heat up the air up to body temperature. To also analyse the impact of the flow on the tissue, i.e., to quantify the forces acting on the tissue, which might lead to irritations or even to inflammations, the wall-shear stress can be evaluated. Figure 6.18b shows a visualisation of the wall-shear stress mapped to the nasal cavity surface of case N_2. Obviously, the swollen turbinates and the bent septum lead to strong wall-shear stresses. Similar to the wall-shear stress, the heat flux can be evaluated by considering $\dot{q} = \kappa \partial T / \partial x_w$ with heat conduction coefficient κ. A result for case N_2 is shown in Fig. 6.18c. From both the wall-shear stress and the heat-flux it is clear that in narrowed geometries the less tempered inhaled flow passes close to the tissue and is thereby heated up quite well.

Fig. 6.17 Total pressure loss and temperature increase for three different cases N_1, N_2, and N_3 for the left (subscript l) and right (subscript r) cavities [30]

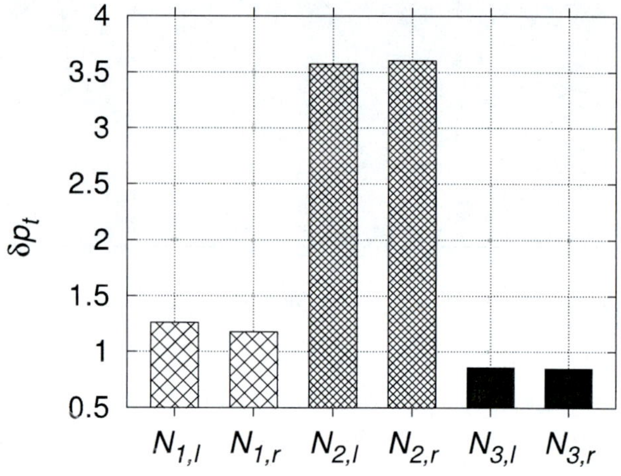

(a) Non-dimensional total pressure loss across the nasal cavities.

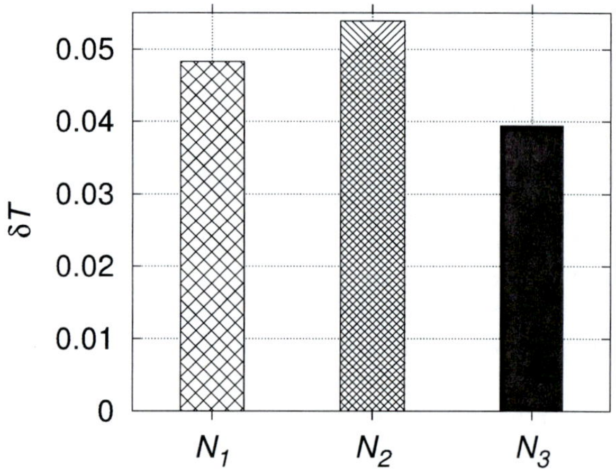

(b) Non-dimensional temperature increase across the nasal cavities.

Further in-depth details on respiratory flows can be obtained by analysing vortical structures, e.g., by means of visualisation techniques (Δ-, Q-criterion, or λ_2 contours), the Reynolds stress tensor, the turbulent kinetic energy, root-mean square values, cross- and autocorrelations, eddy tunraround times, and power spectral densities. These are quantities that directly have an impact on the important parameters such as the pressure loss, the temperature increase, the wall-shear stress, and the

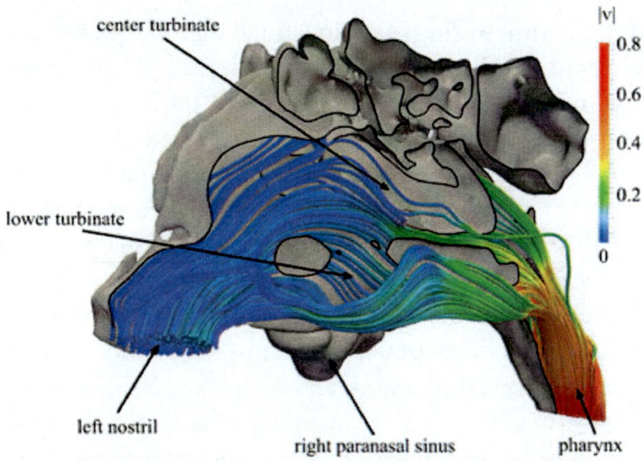

(a) Streamlines in the nasal cavity coloured by the non-dimensional velocity magnitude [33].

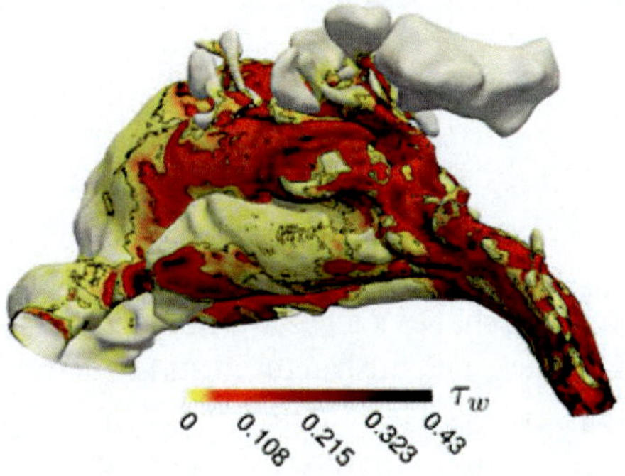

(b) Non-dimensional wall-shear stress mapped to the surface of a nasal cavity [30].

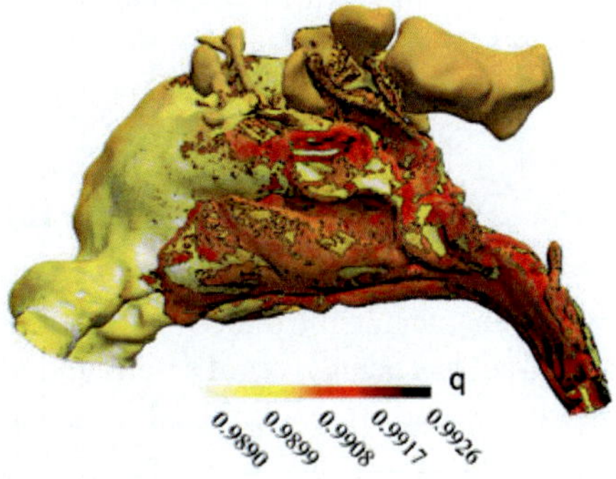

(c) Non-dimensional heat flux mapped to the surface of a nasal cavity [30].

Fig. 6.18 Results of highly resolved simulations from Lintermann [30, 33]

heat flux but are not necessarily crucial to classify and evaluate nasal cavities in preparation of a surgery. Further details can be found in [28–30, 32, 55].

6.6.3 Recapitulation: Using Highly-Resolved Meshes

Simulations in the nasal cavity can be run with varying grades of detail with DNS delivering results with the highest accuracy and RANS with the lowest. Although DNS are quite expensive in terms of computational costs they deliver an extraordinary variety of data that is to both the fluid-mechanics expert and the medical doctor of interest. Concerning the physics of respiration, a doctor might be interested in phenomena causing certain functional degradations, e.g., local constrictions in the airways, which lead to strenuous respiration. The pressure loss and the wall-shear stress are good indicators to find such locations that are potential candidates for a surgery. The latter might also help to find locations with an increased risk of dry-out, irritation, or inflammation. The capability of the nasal cavity to increase the temperature contributes to the well-being of the patient and is hence another key component of the fluid mechanics of respiration. While the global increase along the airways delivers rather superficial information on the heating capability, an analysis of the local heat flux distribution allows to locate regions with low heat flux and hence regions that have a low impact on the temperature increase.

The expert in fluid mechanics is furthermore interested in the finest details of the flow, i.e., in the formation of recirculation zones, the energy distribution, frequency analyses of fluctuating flow, Reynolds stresses, and so forth, to advance fundamental research. Obviously, experts from the medical community, engineering, and HPC need to work together in an interdisciplinary context to extract the maximum of information from highly resolved simulations that are important to treat patients and to advance the fluid-mechanical understanding of respiration and computer science research.

6.7 Summary and Conclusion

In this chapter, an introduction to meshing complex shapes such as the human respiratory tract has been given. Different kinds of meshes, i.e., unstructured and structured meshes with different kinds of mesh elements have been considered and their advantages and disadvantages have been discussed. Using structured meshes, although they allow for an efficient computation and require only a small amount of memory, has shown not to be feasible for complex geometries. Despite their disadvantages from an efficiency point of view, unstructured meshes can be generated fully automatically and are applicable to intricate shapes. However, in this case one has to take care that the quality of the mesh elements is guaranteed. This can be realised by avoiding skewed cells or elements with a too high or too low aspect ratio, and by guaranteeing a sufficient resolution. The latter can be analysed with the help of mesh independence studies.

In addition to fundamentals in meshing, a parallel grid generator has been presented. It is capable of creating large scale hierarchical Cartesian meshes fully automatically on HPC systems in a short amount of time. The grid generator scales up to several hundred thousands of processes. Boundary and patch refinement methods enable high resolutions where necessary to capture of all key flow features. The parallel meshing algorithm is hence a reasonable tool to construct meshes for highly-resolved flow simulations, e.g., in the human airways. Other than that, there also exist other solutions to generate unstructured meshes. They are frequently part of commercial software packages or are free to use and can be used in combination with commercial solvers.

Furthermore, the lattice-Boltzmann method to simulate complex flows in intricate geometries has been presented. It shows high flexibility and high scalability on state-of-the-art supercomputers and is hence well suited to run fully-resolved and highly accurate simulations in the human respiratory tract. This method has been applied to different nasal cavities and a mesh independence study has been presented. It shows to be a suitable tool to evaluate nasal cavities from a fluid-mechanics point of view. Analyses of the pressure loss, the heating capability, the wall-shear stress, and the heat flux are in line with clinical findings.

References

1. K. Bass, P. Worth Longest, Recommendations for simulating microparticle deposition at conditions similar to the upper airways with two-equation turbulence models. J. Aerosol Sci. **119**, 31–50 (2018)
2. R. Benzi, S. Succi, M. Vergassola, The lattice Boltzmann equation: theory and applications. Phys. Rep. **222**(3), 145–197 (1992)
3. P.L. Bhatnagar, E.P. Gross, M. Krook, A model for collision processes in gases. I. Small amplitude processes in charged and neutral one-component systems. Phys. Rev. **94**(3), 511–525 (1954)
4. J. Bonet, J. Peraire, An alternating digital tree (ADT) algorithm for 3D geometric searching and intersection problems. Int. J. Numer. Methods Eng. **31**(1), 1–17 (1991)
5. M. Bouzidi, M. Firdaouss, P. Lallemand, Momentum transfer of a Boltzmann-lattice fluid with boundaries. Phys. Fluids **13**(11), 3452–3459 (2001)
6. C. Burstedde, L.C. Wilcox, O. Ghattas, p4est: scalable algorithms for parallel adaptive mesh refinement on forests of octrees. SIAM J. Sci. Comput. **33**(3), 1103–1133 (2011)
7. M.O. Cetin, V. Pauz, M. Meinke, W. Schröder, Computational analysis of nozzle geometry variations for subsonic turbulent jets. Comput. Fluids **136**, 467–484 (2016)
8. D. D'Humières, I. Ginzburg, M. Krafczyk, P. Lallemand, L.-S. Luo, Multiple-relaxation-time lattice Boltzmann models in three dimensions. Philos. Trans. Ser. A, Math., Phys., Eng. Sci. **360**(1792), 437–451 (2002)
9. A. Dupuis, B. Chopard, Theory and applications of an alternative lattice Boltzmann grid refinement algorithm. Phys. Rev. E **67**(6), 1–7 (2003)
10. N. Filipovic, *Chapter 8 - Computational modeling of dry-powder inhalers for pulmonary drug delivery* (Academic Press, Cambridge, 2020)
11. M. Folk, E. Pourmal, Balancing performance and preservation lessons learned with HDF5, in *Proceedings of the 2010 Roadmap for Digital Preservation Interoperability Framework Workshop on - US-DPIF '10* (2010), pp. 1–8
12. D.O. Frank-Ito, M. Wofford, J.D. Schroeter, J.S. Kimbell, Influence of mesh density on airflow and particle deposition in sinonasal airway modeling. J. Aerosol. Med. Pulm. Drug Deliv. (2015)
13. M. Germano, U. Piomelli, P. Moin, W.H. Cabot, A dynamic subgridâscaleeddy viscosity model. Phys. Fluids A: Fluid Dyn. **3**(7), 1760–1765 (1991)
14. I. Ginzburg, D. D'Humières, Multireflection boundary conditions for lattice Boltzmann models. Phys. Rev. E **68**(6), 066614 (2003)
15. H. Grotjans, F. Menter, Wall functions for industrial applications, in *Computational Fluid Dynamics '98, ECCOMAS (1998)*, ed. by K. Papailiou (Wiley, Hoboken, 1998), pp. 1112–1117
16. Z. Guo, B. Shi, C. Zheng, A coupled lattice BGK model for the Boussinesq equations. Int. J. Numer. Methods Fluids **39**(4), 325–342 (2002)
17. I. Hörschler, M. Meinke, W. Schröder, Numerical simulation of the flow field in a model of the nasal cavity. Comput. Fluids **32**(1), 39–45 (2003)

18. I. Hörschler, W. Schröder, M. Meinke, On the assumption of steadiness of nasal cavity flow. J. Biomech. **43**(6), 1081–5 (2010)
19. S. Hou, J. Sterling, S. Chen, G.D. Doolen, A lattice Boltzmann subgrid model for high Reynolds number flows. Pattern Form. Lattice Gas Autom. **6**, 1–18 (1994)
20. K. Inthavong, A. Chetty, Y. Shang, J. Tu, Examining mesh independence for flow dynamics in the human nasal cavity. Comput. Biol. Med. **102**, 40–50 (2018)
21. T. Isaac, C. Burstedde, O. Ghattas, Low-cost parallel algorithms for 2:1 octree balance, in *2012 IEEE 26th International Parallel and Distributed Processing Symposium* (IEEE, 2012), pp. 426–437
22. T. Ishida, S. Takahashi, K. Nakahashi, Efficient and robust cartesian mesh generation for building-cube method. J. Comput. Sci. Technol. **2**(4), 435–446 (2008)
23. Y. Kuwata, K. Suga, Anomaly of the lattice Boltzmann methods in three-dimensional cylindrical flows. J. Comput. Phys. **280**, 563–569 (2015)
24. P. Lallemand, L.-S. Luo, Theory of the lattice Boltzmann method: dispersion, dissipation, isotropy, Galilean invariance, and stability. Phys. Rev. E **61**(6), 6546–6562 (2000)
25. B.E. Launder, D.B. Spalding, The numerical computation of turbulent flows. Comput. Methods Appl. Mech. Eng. **3**(2), 269–289 (1974)
26. J. Li, M. Zingale, W.-k. Liao, A. Choudhary, R. Ross, R. Thakur, W. Gropp, R. Latham, A. Siegel, B. Gallagher, Parallel netCDF: a high-performance scientific I/O interface, in *Proceedings of the 2003 ACM/IEEE conference on Supercomputing - SC '03* (ACM Press, New York, USA, 2003), p. 39
27. A. Lintermann, Efficient parallel geometry distribution for the simulation of complex flows, in *Proceedings of the VII ECCOMAS Congress 2016* (Athens, 2016). Technical University of Athens (NTUA) Greece, pp. 1277–1293
28. A. Lintermann, G. Eitel-Amor, M. Meinke, W. Schröder, Lattice-Boltzmann solutions with local grid refinement for nasal cavity flows. *New results in numerical and experimental fluid mechanics VIII* (Springer, Berlin, 2013), pp. 583–590
29. A. Lintermann, M. Meinke, W. Schröder, Investigations of the inspiration and heating capability of the human nasal cavity based on a lattice-boltzmann method, in *Proceedings of the ECCOMAS Thematic International Conference on Simulation and Modeling of Biological Flows (SIMBIO 2011)* (Belgium, Brussels, 2011)
30. A. Lintermann, M. Meinke, W. Schröder, Fluid mechanics based classification of the respiratory efficiency of several nasal cavities. Comput. Biol. Med. **43**(11), 1833–1852 (2013)
31. A. Lintermann, M. Meinke, W. Schröder, Zonal flow solver (ZFS): a highly efficient multi-physics simulation framework. Int. J. Comput. Fluid Dyn., 1–28 (2020)
32. A. Lintermann, D. Pleiter, W. Schröder, Performance of ODROID-MC1 for scientific flow problems. Futur. Gener. Comput. Syst. **95**, 149–162 (2019)
33. A. Lintermann, S. Schlimpert, J. Grimmen, C. Günther, M. Meinke, W. Schröder, Massively parallel grid generation on HPC systems. Comput. Methods Appl. Mech. Eng. **277**, 131–153 (2014)
34. A. Lintermann, W. Schröder, A hierarchical numerical journey through the nasal cavity: from nose-like models to real anatomies. Flow, Turbul. Combust. (2017)
35. A. Lintermann, W. Schröder, Simulation of aerosol particle deposition in the upper human tracheobronchial tract. Eur. J. Mech.-B/Fluids **63**, 73–89 (2017)
36. A. Loseille, V. Menier, F. Alauzet, Parallel generation of large-size adapted meshes. Procedia Eng. **124**, 57–69 (2015)
37. F. Menter, Zonal two equation k-ω turbulence models for aerodynamic flows, in *24th AIAA Fluid Dynamics Conference* (Orlando, FL, USA, 1993), pp. AIAA paper 93–2906
38. T. Nakashima, M. Tsubokura, M. Vázquez, H.C. Owen, Y. Doi, Coupled analysis of unsteady aerodynamics and vehicle motion of a heavy-duty truck in wind gusts. Comput. Fluids **80**, 1–9 (2012)
39. H. Nishikawa, B. Diskin, Development and application of parallel agglomerated multigrid methods for complex geometries, in *20th AIAA Computational Fluid Dynamics Conference* (Hawaii, Honolulu, 2011), pp. 27–30

40. M. Peric, Flow simulation using control volumes of arbitrary polyhedral shape. ERCOFTAC Bulletin, No. 62 (2004)
41. A. Pogorelov, M. Meinke, W. Schröder, large-eddy simulation of the unsteady full 3d rim seal flow in a one-stage axial-flow turbine. Flow, Turbul. Combust. (2018)
42. A. Pogorelov, L. Schneiders, M. Meinke, W. Schröder, An adaptive cartesian mesh based method to simulate turbulent flows of multiple rotating surfaces. Flow Turbul. Combust. **100**(1), 19–38 (2018)
43. Y.H. Qian, D. D'Humières, P. Lallemand, Lattice BGK models for Navier-Stokes equation. Europhys. Lett. (EPL) **17**(6), 479–484 (1992)
44. L. Richardson, On the approximate arithmetical solution by finite differences of physical problems involving differential equations, with an application to the stresses in a masonry dam. Proc. R. Soc. London. Ser. A **83**, 335–336 (1910)
45. P. Roache, Perspective: a method for uniform reporting of grid refinement studies. J. Fluids Eng. **116**, 405–413 (1994)
46. H. Sagan, *Space-filling curves*, 1st edn. (Universitext. Springer,New York, 1994)
47. B. Saint-Vernant, L. Wantzel, Mémoire et expérience sur l'écoulement déterminé par des différences de pressions considérables. Journal de l'École Polytechnique H **27**, 85ff (1839)
48. L. Schneiders, J.H. Grimmen, M. Meinke, W. Schröder, An efficient numerical method for fully-resolved particle simulations on high-performance computers, in *Proceedings in Applied Mathematics and Mechanics* (Lecce, Italy, 2015), GAMM, Ed., Wiley-VCH
49. L. Schneiders, C. Günther, M. Meinke, W. Schröder, An efficient conservative cut-cell method for rigid bodies interacting with viscous compressible flows. J. Comput. Phys. **311**, 62–86 (2016)
50. L. Schneiders, M. Meinke, W. Schröder, Direct particleâfluid simulation ofKolmogorov-length-scale size particles in decaying isotropicturbulence. J. Fluid Mech. **819**, 188–227 (2017)
51. J. Smagorinsky, General circulation experiments with the primitive equations. Mon. Weather Rev. **91**(3), 99–164 (1963)
52. J.F. Thompson, B.K. Soni, N.P. Weatherill, *Handbook of grid generation* (Taylor & Francis Inc., CRC Press, 1998)
53. J.F. Thompson, Z. Warsi, C.W. Mastin, *Numerical grid generation: foundations and applications* (Elsevier Science Pub. Co., New York, 1985)
54. S. Vinchurkar, P.W. Longest, Evaluation of hexahedral, prismatic and hybrid mesh styles for simulating respiratory aerosol dynamics. Comput. Fluids **37**, 317–331 (2008)
55. M. Waldmann, A. Lintermann, Y.J. Choi, W. Schröder, Analysis of the effects of MARME treatment on respiratory flow using the lattice-Boltzmann method. *New results in numerical and experimental fluid mechanics XII* (2020), pp. 853–863
56. A.T. White, C.K. Chong, Rotational invariance in the three-dimensional lattice Boltzmann method is dependent on the choice of lattice. J. Comput. Phys. **230**(16), 6367–6378 (2011)
57. D. Yu, R. Mei, L.-S. Luo, W. Shyy, Viscous flow computations with the method of lattice Boltzmann equation. Prog. Aerosp. Sci. **39**(5), 329–367 (2003)

Chapter 7
Fundamentals of Fluid Dynamics

Alister J. Bates

Abstract The purpose of CFD is to map and quantify fluid flows in space and time. CFD simulations achieve this by solving the equations that govern flow: conservation of mass; balance of momentum; and conservation of energy. This chapter describes what these equations conceptually mean and how they are derived by applying basic physics statements such as Newton's second law and the first law of thermodynamics to fluids. Fluid flow phenomena including boundary layers, turbulence, and unsteadiness are introduced and techniques to model them via CFD simulations are discussed. Turbulence models, which allow CFD simulations to achieve accurate results without having to calculate the smallest velocity fluctuations in a flow, thereby accelerating simulations, are also discussed and the most common models used in respiratory airflow simulations are compared.

7.1 Introduction

7.1.1 CFD Fundamentals

The purpose of computational fluid dynamics (CFD) is to map and quantify fluid flows in space and time. CFD simulations calculate how a fluid will behave given certain constraints, forces, and previous behaviours. Just three basic principles are required to achieve this:

- Conservation of Mass
- Newton's Second Law of Motion
- Conservation of Energy

A. J. Bates (✉)
Cincinnati Children's Hospital, Cincinnati, USA
e-mail: alister.bates@cchmc.org

K. Inthavong et al. (eds.), *Clinical and Biomedical Engineering in the Human Nose*, Biological and Medical Physics, Biomedical Engineering,
https://doi.org/10.1007/978-981-15-6716-2_7

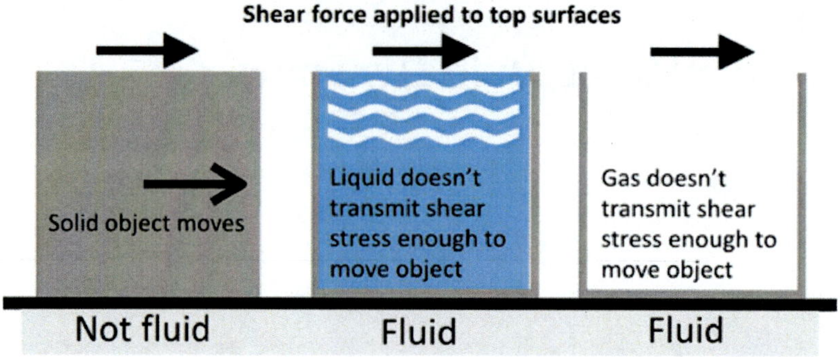

Fig. 7.1 Transmission of shear stress. The first container is solid, so transmits shear stress well. Pushing sideways on the solid container moves the whole container sideways. The other two containers are full of water and air respectively, which are fluids and don't transmit shear stresses well. We can't move these open containers by applying a sideways force to the top surface of the fluids

This chapter will review each of these principles in detail in Sect. 7.2, first conceptually, and then, in Sect. 7.2.5, mathematically. Before delving into these principles, we will begin by defining some necessary concepts and asking why we need these principles.

7.1.2 CFD Solver Basics

Fluids

The first question to ask is what do we mean by a fluid? This becomes especially relevant in CFD of airflow in the respiratory system, because clinicians use the term fluid to mean liquid or mucus in the airways, a common problem in patients with certain respiratory diseases. In engineering, we often assess whether a material can be treated as a fluid based on how well it transmits shear stress. Fluids do not transmit shear stresses as well as solids. This idea is perhaps best illustrated with an example. Imagine two containers on a table. The first is a solid box. If we push sideways on the top surface, we can move the entire box sideways as shown in Fig. 7.1. We have a second container, with the same size and weight as the first, but this time the container has no upper surface and is full of water. If we press sideways on the surface of the water, the container does not move. No matter how hard we press sideways on the surface of the water, the box containing the water will not move. The reason is that water does not transmit shear stress well, so we call it a fluid. Similarly, air does not transmit shear stress well. If we had a third container with no upper surface that was full of air, we could not move the container just by moving our hands sideways over the open top of the container. Therefore, air (in fact all gases) are fluids to an engineer, and we can use CFD simulations to map their behaviour.

Basic CFD Concepts

Boundary Conditions: To set up a CFD simulation, we create a region of interest surrounded by boundary conditions. This means we define what is happening at the edge of our region of interest. For example, in respiratory CFD, our region of interest may be the airway from the nose to the carina, or it may descend further into the lungs, or include a region of air outside the nose. Whatever our region of interest is, we need to set rules about what happens on the boundary of that region. For example, how is the air entering our region of interest, and how is it leaving? Are the walls hot and therefore warming the air that rushes past it? Are the walls stationary? Air entering and leaving our region of interest is usually defined in terms of pressure and flow rates. Differences in pressure between locations in the airway cause flow to move between the locations. For example, air tends to flow towards regions of low pressure. Therefore in respiratory CFD, we usually either set the pressure differential and measure how much flow moves in response or we set the amount of airflow entering the airway and measure what pressure difference between the inlet and the outlet was required to drive that flow.

Once we have set the boundary conditions, we can use CFD simulations to calculate the subsequent course of air within the airways. This works because air will always behave according to certain rules, which are described in Sect. 7.2. What makes air behave differently in one airway from another, or indeed as air flows along the side of an aeroplane, is purely due to the boundary conditions.

Control Volumes: It is often convenient to think about fluids in terms of a specific region of the flow, i.e. define a specific region of interest (a 'control volume') and then consider what is happening at the edges of that region. How much flow is going in and out? What are the pressure forces at the edges? We don't usually need to worry about what happens inside the control volume. Depending on what we use the control volume for, control volumes can be large, but often they are very small as this allows us to make assumptions about what is going on inside more easily. Control volumes may be stationary, with flow passing through them, or may move along with the flow. In Sect. 7.2.5, we will use control volumes to define the governing equations of fluid flow. We will use a static control volume to derive the continuity equation and a moving control volume to derive the momentum and energy equations. This choice is purely for convenience and we could use either type of control volume to derive any of the equations. When we use a stationary control volume, the equation we derive is known as the conservation form of the equation. A moving control volume produces the non-conservation form. Each cell in a CFD mesh (see Chap. 6) can be thought of as its own small control volume, with flow moving through it and pressure acting on the faces of the cell.

Non-dimensional Numbers: Non-dimensional numbers are a way of describing flows without worrying about specific details of an individual case. They are composed of ratios of variables—such as different types of forces acting within a flow, or a flow's ability to transfer quantities such as heat or momentum. The numbers are non-dimensional as they have no net units (i.e. the units in the top and bottom of the equations cancel out). Non-dimensional numbers allow us to characterise flows and compare them to other flows with similar behaviour but different absolute values.

Why We Need the Navier–Stokes and Conservation Equations
Let's suppose we are interested in determining the pressure, velocity, and temperature of the airflow throughout our airway. That means we have five unknowns to determine—pressure and temperature, and also the three components of the 3-dimensional velocity vector. Pressure and temperature are scalars, so don't have directions associated with them. When there are five unknowns, we need five independent equations to obtain their solutions. Section 7.2 describes how we obtain these equations. The fundamental equation of fluid dynamics is simply Newton's 2nd law of motion, $\mathbf{F} = m\mathbf{a}$, applied to a fluid. This is known as the Navier–Stokes equation. This is a vector equation, so it really counts as 3 equations, because we can write it as a separate equation in each of the x, y, and z directions. The concepts of conservation of mass and energy, combined with the Navier–Stokes equations, provide us with the required five equations. In fluid dynamics these equations are always used together and the term Navier–Stokes equations is often applied to the whole group of equations described below.

CFD simulations produce many more outputs than the pressure, velocity, and temperature, but many variables, such as wall shear stress or vorticity, are derived from these primary variables.

7.2 Fluid Dynamics and Governing Equations

7.2.1 Conservation of Mass

Conservation of mass means simply that fluid mass cannot be created or destroyed. Any fluid that enters our area of interest must go somewhere. Depending on the situation, this new fluid entering our flow domain may either (1) cause an equal mass of fluid to exit the domain, or (2) the density of the fluid may increase as it absorbs the new mass, or (3) the domain may change in size to accommodate this new mass.

To illustrate this, let's make some assumptions about our area of interest and our fluid. First, let's consider that our area of interest is a pipe of a fixed size. As it can't change in size to accommodate fluid flowing into it, option (3) above is not permitted in this case. Secondly, let's assume our fluid is incompressible. That means the fluid has a fixed, constant density, i.e. a given volume of the fluid will always have the same mass. As this assumption prevents the density of the fluid increasing as new fluid enters our pipe, option (2) above can't happen and we are left with option (1)—fluid flowing into a domain will cause an equal mass of fluid to exit the domain. Therefore, if we push fluid into our rigid pipe at a flow rate of 100 ml/s, we will get 100 ml/s out

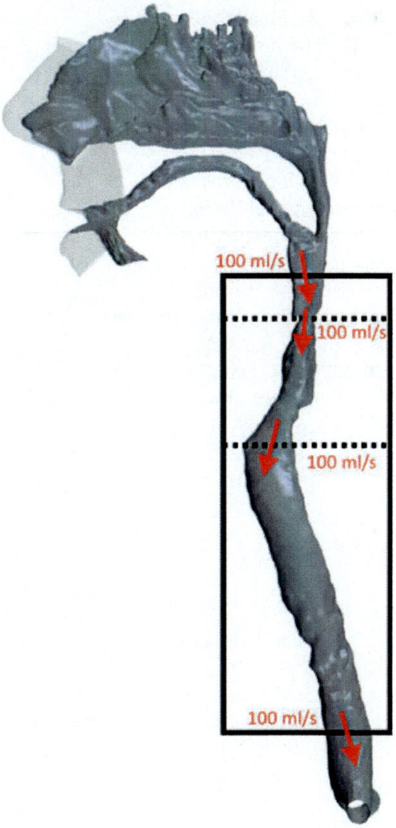

Fig. 7.2 An upper airway model obtained from an adult. Various regions of interest are shown via the black boxes. Conservation of mass tells us that if the flow is incompressible and the airway walls are rigid, whatever flows into the top of our region of interest must flow out at the other end of the region of interest

of the other end. Notice, that we didn't make any assumptions about the length of our pipe. Whether the pipe is 1 cm or 1 mile long, this still holds true (assuming we are driving the flow with a pump which is powerful enough to maintain our desired flow rate).

Let's consider how this affects airflow in the airway. We will choose a section of the airway with only one path through (i.e. not the nasal passages, where some airflow goes into the left nostril and some into the right). Let's choose the airway from the level of the back of the tongue (the oropharynx) and after the oral and nasal airways have joined together, and extending to just above the carina, where the airway splits into the two main bronchi. We will use the same assumptions as above—that the airway is rigid and won't expand or contract, and that air is incompressible.

At the inlet to our domain (Fig. 7.2), the flow rate is 100 ml/s. We know therefore that the flow rate out of our domain, at the end of the trachea must also be 100 ml/s. However, our choice of domain was arbitrary and we could have chosen our domain to finish somewhere else in the airway. Therefore, if we choose our domain to end at the vocal folds, the airflow through the vocal folds must also be 100 ml/s. Similarly at the level of the epiglottis, the flow rate is 100 ml/s as it is everywhere throughout

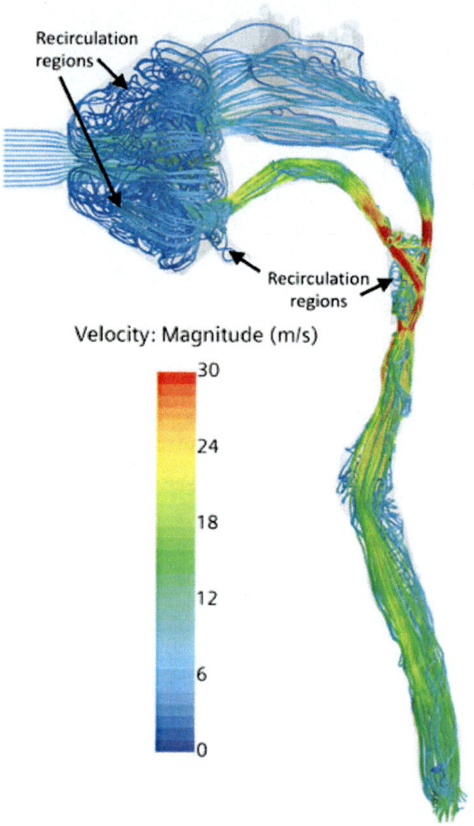

Fig. 7.3 An adult's upper airway with streamlines showing the path of airflow from the inlet to a mask worn by the subject and through the airway. There is a recirculation region, where the flow travels in the opposite direction to the main flow in the mask and oropharynx. However, conservation of mass tells us that the net mass flow through this region must still equal the mass flow elsewhere in the airway

our domain. We know that the cross-sectional area of the airway lumen is different at each of these locations within the airway. In this case, it is smallest through the vocal folds and largest in the trachea. To get the same flow rate through a narrow section as through a large section, the velocity of the airflow must change. Therefore, anywhere that the airway is narrow, such as the vocal folds, the airflow must travel faster.

Figure 7.3 shows airflow streamlines from a CFD simulation of inspiratory airflow. The streamlines do not follow a uniform path through the airway but form recirculation regions, such as in the mask and oropharynx where the flow doubles back on itself. If our region of interest ends in a region where the flow is moving in multiple directions, conservation of mass tells us that there must still be a flow exiting the region equal to that entering. Therefore, it is the net flow of air leaving our region that is important, and any flow re-entering the domain must be countered by an increased flow rate leaving the domain.

Bronchoscopy tells us that the airway is not in fact rigid. The tissue structures surrounding the large airways, such as the tongue, move as we breathe, and the trachea dilates as we breathe in and contracts as we breathe out due to the pleural pressure acting on the trachea's exterior. This means our first assumption above (that the airway is rigid) is not physiological. To account for the change in volume of the airway, if the volume of the airway is increasing, the flow rate out of the domain must be decreased to account for the amount of fluid it takes to fill the extra volume of the airway. Conversely, if the volume of the airway decreases, the flow rate out of the domain will increase due to the air displaced by the shrinking airway. The extreme case of volume change is inflating a balloon, where there is no flow out of the balloon, but the volume of the balloon changes to account for the air flowing into it. The expansion of the lungs or terminal airways can be modelled in this way.

Air is not really incompressible, so our second assumption above is also not realistic. Air can become compressible with high enough velocity. However, the change in air density due to velocity does not become a significant factor until the airspeed reaches a Mach number of around 0.3 (0.3 times the speed of sound—around 100 m/s), much faster than the velocity of air within the respiratory system, which usually doesn't reach more than 30 or 40 m/s.

Air can also change in density due to pressure, but again this does not usually occur within the airways. The pressures within the respiratory system are not high enough to change the air density significantly. The ideal gas law tells us that a change in density of air by 10% from its density at atmospheric pressure would require an airway pressure gradient of approximately 78 cmH_2O ($\approx$7,700 Pa). That value is much higher than the pressure differences usually found in the airway.

In the airway, the most likely factor to change the air's density is the heating and humidification of air. As we inhale air, we change the temperature and humidity of the air to approximately body temperature and 100% humidity. Inhaling air at 20° C and 50% humidity would cause a change of density of 7% as the air is warmed and humidified in the nose. Therefore, CFD simulations designed to include these effects should account for the change in density. In this case, it is necessary to ensure that mass flow rates in and out of an area of interest are conserved, even though the volumetric flow rates will differ proportionally to the difference in density.

7.2.2 Newton's Second Law

Newton's Second Law tells us that the sum of forces acting on an object of constant mass are equal to the object's rate of change of momentum. Commonly written $\mathbf{F} = m\mathbf{a}$, where $\mathbf{F}$ is the sum of forces acting on the object, m is the object's mass, and $\mathbf{a}$ its acceleration. In fluid dynamics, we cannot consider a single molecule of our fluid (it would be computationally impossible to track every molecule of our fluid), but instead consider how forces act on a small region of our fluid. This small region is similar to the area of interest we considered in Sect. 7.2.1, but not necessarily extending across the whole width of the airway and formally known as a control volume, as described in Sect. 7.1.2. For convenience, our control volume will move along with the flow in this example.

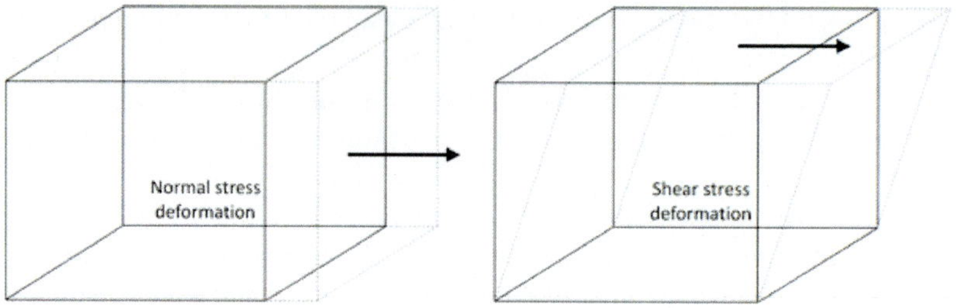

Fig. 7.4 Normal and shear forces in the x-direction acting on a control volume moving with the flow

To apply $F = ma$ to our control volume, let's first consider the forces acting on our control volume. We can divide these forces into several groups. First we have body forces, that is forces remotely acting on the fluid as a whole, such as gravity. Next, we have surface forces—forces that act on the surface of our control volume. Surface forces can be further divided into **pressure forces** that act directly on our control volume and **shear stress** forces that act along the face of our control volumes, distorting its shape (Fig. 7.4).

Pressure is generally thought of as a gradient rather than in absolute terms. As we are surrounded by the atmosphere, which has its own pressure fluctuations, it is the difference in pressure between two locations that is important. Flow always wants to move away from high pressure and towards low pressure. When we want to inhale, we make our lungs expand, causing the pressure within them to drop below atmospheric pressure. The airflow rushes in along our airway to the lungs. When we want to breathe out, we contract our lungs, causing the pressure to rise. The air then escapes along the airway to the comparatively low pressure of the outside atmosphere.

Shear stress forces acting on a fluid are occur when there is a gradient of velocity in one or more direction across our control volume. As air flows through our airway, it does not move at the same speed. Air adjacent to the surface of the airway does not move due to the friction between the air molecules and the molecules of the surface (this is known as the **no-slip condition** and occurs on most surfaces, not only in the airway). However, air that is far from the airway surface will move quickly. Between the fast moving air in the middle of the airway and the stationary air at the airway wall, the air's velocity gradually changes from fast to slow (Fig. 7.5).

As air molecules move past each other, frictional forces exchange momentum between air at different speeds, smoothing out the gradient of velocity. The region of slow moving air close to a surface is called the boundary layer, and it is vital to model this layer accurately in CFD simulations (see Chap. 6). The friction between air moving at different speeds is known as the fluid's viscosity. Fluids with high viscosity (e.g. treacle or molasses) will have more friction between layers of fluid than those with low viscosity (e.g. air) as the high friction exchanges momentum between fluid regions. In fluids, the exchange of a property (in this case momentum) between regions of the fluid is known as **diffusion**.

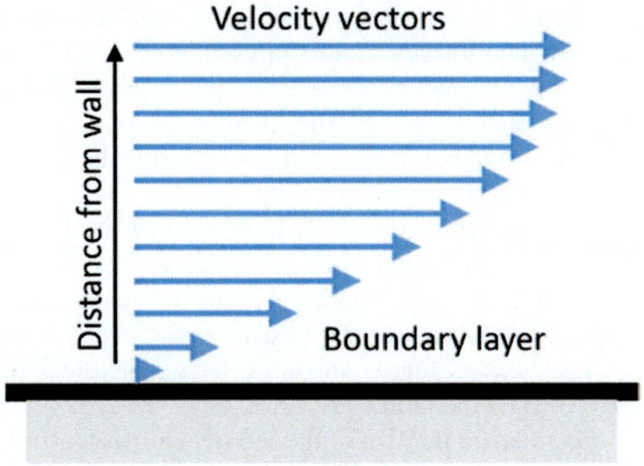

Fig. 7.5 Velocity profile in a boundary layer. Flow is stationary at the wall and fast in the midstream. Viscosity is friction between layers of flow, which causes momentum transfer between layers of flow and a smooth velocity gradient to form from the wall to the mid-stream

Just as we divided the force term in Newton's Second Law into groups, our control volume's rate of change of momentum can be divided into two groups. The first is a change of momentum with respect to time. For example, we don't inhale at the same speed all day, but breathe in, slow our inhalation down to a stop, then breathe out. Air within the airway will therefore be accelerated into the airway as we inhale, then decelerate as we stop inhaling. It will accelerate in the opposite direction as we exhale and then decelerate again.

The second change of momentum is due to the change in position of our region of interest. As a piece of fluid moves from a slow flowing region to a fast flowing region, it will accelerate. This is analogous to a car moving quicker on a clear section of highway than in a congested section of highway—the car's velocity has changed due to a change to its position. In fluids, this effect is known as **convection**.

We can now write out a word equation representing $F = ma$ for a fluid and replace the force and rate of change of momentum terms with the mechanisms responsible for these phenomena.

$$\begin{aligned}&\text{Body Forces} - \text{Pressure gradient} + \text{Diffusion}\\&\qquad = \text{Change in momentum with time} + \text{Convection}\end{aligned} \tag{7.1}$$

The right hand side of the equation represents acceleration of the flow. Flow is accelerated when there is a negative pressure gradient and therefore, the pressure gradient term here is negative. In other words, flow tends to move from high to low pressure.

Reynolds Number

Now that we have separated the balance of forces and changes in momentum in our fluid into various terms, considering the relative importance of each of these terms can help us understand the behaviour of our flow.

The rate of change of momentum part of Newton's second law, ma, can be thought of as **inertial forces** which represent how hard it is to accelerate the fluid due to its mass. These forces are represented by the product of the fluid's density, ρ, its velocity, U, and a length scale, D. The choice of length scale depends on the conduit the flow is passing through. For flow within a cylindrical pipe, the pipe's diameter is often used, or in a non-circular pipe, the hydraulic diameter, D_h (the ratio of 4 times the pipe's cross-sectional area to its perimeter) can be used. There is a degree of subjectivity in the selection of length scale, particularly in respiratory airflow, as the airway changes size and shape change abruptly along its length. Therefore, care needs to be taken to ensure that the chosen length scale is suitable.

The **viscous forces** are those frictional forces between fluid layers which are governed by the fluid's viscosity, μ.

To compare the size of the inertial and viscous forces within a fluid flow, we can take the ratio of them. This ratio is known as the flow's **Reynolds Number**, which is often abbreviated to Re.

$$\text{Reynolds Number} = Re = \frac{\text{Inertial Forces}}{\text{Viscous Forces}} = \frac{\rho U D}{\mu} \tag{7.2}$$

Reynolds Number is important because just four parameters can tell us a lot about a flow. It consists of two parameters that concern the type of fluid flowing (the density and viscosity), one parameter (the characteristic distance) that tells us about the geometry through which the fluid is flowing, and the velocity tells us about the flow itself.

Flows with a low Reynolds Number are dominated by viscosity, meaning frictional forces keep each parcel of fluid moving smoothly with surrounding fluid parcels. This results in smooth layers of flow, known as **laminar** flow.

In flows with high Reynolds Numbers, the inertial forces are much higher than the viscous forces, therefore each parcel of fluid is more likely to move according to its own inertial forces. This results in more chaotic flows, with individual flow parcels moving at different speeds and in different directions, often in contrary directions to the main fluid flow. This is known as a **turbulent** flow.

Consider pouring water out of a bottle into a bowl and then pouring thick honey into a bowl at the same speed. The water will flow quickly, and there will be chaotic mixing in the bowl. This flow of water is dominated by the inertial forces. The thick honey will flow slowly, and will layer smoothly in the bottom of the bowl. The difference in behaviour is because the viscosity of the honey is much higher than the water, although the other three parameters are approximately the same between the fluids, and therefore the Reynolds Number is much lower for the flow of honey than the flow of water.

In long, straight, regular circular pipes with steady incoming flow, laminar flow will persist up to a Reynolds number of approximately 2,300. At a Reynolds number above 4000, flow will be turbulent. Between these values, the flow will be **transitional** (a mixture of laminar and turbulent flow). While these numbers can be useful references, they should not be used as determinants of laminar or turbulent flow in the airways, as the airways change cross-sectional shape and area abruptly, include corners, bifurcations, moving structures and respiratory airflow is not steady in one direction. These factors may cause turbulent flow to occur at lower Reynolds numbers than would be expected in a circular pipe (see Sect. 7.3.1 below).

In respiratory airflow, the air is largely uniform, with only a small change in density or viscosity as we breathe. However, the other two parameters, flow velocity and the airway length scale, vary significantly through the airway, causing different airflow characteristics at different levels. In the trachea, the air is moving quickly and the trachea is relatively large, around 20 mm [1, 2] in diameter in adults. The Reynolds number can reach values of 14,000 depending on the breathing condition [4, 8, 9]. Therefore, inertial forces are much larger than viscous forces within the trachea. These studies have shown that the airflow is often turbulent in the trachea due to the rapid expansion of the airway cross-section as air moves from the constriction of the vocal folds into the trachea.

As the airway tree bifurcates as it progresses through the lung, the airways narrow and the airflow velocity reduces. Therefore the Reynolds number reduces and the flow is more likely to be laminar. The airflow slows down because although each individual airway narrows at each level of the airway tree, the flow is divided between more airways at each level, so the total area available to the flow increases at each level of the airway tree. Conservation of mass tells us that as cross-sectional area increases, the flow must slow down. Therefore, flow further down the airway tree is much more likely to be laminar than flow in the trachea.

As both airflow velocity, U and length-scale, D, in the alveolar ducts are much lower than in the trachea, (alveolar duct diameter $\approx$1 mm [26, 27]) the inertial forces are also much lower. In these small airways, the Reynolds number is as low as 0.01 to 1 [15]. At these low Reynolds number, the flow is dominated by viscous forces, and transport occurs due to diffusion rather than bulk flow.

7.2.3 Conservation of Energy

Conservation of energy states that energy cannot be created or destroyed. Let's consider our control volume of fluid, moving along within the flow. We can add energy to our control volume in two ways, by transferring heat into our control volume or by doing work on the fluid.

The net rate of heat added to our control volume is the heat that flows in minus the heat that flows out in a given time. Similarly, the net rate of work done by our fluid is the work done on our fluid minus the work done by our fluid on the surroundings.

Conservation of energy tells us that the rate of change of energy inside our control volume equals the net heat transfers into our control volume plus the net work done:

$$\text{Rate of change in energy} = \text{Net rate of heat transfer} + \text{Net rate of work done} \tag{7.3}$$

Heat Transfer

Heat can flow in or out of our control volume. The flow of heat into a control volume of fluid moving along with the flow changes due to several different mechanisms. The temperature anywhere within the control volume can change due to changes in temperature with time, for example as a person's body temperature varies during the day. The temperature of our control volume will also change due to changes in the position of our control volume, for example as our control volume moves from the cool region of air within the nostrils to further down the airway, where the air temperature is higher. Finally, heat can diffuse into our control volume via diffusion, that is heat energy transferring from fluid molecule to adjacent fluid molecules based on a property of the fluid itself; the thermal conductivity, k.

Finally, we have external sources of heat. For example, our fluid may receive thermal energy from an external source via radiation. We can write a word equation for the transfer of heat in and out of our control volume.

$$\text{Change in temperature with time} + \text{Thermal convection} = \text{Thermal diffusion} + \text{Heat sources} \quad (7.4)$$

The relative influence of these terms depends on the flow of interest. For example, when the flow velocity in the nose is low, heat transfers from the warm tissue surrounding the airway into the cool inhaled air by thermal diffusion, which is also known as conduction. However, if we inhale quickly, the cool in rushing air carries the thermal energy away in convection. The relative importance of the two mechanisms is captured by another non-dimensional number, the thermal Péclet number, Pe.

$$\text{Péclet Number} = Pe = \frac{\text{Heat convected by flow}}{\text{Heat conducted}} = \frac{\rho U D c_p}{k} \quad (7.5)$$

Where c_p is the ability of the fluid to store heat, its specific heat capacity. The Péclet number looks a lot like the Reynolds number, and contains the same parameters ρ, u, and D. However, the Péclet Number's top line represents the massflow of fluid multiplied by the fluid's ability to carry heat with it. The bottom line is now the thermal conductivity, k, rather than the viscosity, or in other words, the ability of the fluid molecules to exchange heat, rather than momentum.

Work Done on the Fluid

We can do work on our control volume of fluid via body and surface forces. Surface forces include pressure forces compressing our control volume, or shear forces deforming it. To calculate how much work the surface forces do to one face of our control volume, we use the basic relationship:

$$\text{Work} = \text{Force} \cdot \text{Distance Moved} \quad (7.6)$$

The force acting on one face of our control volume due to the pressure is the pressure multiplied by the area of the face. The distance it moves in a given time is the fluid velocity; multiplying these two values together gives us the rate of work done. However, the pressure force and velocity at either end of our control volume are very similar, so nearly all the work done at one end of our control volume is cancelled out at the other end. Often, the change in energy in our control volume due to work done is negligibly small and can therefore be assumed to equal zero in respiratory airflow.

Summary

Solving the energy equation in respiratory CFD allows us to calculate the change of air temperature during inhalation. The nose is a very good heat exchanger and even in extreme outside temperatures, air is heated to close to body temperature by the time it reaches the back of the nose [10–13, 16]. As air changes temperature, it also changes density. While this change in density is small, the change may be significant depending on the quantities of interest for a simulation. In this case, an extra equation is needed to map the change in air density with changing temperature. A good approximation for this relationship is the **ideal gas law**, which relates the pressure, density, and temperature of a gas together based on a property of the gas known as the specific gas constant. The ideal gas law needs to be solved at each iteration of the CFD simulation, posing some extra computational cost.

In practice, respiratory CFD simulations often neglect the energy equation. If it can be assumed that the flow is the same temperature throughout and the same temperature as the airway walls (i.e. the heating of inhaled air is not of interest), or change in density of the air is assumed to be insignificant, then simulations can be treated as adiabatic. This means all net heat transfer is zero and the energy equation does not need to be solved, reducing the computational cost of the CFD simulation.

7.2.4 Transport

The word equations for momentum and heat transfer above ((7.1) and (7.3), respectively) are very similar in that they both contain terms representing convection, diffusion, sources, and rate of change with time. Both momentum and heat can be thought of as properties being carried by the fluid. In fact, a similar equation can be written for any quantity of interest being carried by the fluid. Imagine we hold a source of red smoke in front of a subject and want to track the path of that red smoke as they inhaled. The red smoke would be transported by convection (i.e. carried along by the moving fluid) and diffusion (i.e. even when the airflow is stationary, the smoke would spread out). The concentration of smoke in the nose would change with time if we varied the amount of smoke we released, and smoke could be absorbed or created by external sources, such as a second smoke source.

In general any property of the flow of interest can be modelled with the same transport equation, by simply replacing the quantity of interest in the transport equation and understanding what this means physically.

$$\text{Change with time} + \text{Convection} = \text{Diffusion} + \text{New Sources} \tag{7.7}$$

As a practical example, we could model the humidification of air by considering how the amount of water vapour changes with time, how air convects water vapour, how water vapour diffuses in air, and the sources of water vapour.

7.2.5 Derivation of the Navier–Stokes and Continuity Equations

Conservation of Mass

If we have a stationary control volume, the concept of conservation of mass tells us that the change in mass inside our control volume over a given time, $\frac{\partial m}{\partial t}$ must equal the difference between the sum of all the mass flowing into the control volume, $\sum \dot{m}_{in}$ and mass flowing out of the control volume, $\sum \dot{m}_{out}$.

$$\frac{\partial m}{\partial t} = \sum \dot{m}_{in} - \sum \dot{m}_{out} \tag{7.8}$$

Figure 7.6 shows mass flowing through a control volume of infinitesimal size. The cube's dimensions are dx, dy, and dz. The velocity of flow in each of the x, y, and, z directions is u, v, and, w, respectively.

Let's consider mass moving in and out of the faces of our cube in the x direction. The massflow entering in the x direction $\dot{m}_{in,x}$, has a density ρ, travels at a velocity u and enters through a surface of area A.

$$\dot{m}_{in,x} = \rho u A \tag{7.9}$$

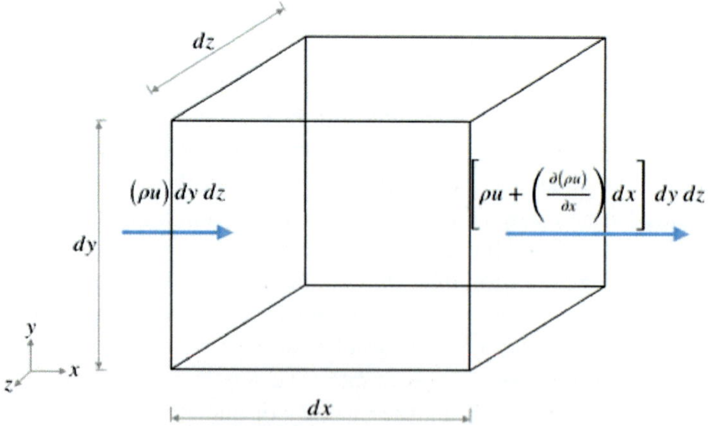

Fig. 7.6 Mass flowing in and out of a stationary control volume in the x-direction

We know the dimensions of our cube and the area A is the multiple of the edge lengths dy and dz. Therefore:

$$\dot{m}_{in,x} = \rho u dy\, dz \tag{7.10}$$

The massflow out of our domain may be traveling at a different velocity because it has travelled a distance dx across our control volume. Its new velocity is the old velocity, plus the change in velocity due to the change in x position, multiplied by the gradient of velocity in that direction. Similarly, the density of the fluid may have changed due to the change in x position. The term $\frac{\partial(\rho u)}{\partial x}$ is the gradient in density and velocity with change in x position and dx is the total distance moved.

$$\dot{m}_{out,x} = \left(\rho u + \frac{\partial(\rho u)}{\partial x} dx\right) dy\, dz \tag{7.11}$$

So the net massflow, $\dot{m}_{in,x} - \dot{m}_{out,x}$ in the x direction is

$$-\frac{\partial(\rho u)}{\partial x} dx\, dy\, dz \tag{7.12}$$

We can repeat calculation of the net massflow in the y and z directions to get the total net massflow in our control volume:

$$\sum \dot{m}_{in} - \sum \dot{m}_{out} = -\frac{\partial(\rho u)}{\partial x} dx\, dy\, dz - \frac{\partial(\rho v)}{\partial y} dx\, dy\, dz - \frac{\partial(\rho w)}{\partial z} dx\, dy\, dz \tag{7.13}$$

We must also consider how the total mass of our fluid changes with time. The mass in our control volume is the density multiplied by the volume, $\rho dx\, dy\, dz$. The volume of our control volume is fixed, so the increase in mass with time is

$$\frac{\partial m}{\partial t} = \frac{\partial \rho}{\partial t} dx\, dy\, dz \tag{7.14}$$

We can substitute our equation for the total net massflow into our control volume, (7.13), and the change in mass with time, (7.14), into our statement of conservation of mass, (7.8)

$$\frac{\partial \rho}{\partial t} dx\, dy\, dz = -\frac{\partial(\rho u)}{\partial x} dx\, dy\, dz - \frac{\partial(\rho v)}{\partial y} dx\, dy\, dz - \frac{\partial(\rho w)}{\partial z} dx\, dy\, dz \tag{7.15}$$

Cancelling the volume term, $dx\, dy\, dz$, we get:

$$\frac{\partial \rho}{\partial t} = -\frac{\partial(\rho u)}{\partial x} - \frac{\partial(\rho v)}{\partial y} - \frac{\partial(\rho w)}{\partial z} \tag{7.16}$$

If we move all the terms to the left hand side and rewrite the equation in vector form, where our velocity vector $\mathbf{u} = (u, v, w)$, we get

$$\frac{\partial \rho}{\partial t} + \nabla \cdot \rho \mathbf{u} = 0 \tag{7.17}$$

Equation (7.17) is the vector form of our statement of conservation of mass. We call this the conservative form of the equation because we derived it from a control volume that was not moving.

Finally, we can consider the special case of a fluid that is incompressible, that is the density, ρ does not change in space or time. In this case, our statement of conservation of mass simplifies to

$$\nabla \cdot \mathbf{u} = 0 \tag{7.18}$$

i.e. for incompressible flows, the divergence of velocity is zero.

Newton's Second Law

To derive the momentum equation equivalent to (7.1), we will apply Newton's Second Law, $F = ma$, to an infinitesimal control volume of fixed mass. For convenience, this time our control volume will move along with the fluid.

The mass of fluid within the control, m, is simply its density multiplied by the volume of the control volume, which equals $dx\, dy\, dz$ as in Sect. 7.2.5.

$$m = \rho dx\, dy\, dz \tag{7.19}$$

The acceleration of the fluid, a, can occur for two reasons, as shown in Fig. 7.7. The velocity can change over time, $\frac{\partial u}{\partial t}$, or because our control volume is moving. As the control volume moves along in the x-direction, the rate at which its x-component of velocity changes is $\frac{\partial u}{\partial x}$. We now need to know how far our control volume will move. In a period of time, dt, the distance our control volume will move in the x-direction is $\frac{dx}{dt}$, but of course a change in position with time is just a velocity, i.e. $\frac{dx}{dt} = u$. Therefore, the change in our control volume's x-velocity due to movement in the x-direction is $u\frac{\partial u}{\partial x}$. The velocity can change as our control volume moves position in each of the x, y, and z directions. In total, the change in the x-component of velocity, u, for our moving control volume in a period dt is

$$a_x = \frac{\partial u}{\partial t} + u\frac{\partial u}{\partial x} + v\frac{\partial u}{\partial y} + w\frac{\partial u}{\partial z} = \frac{Du}{Dt} \tag{7.20}$$

Where $\frac{Du}{Dt}$ is a shorthand way of writing $\frac{\partial u}{\partial t} + u\frac{\partial u}{\partial x} + v\frac{\partial u}{\partial y} + w\frac{\partial u}{\partial z}$ and is known as the material or substantive derivative. In cartesian vector notation, we can write this as

$$\frac{Du}{Dt} = \frac{\partial u}{\partial t} + \mathbf{u} \cdot \nabla u \tag{7.21}$$

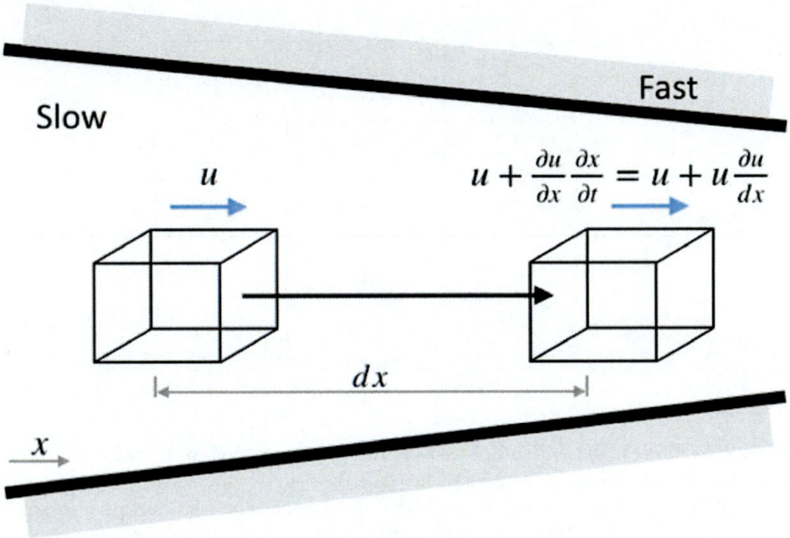

Fig. 7.7 The acceleration of a control volume of fluid as it moves through the flow. In the period of time, dt, the volume has moved a distance, dx. The geometry surrounding the flow converges, creating a velocity gradient. As our volume moves to the right it will accelerate. The amount of acceleration due to this motion is the gradient of velocity with respect to position, $\frac{\partial u}{\partial x}$, multiplied by the distance the volume has moved in the given time, $\frac{dx}{dt}$. However, $\frac{dx}{dt} = u$, so the total acceleration due to the motion of the volume is then $u\frac{\partial u}{\partial x}$

So far, we have only considered acceleration of our x-component of velocity (i.e. u), but the same is true of the y and z components of velocity. We can use the material derivative for our velocity vector $\mathbf{u} = (u, v, w)$ too

$$\frac{D\mathbf{u}}{Dt} = \frac{\partial \mathbf{u}}{\partial t} + (\mathbf{u} \cdot \nabla)\,\mathbf{u} \tag{7.22}$$

Now we must consider the forces acting on our moving control volume. Firstly, we will separate our forces, F, into body forces which act over the entire fluid, and surface forces which act only on the faces of our control volume. Body forces tend to be forces that act from a distance and include gravity forces, Coriolis forces, and electromagnetic forces. We can state that the body force per unit mass of our fluid is in the x-direction is $f_{Bx}\rho dx\, dy\, dz$ and the total body force in vector form is $\mathbf{F_B}\rho dx\, dy\, dz$.

The surface forces acting on the faces of our cube can also be divided into various types. For example, pressure forces imposed by the fluid outside our control volume, which act perpendicular to the surface of our control volume. We also have both normal and shear stresses due to friction. The normal stress comes from the change in volume of our fluid element over time. The shear stress comes from the shear deformation of our fluid element. We will denote stresses with the symbol τ,

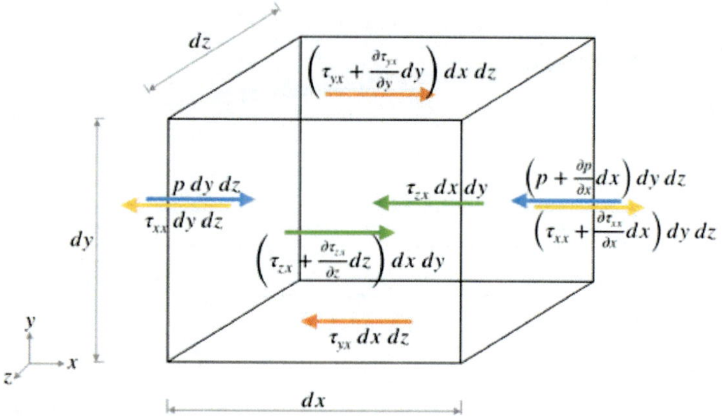

Fig. 7.8 Forces acting in the x-direction on a control volume moving with the flow. Pressure forces (i.e. pressures, p, multiplied by the area of the face they act on) are shown with blue arrows and act on the left and right faces. Normal stresses produce forces due friction with surrounding fluid elements in the normal direction and are shown in yellow. These normal stresses also act on the left and right face of our volume. Shear stresses due to friction from surrounding forces acting perpendicular to the faces of our control volume act on the front and back faces (green) and the top and bottom faces (orange)

subscripted first with the plane they act perpendicular to and secondly with direction in which they act. For example τ_{zx} acts in the x-direction on a plane perpendicular to the z-axis. Figure 7.8 shows the forces acting on a control volume in the x-direction.

Figure 7.8 shows a pressure p acting on a face of size $dy\,dz$ on the left of the figure, providing a force of $pdy\,dz$. On the opposite side of the cube, the pressure p acts on the same sized face, but the pressure here is different by the gradient of pressure in the x-direction multiplied by the distance the two faces are apart, dx. Given that the two pressures act in opposite directions, we get a net pressure of $\left[p-\left(p+\frac{\partial p}{\partial x}dx\right)\right]dy\,dz$. Repeating this analysis for the stresses, τ, acting on our volume and summing all the surface forces in the x-direction, we get

$$\left[p-\left(p+\frac{\partial p}{\partial x}dx\right)\right]dy\,dz+\left[\left(\tau_{xx}+\frac{\partial\tau_{xx}}{\partial x}dx\right)-\tau_{xx}\right]dy\,dz+$$
$$\left[\left(\tau_{yx}+\frac{\partial\tau_{yx}}{\partial y}dy\right)-\tau_{yx}\right]dx\,dz+\left[\left(\tau_{zx}+\frac{\partial\tau_{zx}}{\partial z}dz\right)-\tau_{zx}\right]dx\,dy \quad (7.23)$$

Summing surface forces in the x-direction with body forces in the x-direction and simplifying, we get the total force in the x-direction, F_x

$$F_x=\left(-\frac{\partial p}{\partial x}+\frac{\partial\tau_{xx}}{\partial x}+\frac{\partial\tau_{yx}}{\partial y}+\frac{\partial\tau_{zx}}{\partial z}\right)dx\,dy\,dz+f_{Bx}\rho dx\,dy\,dz \quad (7.24)$$

We now have, in the x-direction, our force, F_x (7.24), our mass, m (7.19), and our acceleration, a (7.20). We can now write our momentum balance, $F = ma$ for our fluid element in the x-direction

$$\left(-\frac{\partial p}{\partial x} + \frac{\partial \tau_{xx}}{\partial x} + \frac{\partial \tau_{yx}}{\partial y} + \frac{\partial \tau_{zx}}{\partial z}\right) dx\, dy\, dz + f_{Bx}\rho dx\, dy\, dz = \rho dx\, dy\, dz \frac{Du}{Dt} \tag{7.25}$$

We can simplify this equation by cancelling through the volume of our fluid element $dx\, dy\, dz$.

$$-\frac{\partial p}{\partial x} + \frac{\partial \tau_{xx}}{\partial x} + \frac{\partial \tau_{yx}}{\partial y} + \frac{\partial \tau_{zx}}{\partial z} + f_{Bx}\rho = \rho \frac{Du}{Dt} \tag{7.26}$$

We could also write equivalent equations in the y and z-directions, or in vector form, with our velocity vector $\mathbf{u} = (u, v, w)$, and stress tensor $\boldsymbol{\tau} = \begin{bmatrix} \tau_{xx} & \tau_{xy} & \tau_{xz} \\ \tau_{yx} & \tau_{yy} & \tau_{yz} \\ \tau_{zx} & \tau_{zy} & \tau_{zz} \end{bmatrix}$, this equation becomes

$$-\nabla p + \nabla \cdot \boldsymbol{\tau} + \rho \mathbf{F_B} = \rho \frac{D\mathbf{u}}{Dt} \tag{7.27}$$

Equation (7.26) and its equivalents in the y and z-directions are the Navier–Stokes equations in non-conservation form (as our fluid element was moving with the flow rather than stationary). However, we can convert them into a more usable form by considering the stress-strain relationships in our fluid. Air is a Newtonian fluid, which means the shear stress is proportional to the rate of strain. The strains in our fluid are the gradients of velocity, i.e. how one part of the fluid is moving differently from the part adjacent to it. In a Newtonian fluid with a molecular or dynamic viscosity of μ, Stokes suggested the following relationships between stresses and strains in fluids

$$\tau_{xx} = -\frac{2}{3}(\nabla \cdot \mathbf{u}) + 2\mu \frac{\partial u}{\partial x} \text{ and } \tau_{yx} = \tau_{xy} = \mu\left(\frac{\partial v}{\partial x} + \frac{\partial u}{\partial y}\right) \tag{7.28}$$

The derivation of this result is beyond the scope of this chapter, but is well described by Kundu and Cohen [17]. Similar relations are found for stresses in other directions. Substituting these relationships into (7.27) gives us

$$-\nabla p + \mu \nabla^2 \mathbf{u} + \frac{1}{3}\mu \nabla (\nabla \cdot \mathbf{u}) + \rho \mathbf{F_B} = \rho \frac{D\mathbf{u}}{Dt} \tag{7.29}$$

If we consider our fluid to be incompressible, the strain tensor, $\boldsymbol{\tau}$, is simpler

$$\boldsymbol{\tau} = \mu \left(\nabla \mathbf{u} + (\nabla \mathbf{u})^{\mathbf{T}}\right) \tag{7.30}$$

The incompressible form of the Navier–Stokes equation in vector notation is then

$$-\nabla p + \mu \nabla^2 \mathbf{u} + \rho \mathbf{F_B} = \rho \frac{Du}{Dt} \tag{7.31}$$

Writing out the material derivative, $\frac{Du}{Dt}$, in full and reordering this equation we find

$$\underbrace{\rho F_B}_{\text{Body forces}} \quad \underbrace{-\nabla p}_{\text{Pressure gradient}} \quad + \underbrace{\mu \nabla^2 \mathbf{u}}_{\text{Diffusion}} = \underbrace{\frac{\partial \mathbf{u}}{\partial t}}_{\text{Change with time}} \quad + \underbrace{(\mathbf{u} \cdot \nabla)\, \mathbf{u}}_{\text{Convection}} \tag{7.32}$$

This form of the equation is exactly analogous to the word equation (7.1) with the terms representing body forces, the pressure gradient, diffusion, acceleration with respect to time, and convection, respectively.

Conservation of Energy

We can state our conservation of energy rule, (7.3), in terms of the energy in the fluid per unit mass, E, the rate at which heat is added $\dot{Q}$ and the rate at which work is done on the fluid, $\dot{W}$, as follows

$$\frac{DE}{Dt} = \sum \dot{Q} + \sum \dot{W} \tag{7.33}$$

The energy per unit mass, E, is just the temperature of the fluid T multiplied by a property of the fluid—its specific heat capacity, c_p.

$$E = c_p T \tag{7.34}$$

Substituting this relationship into the material derivative for the energy per unit mass, E, we find

$$\frac{DE}{Dt} = c_p \left(\frac{\partial T}{\partial t} + u \frac{\partial T}{\partial x} + v \frac{\partial T}{\partial y} + w \frac{\partial T}{\partial z} \right) \tag{7.35}$$

The total amount of energy in a fluid element per unit mass is then $\frac{DE}{Dt} \rho \, dx \, dy \, dz$.

The term $\sum \dot{Q}$ is the sum of heat added to or lost from the control volume. The volumetric heating of our fluid element (due to radiation for instance) is the heat flux $\dot{q}$ multiplied by the mass of the control volume, $\rho dx \, dy \, dz$ to give

$$\rho \dot{q} dx \, dy \, dz \tag{7.36}$$

Figure 7.9 shows the heat flux due to conduction into a infinitesimal control volume moving with the flow. Just as we saw for mass and momentum, the difference between the heat flux into one side of the control volume and out of the other side is related simply to the size of the control volume and the gradient in that direction. In the x-direction, the net heat flux through the face of size $dy \, dz$ is

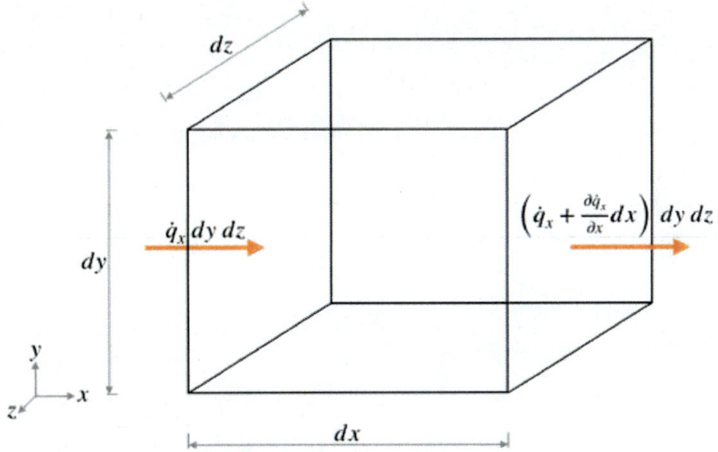

Fig. 7.9 Heat flux in and out of a control volume in the x-direction. Energy flux due to work done on the fluid is assumed to be negligibly small

$$-\left(\frac{\partial \dot{q}_x}{\partial x} dx\right) dy\,dz \tag{7.37}$$

Summing equivalent terms tells us that the new heat flux into our cube via conduction is

$$\sum \dot{Q} = \left[\rho \dot{q} - \left(\frac{\partial \dot{q}_x}{\partial x} + \frac{\partial \dot{q}_y}{\partial y} + \frac{\partial \dot{q}_z}{\partial z}\right)\right] dx\,dy\,dz \tag{7.38}$$

Fourier's law of heat conduction tells us that the heat flux per unit area $\dot{q}$ is related to the temperature gradient by the thermal conductivity, k

$$\dot{q}_x = -k \frac{\partial T}{\partial x} \tag{7.39}$$

Substituting this into (7.38), the net heat flux is

$$\sum \dot{Q} = \left[\rho \dot{q} - k\left(\frac{\partial^2 T}{dx^2} + \frac{\partial^2 T}{dy^2} + \frac{\partial^2 T}{dz^2}\right)\right] dx\,dy\,dz \tag{7.40}$$

We must now consider the work done on the fluid, $\sum \dot{W}$. We know that the rate of work done is the net force multiplied by the fluid velocity. The forces are the body and surface forces we saw in Sect. 7.2.5. However in effect, in respiratory fluid mechanics, the forces are not sufficient to generate meaningful work. Therefore, it is usually reasonable to assume $\sum \dot{W} \approx 0$. With this assumption, we find the resultant energy balance from substituting (7.35) and (7.40) into (7.33).

$$\rho c_p\left(\frac{\partial T}{\partial t}+u\frac{\partial T}{\partial x}+v\frac{\partial T}{\partial y}+w\frac{\partial T}{\partial z}\right)dx\,dy\,dz = \left[\rho\dot{q}-k\left(\frac{\partial^2 T}{dx^2}+\frac{\partial^2 T}{dy^2}+\frac{\partial^2 T}{dz^2}\right)\right]dx\,dy\,dz \quad (7.41)$$

Rearranging this equation and cancelling through the volume of the control volume, we get

$$\underbrace{\frac{\partial T}{\partial t}}_{\text{Change with time}}+\underbrace{\left(u\frac{\partial T}{\partial x}+v\frac{\partial T}{\partial y}+w\frac{\partial T}{\partial z}\right)}_{\text{Convection}} = \underbrace{\frac{\dot{q}}{c_p}}_{\text{External source}} \underbrace{-\frac{k}{\rho c_p}\left(\frac{\partial^2 T}{dx^2}+\frac{\partial^2 T}{dy^2}+\frac{\partial^2 T}{dz^2}\right)}_{\text{Diffusion}} \quad (7.42)$$

Conceptually, these terms can be thought of as change of temperature with time, convection of heat, source terms, and thermal diffusion, respectively.

Transport Equation

In the previous sections, we followed the transport of mass, momentum, and energy through control volumes of fluids. The same methods can be employed for a generic variable which we will call ϕ. The variable ϕ may change with time; may be convected by the flow, may diffuse through the flow, depending on its diffusion coefficient, Γ, and we may have source terms, S_ϕ to add or subtract ϕ from our domain. In general, the transport equation for our variable ϕ is

$$\underbrace{\frac{\partial \rho\phi}{\partial t}}_{\text{Change with time}}+\underbrace{\nabla\cdot(\rho\mathbf{u}\phi)}_{\text{Convection}} = \underbrace{\nabla\cdot(\Gamma\nabla\phi)}_{\text{Diffusion}}+\underbrace{S_\phi}_{\text{Sources}} \quad (7.43)$$

Replacing the term ϕ with a useful concept, such as water vapour, provides us with an equation to model the transport of water vapour through our flow.

7.3 Turbulent Flow

7.3.1 Flow Regimes—Laminar and Turbulent Flow

As we have already seen in our discussion of Reynolds Number (Sect. 7.2.2), fluid flow is classified into three regimes; laminar, turbulent, and transitional. Transitional flow is a combination of the first two regimes, where the flow exhibits characteristics of both laminar and turbulent flow in different regions. Therefore, we will concentrate

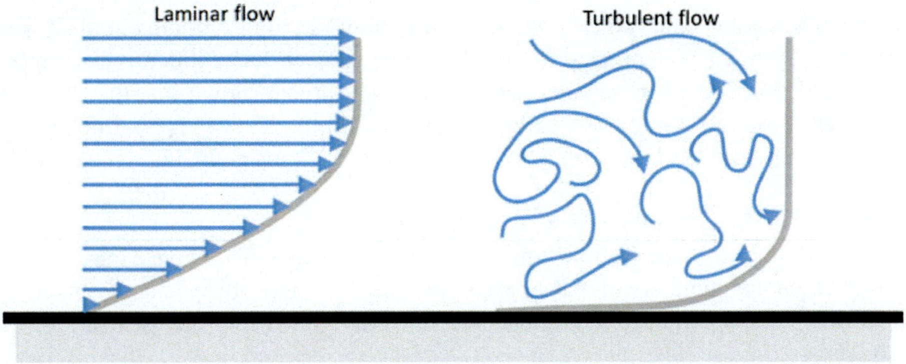

Fig. 7.10 Velocity profiles in laminar and turbulent flows. The laminar flow has a larger boundary layer as turbulent mixing exchanges momentum quicker in the turbulent flow, reducing the boundary layer size

on laminar and turbulent regimes in this chapter. Transitional flow often occurs in regions where laminar flow becomes turbulent in response to a disturbance. The main differences between laminar and turbulent flow regimes are the distribution of local flow velocities and the length scales of features within the flow. These factors are important as they have many consequences for the flow including how much energy is required to drive the flow and the transport of momentum and particles within the flow. Let's consider flow in a pipe. The pipe is smooth, straight, and has a round cross-section. The flow is of low velocity and has travelled some distance since the beginning of the pipe. If we look at the velocity at various points throughout the flow, they all point in the same direction—along the pipe. Nowhere within the flow are particles moving in any direction other than along the pipe. We know that fluid particles right along the wall are fixed to the wall by friction, but the flow in the middle of the pipe is moving. The viscosity of the fluid causes diffusion of momentum through the fluid until an equilibrium is reached: a transition of velocity from stationary at the wall to a maximum velocity in the centre of the pipe (laminar flow, Fig. 7.10).

As our fluid travels along the pipe, let's consider what happens when the flow travels faster, so that the Reynolds number is higher and inertial forces dominate viscous forces. If a parcel of fluid is disturbed, perhaps by irregularities in the pipe wall, it will be thrown off course from the direction it was travelling. The fluid parcel's own momentum will cause it to keep travelling in this new direction and the friction from the viscosity of other, surrounding fluid parcels is insufficient to damp out this new motion. These disturbances spread quickly throughout a flow causing local velocities to point in a variety of different directions, as the flow forms vortices and eddies (turbulent flow, Fig. 7.10, right). In addition to surface irregularities, other disturbances that can cause turbulence include changes in pipe cross sectional shape or area, pipe curvature, vibrations, and motion. All of these types of disturbances happen in the airway, so the flow in the airway may become turbulent in the airways at lower Reynolds than is the case in ideal pipe-flows. In long, smooth circular pipes, we expect flow to be laminar at Reynolds numbers lower than 2400, turbulent above 4000 and transitional in between. However, studies have shown that theoretically, turbulent flow can occur at Reynolds numbers as low as 81 [14, 24] before viscous forces damp out oscillations in local flow velocities. Conversely, in well controlled experiments, laminar flow has been maintained at Reynolds numbers as high as 100,000 [21, 24]. While these are extreme values, it should be remembered that the standard values for pipe flow are empirical approximations. Turbulence may occur at lower Reynolds numbers in real airways particularly where abrupt changes in cross-sectional area occur, such as the opening from the glottis into the trachea [1–3, 9, 18, 25].

7.3.2 Features of Turbulent Flow

If we recorded the velocity of passing air at a particular point within our pipe, we would get a constant value in a laminar flow, as we would always sample from the same layer of flow. However, in a turbulent flow, we would see that the velocity fluctuates as the flow passes and we record the velocity of flow in eddies which may be rotating with or against the mean flow direction. This happens if we measure any component of velocity, that is whether we consider velocity in the direction of the flow, or moving at 90° to the main flow.

Having pockets of flow moving in all directions causes a great deal of mixing in the flow. If we inject a stream of ink into turbulent flow, it will quickly become dispersed across the width of the pipe, whereas in a laminar flow, the stream of ink would simply progress downstream in a line. Turbulent mixing affects more than just ink however. Momentum is also shared much more quickly between different parts of the flow in turbulent flow than in laminar flow. In laminar flow, the slow diffusion of momentum between layers causes large boundary layers to form (Fig. 7.10, left). However, in turbulent flow, momentum is mixed more quickly resulting in a much smaller boundary layer (Fig. 7.10, right).

In addition to momentum, inhaled particles, heat, and water vapour are all mixed more quickly by a turbulent flow than a laminar one. This has important consequences for CFD simulations, which have to model this turbulent mixing. Let's consider the size of the flow structures that cause mixing—the eddies within the flow. As our flow passes an obstacle—for instance the epiglottis protruding into the upper airway—the eddy that forms in the wake behind the obstruction will be approximately the same size as the obstruction itself. This is easy for CFD simulations to handle, as we have already designed our mesh to be capable of capturing the large features of the flow (see Chap. 6). However, we must consider what happens to that large eddy as it moves along with the flow. The large eddy is unstable and will break up into smaller eddies as it moves along with the flow. These smaller eddies are also unstable and will consequently break into even smaller eddies. This process is repeated over and over again to smaller and smaller length scales (Fig. 7.11). For a flow to be considered

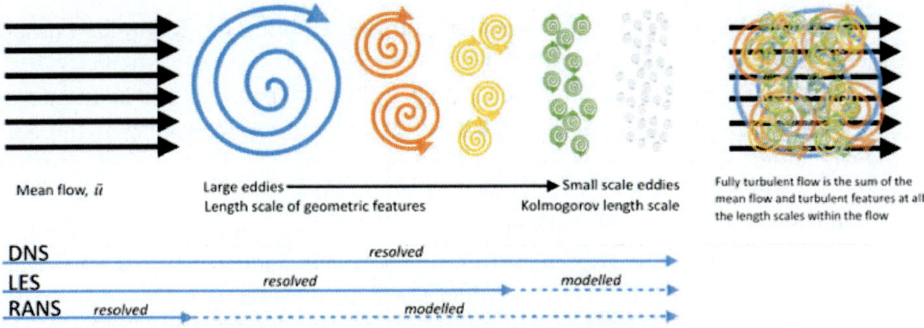

Fig. 7.11 The variety of length scales within the flow. We have a mean flow field filling our domain. The largest turbulent structures can be of the same length scale as geometric features of the flow domain. The large eddies break down into progressively smaller sizes (orange, yellow, green, and grey) until they dissipate into heat due to viscosity at the Kolmogorov length scale. In a fully turbulent flow, all these scales occur within the flow at the same time and place. The flow appears turbulent at any length scale between the scale of the geometry and the Kolmogorov length scale. This complicated pattern of velocity fluctuations at different length scales can be represented by measuring a mean velocity in addition to a fluctuating component. Lower—Various turbulence models calculate different length scales of the flow. DNS simulations calculate all length scales within the flow, LES simulations calculate large length scales and model the effect of smaller length scales. RANS models only accurately calculate the mean flow pattern

fully turbulent it needs to contain eddies of a range of sizes throughout this length scale overlapping each other (Fig. 7.11). In order to accurately model these smallest of eddies, we need to consider what the size of the smallest possible eddy would be. Just as we characterised the flow by the Reynolds number (7.2)—the ratio or inertial to viscous forces, we can characterise each eddy its own Reynolds number. Instead of using values of velocity and distance from the main flow, we can use values that represent the velocity and size of the eddy.

$$\text{Eddy Reynolds Number} = Re_{eddy} = \frac{\text{Inertial Forces}}{\text{Viscous Forces}} = \frac{\rho U_{eddy} D_{eddy}}{\mu} \qquad (7.44)$$

From this equation we can see that as smaller and smaller eddies are produced with smaller values of D_{eddy}, the size of the eddy, then the inertial forces of the eddy to keep it rotating get smaller and smaller. The viscous forces that damp out velocity fluctuations get relatively larger and larger, until we eventually reach a point where the eddy can no longer rotate (at approximately $Re_{eddy} = 1$). At this point, instead of breaking down into even smaller eddies, the energy is dissipated by friction as heat. Therefore, viscosity is responsible for the size of the smallest feature within a flow. We should also consider the velocity of the eddy, U_{eddy}. Conservation of energy tells us that the turbulent kinetic energy fed into the initial large eddy must equal the total turbulent kinetic energy in each subsequent level of smaller eddies and also equal the energy dissipated as heat by the smallest level of eddies. Mathematical analysis of this kinetic energy cascade tells us that the eddy velocity U_{eddy} gets lower as the eddies get smaller (i.e. D_{eddy} gets smaller), further reducing the Reynolds number of successive eddies, Re_{eddy}.

The smallest length scale of eddies within a flow is known as the Kolmogorov length scale, after the mathematician who derived it. In an adult airway, the Kolmogorov length scale has been calculated to be as small as 32 μm while sniffing [8]. For a CFD mesh to be fine enough to resolve these flow features everywhere within the airway takes many millions or even billions of elements and is therefore extremely computationally expensive to run. Therefore, another approach has been developed—to model the effect of turbulence (e.g. increased mixing in a flow) without having to fully resolve every eddy within the flow. This approach is known as turbulence modelling.

7.3.3 Introduction to Turbulence Modelling

Running CFD simulations on a mesh capable of resolving all the features in a flow (known as Direct Numerical Simulation or DNS) is often computationally prohibitive. An alternative approach is to create a CFD mesh capable of resolving large flow features, and then to model the effect of turbulent flow features smaller than the mesh size by adding correction terms to your governing equations.

Reynolds-Averaged Navier Stokes (RANS) Turbulence Models

In Sect. 7.3.2, we saw that turbulent flows are characterised by fluctuations in the local velocity. If we measure the velocity at a point, it will fluctuate as eddies of different size and strength pass through our point and will have a mean value representing the average flow pattern. Turbulence modelling separates these velocities into average and fluctuating components.

$$\text{velocity} = \text{mean velocity} + \text{fluctuating velocity} \tag{7.45}$$

This technique is known as Reynolds decomposition. We can revisit the equations we derived in Sect. 7.2 and replace any terms that include velocity with separate mean and fluctuating velocity terms. In a fluid flow, the local velocity of the fluid is extremely important, as each of the conservation equations for mass, momentum balance, and energy contain velocity terms. A fluid's velocity governs the flow of mass in the conservation of mass equation (7.17). In the momentum equation, velocity appears in the acceleration, convection, and diffusion terms (7.27). In the energy equation, fluid velocity appears in the convection term (7.42).

One approach to solving the equations in this form is to only solve for the time-averaged solution, and not worry about the small fluctuations to yield what are known as the Reynolds-Averaged Navier Stokes (RANS) Equations. Taking the time averaged solution is useful because by definition, the fluctuating velocity components all have an average value of zero and can be discounted from the equation. However, the convection term in the Navier–Stokes equations produces terms that multiply two fluctuating velocity components together (e.g. the fluctuating x-velocity multiplied by the fluctuating y-velocity). While the mean of either fluctuating velocity on its own is zero, the mean of the product of two fluctuating velocities is not zero and we end up with a term including fluctuating velocities in our time-averaged Navier–Stokes Equations, known as the Reynolds Stress terms. In three-dimensions, six Reynolds Stress terms appear due to the combination of fluctuating terms in each of the three directions. Now we have the extra Reynolds Stress unknowns, we need some extra equations to solve the governing equations. There are several different approaches to solving these Reynolds Stress terms.

We know that the effect of turbulence is increased mixing. An approximation known as the Boussinesq approximation creates a term called eddy viscosity, which relates this extra mixing due to turbulence to the mean flow field and can be added to the normal molecular viscosity. The effect of this technique is to model the fluid as more viscous than it really is to account for the effect of turbulent mixing. While the simplest approach is to assume that this eddy viscosity is an empirically determined constant value, there are several more advanced approaches to determine the value of the eddy viscosity in RANS turbulence models.

k-ω RANS Turbulence Models

k-ω (k-omega) turbulence models calculate the local eddy viscosity by solving two equations in addition to the conservation of mass, momentum, and energy equations. These two equations concern the transport of the kinetic energy of turbulence, k, and the dissipation rate per unit volume and time, ω. While solving these two additional equations adds computational expense, the benefit is a more accurate calculation of the eddy viscosity, and hence more accurate simulations for most problems.

The turbulent kinetic energy is simply how much energy a parcel of fluid has due to its turbulent motion and is related to the square of the fluctuating velocity terms. The turbulent kinetic energy is transported through the flow by the transport equation just like any other property.

The dissipation rate per unit mass, ω, tells us how quickly the turbulent kinetic energy dissipates as heat as it progresses through the length scales of turbulence. It is found by dividing gradients in the local fluctuating velocity field by the turbulent kinetic energy. The dissipation rate is also transported through the flow by the transport equations[22].

Once the turbulent kinetic energy and dissipation rate have been calculated, the eddy viscosity is simply the turbulent kinetic energy divided by the dissipation. Some more complex versions of the k-ω turbulence model multiply this ratio by additional terms to account for damping of the eddy viscosity.

The k-ω turbulence model has been widely used in respiratory CFD simulations because it is particularly suited to low Reynolds number flows which are in the transitional or mildly turbulent regimes. It predicts flow behaviour well in terms of flow separation from smooth surfaces, however it can over-predict the turbulent kinetic energy in regions where the flow accelerates or stagnates. The k-ω turbulence model can be used everywhere in the airway, including down to the wall, however it requires a very fine mesh near to the wall which increases computational expense.

k-ε RANS Turbulence Models

k-ε (k-epsilon) turbulence models are an alternative approach to calculating the eddy viscosity. The turbulent kinetic energy transport is solved, as in k-ω models, but the transport of turbulent dissipation, ε, forms the second equation, rather than the turbulent dissipation rate per unit volume [23]. This turbulent dissipation, ε, is proportional to the product of k and ω. Essentially, ε represents the scale over which the turbulence acts. However, k-ε turbulence models are widely used in CFD simulations of external flows (i.e. flow over the outside of objects such as planes and cars, as opposed to internal flows inside pipes) and flows that are fully turbulent, rather than transitional. k-ε models do not provide accurate solutions in regions where there is an adverse pressure gradient. Therefore, k-ε models are less commonly used in respiratory CFD than in other industrial CFD applications.

k-ω Shear Stress Transport (SST) Turbulence Models

Shear Stress Transport (SST) turbulence models make use of the relative strengths of the k-ω and k-ε turbulence models by switching between the models depending on which model is most appropriate for that region of the flow [19, 20]. For instance, the k-ω model is used in the inner region of boundary layers, which allows the model to be valid throughout the domain, including at the wall, while the k-ε is used in the free stream regions.

Reynolds Stress RANS Turbulence Models

Reynolds Stress RANS turbulence models offer a significant advantage over other RANS turbulence models. We saw in the introduction to RANS turbulence models that the Reynolds Stress term comes from the product of fluctuating velocity terms. However, both the k-ω and k-ε turbulence models assume that these fluctuating velocity terms are equal in all directions (i.e. the turbulence is isotropic). While this assumption is valid in certain situations, there are many situations where this assumption is not valid. In respiratory CFD, the most significant of these are regions of the airway where the flow path curves significantly, for instance at the 90° turn in the nasopharynx, and regions where the flow separates from the airway surface and recirculation zones form, as occurs when flow passes through the vocal folds as a jet, and recirculates in the upper trachea.

Instead of calculating an eddy viscosity which is insensitive to directions, Reynolds Stress models calculate each of the fluctuating velocity terms separately, allowing for anisotropic velocity fluctuations within the flow. While this approach is much more accurate than the two-equation approaches of the k-ω and k-ε models, it comes at the computational expense of solving extra transport equations for each of the products of fluctuating velocity terms.

Large Eddy Simulation (LES) Turbulence Model

An alternative to RANS turbulence modelling is the large eddy simulation (LES) turbulence model. The major difference between LES and RANS approaches is that the LES model does not take a temporal mean of the simulation, but rather calculates the instantaneous flow field. Therefore, LES simulations are only usable for unsteady simulations (see Sect. 7.4). However, it is still too computationally expensive to solve the flow equations at a resolution that captures flow features all the way down to the Kolmogorov length scale. Therefore, LES simulations spatially and temporally filter the simulation. The large flow features are solved directly, while the effect of small features is modelled. One feature of turbulence that allows LES models to work is that at small scales, turbulent features act uniformly in all directions and do not bear much relationship to the larger eddies which are dominated by the disturbances which formed them and the local geometry. Therefore, the small scales of turbulence can be treated homogeneously, in contrast to the large scales, which depend strongly on local conditions. LES simulations require higher resolution meshes than RANS, but lower resolution than DNS.

The choice of filter size in LES modelling has a large impact on the accuracy of the simulation as it should be small enough that the flow fluctuations below the length scale of the filter size are homogenous. One approach is to use the mesh resolution as the filter size and assume that the numerical scheme will account for turbulence below that value. Other approaches set the filter size explicitly. There are several different approaches to model the effect of flow fluctuations smaller than the filter size. One commonly used approach is known as the Smagorinsky–Lilly method which works on the assumption that energy production and dissipation in the small scales are in equilibrium.

In practice, LES simulations are the most popular option for simulations where the aim is to simulate respiratory airflow with a high degree of accuracy. The results from LES simulations more accurately match experimental flow results than RANS simulations, but at significantly less computational expense than DNS simulations.

Detached Eddy Simulation Turbulence Model

Detached eddy simulation turbulence models combine RANS and LES approaches, so that a RANS approach is used near the wall and LES is used away from the walls. This is achieved by calculation of the local turbulent length scale. Where the turbulent length scale is smaller than the mesh size, the RANS approach is used, and the LES approach is used elsewhere. This technique has the advantage of allowing a coarser grid resolution near the wall. The effect of this is to reduce the computational cost of LES simulations.

7.3.4 Near-Wall Modelling

The flow immediately adjacent to the airway wall does not move relative to the wall due to friction. This effect is known as the **no-slip condition**. However, the flow further from the wall will be moving and a large gradient of velocity forms between the stationary flow and the fast moving mid-stream flow.

The discussion of the Reynolds number in Sect. 7.2.2, showed viscous forces are dominant in slow moving flows which have low Reynolds numbers. Therefore, viscous forces are dominant near the wall, and become less important as we move away from the wall. We call the region where viscous forces dominate the **viscous sub-layer**.

Not all turbulence models can accurately model the viscous effects near the wall (notably for example, the k-ε model) and instead use near-wall modelling to capture the effect of the viscosity near the wall. Near-wall modelling uses empirical relationships to model the velocity near the wall.

Near-wall modelling makes use of non-dimensional numbers to allow empirical flow results to be applied to a variety of flow situations. We can define a non-dimensional number for our velocity near the wall, u^+. It is the ratio of the absolute velocity, u, and the wall friction velocity, u_τ, which is derived from the wall shear stress, τ_w, by the equation $u_\tau = \sqrt{\frac{\tau_w}{\rho}}$. The wall friction velocity is a way of expressing shear stress in the flow in terms of velocity. The wall shear stress is the tangential pressure the flow exerts on the walls as the flow tries to pull the walls of the conduit along with it via friction. The dimensionless velocity u^+ is given by

$$u^+ = \frac{u}{u_\tau} \tag{7.46}$$

Given this value, the non-dimensional distance to the wall is y^+.

$$y^+ = \frac{y u_\tau}{\nu} \tag{7.47}$$

where y is the absolute distance to the wall, and ν is the kinematic viscosity, which is $\nu = \frac{\mu}{\rho}$. In the **viscous sub-layer,** i.e. the region where $y^+ < 5$, the following relationship is found

$$u^+ = y^+ \tag{7.48}$$

This equation tells us that in the region very close to the wall, there is a linear-relationship between non-dimensional velocity and distance from the wall. At the edge of the viscous sub-layer $u^+ = y^+ = 5$ and the velocity u at the edge of the viscous sublayer is u_δ. Substituting these values into (7.46) and (7.47) gives us the relationship for y, the height of the viscous sublayer

$$y = \frac{25\nu}{u_\delta} \tag{7.49}$$

Since u_δ is related to the free stream velocity, this tells us that the faster the flow travels, the smaller the viscous sublayer becomes. Therefore if we are to create a mesh that captures the viscous sublayer, we will need fine meshes at the wall for simulations for fast moving flows.

In respiratory CFD, flow speeds are not particularly high and in general it is possible to obtain a mesh where the first mesh element has a $y^+ < 1$ throughout the mesh. This ensures good coverage of the viscous sublayer and obviates the need for further near wall modelling. However, other empirical relationships do exist for higher y^+ values.

In the region known as the logarithmic sublayer, $30 < y^+ < 400$, a logarithmic relationship is found between u^+ and y^+ from experimental data and is known as the law of the wall

$$u^+ = 2.5 \ln\left(y^+\right) + 5.45 \tag{7.50}$$

Between the viscous sublayer and the logarithmic sublayer, i.e. $5 < y^+ < 30$, a transitional sublayer is found in the region where both turbulence and viscous forces are important. In this region, the relationship between y^+ and u^+ can be modelled as in-between the linear and logarithmic relationships above.

When modelling turbulence at the wall, it is necessary to ensure the employed turbulence model is valid throughout the domain. If the desired turbulence model is not valid at the wall, a hybrid scheme can be used to capture viscous effects down to the wall, along with a mesh fine enough to have mesh elements within the viscous sublayer. Alternatively, wall modelling techniques based on the relationships given above can be used to connect the free stream velocity field with the wall.

7.4 Flow and Time—Steady or Unsteady?

Respiratory airflow is not constant in time. Each breath involves accelerating airflow to peak inhalation, decelerating the airflow and reversing its direction and then exhaling. We must decide how our CFD simulations will deal with the temporal variation in airflow. Not only is respiratory airflow cyclical, but it contains many fluctuations in local velocity patterns that occur at higher frequencies than the respiratory rate [4].

Unsteady flow should not be confused with turbulent flow. A flow can be unsteady and laminar, for example a slow moving flow that moves up and down in response to some disturbance, but remains in smooth layers. Turbulent flow is characterised by oscillations in local velocities with respect to time, but the mean flow field can be relatively unchanging with time, so a steady simulation may suffice (with appropriate modelling of the effect of the turbulent oscillations).

7.4.1 Steady Simulations

The simplest approach to modelling the changes in respiratory airflow throughout a breath is to ignore them. In many cases, it is appropriate to choose one part of the breath that best represents what you are interested in and to model just that part of the breath. For example, if you are interested in the pressure loss along the airway during inhalation, the pressure loss is likely to be highest when the airflow rate is highest (ignoring airway motion). Therefore, rather than simulating a full breath, it takes much less computational effort to only model airflow at peak inhalation. The RANS turbulence models are best suited to this approach, as they yield a temporal average solution in their formulation. LES simulations are time-dependent, so cannot produce steady flow results (see instead Quasi-Steady Simulations below).

The result of a steady simulation is equivalent to the average airflow pattern at constant boundary conditions, i.e. as if a person could inhale at the same rate for an infinitely long time, and we took the average flow field of that inhalation. No fluctuations are apparent in the solution. Some flow fields do not have a steady solution, for example the downstream position of the jet that forms in the vocal folds may oscillate over time. In a steady simulation, movement of this jet will appear as a blurring of the jet's boundary and it may be hard to achieve simulation convergence if no steady solution exists.

The great advantage of steady simulations is that the flow solution only needs to be obtained for one time point, meaning results can be achieved very quickly. However, the flow solution at any time during a breath depends to some degree on what the flow was doing at previous time points, and this information is lost in steady flow simulations.

7.4.2 Unsteady Simulations

Unsteady Simulations with Changing Boundary Conditions

The alternative to steady simulations is to simulate the flow at each point in time over the period of interest. Just as we divide the spatial region of interest up into many small cells during meshing, we must divide our temporal period of interest into many smaller periods, called time steps, to perform unsteady CFD simulations. Just as mesh size must be carefully chosen, so time steps must also be determined based on the period of fluctuations which you wish to resolve, and the rate at which boundary conditions change with time.

The iterative solving of the governing equations must be repeated at each time step. Therefore an unsteady CFD simulation is a series of CFD simulations, one for each time step. While it is computationally expensive to iterate to a solution at each time step, this problem is somewhat mitigated by the fact that the flow does not change much between each time step (assuming the duration of the time step is sufficiently small) so it often does not take many iterations to achieve convergence at each successive time step in an unsteady simulation.

The advantage of unsteady CFD simulations is revealing the temporal dynamics of airflow; for example, how do flow features form during early inhalation, or how does airway resistance vary with time? Particles that enter the airway during early inhalation, when the flow speed is low and is dominated by viscous effects, may follow very different trajectories from particles released at peak inhalation where inertial forces dominate. This effect may be lost in steady simulations.

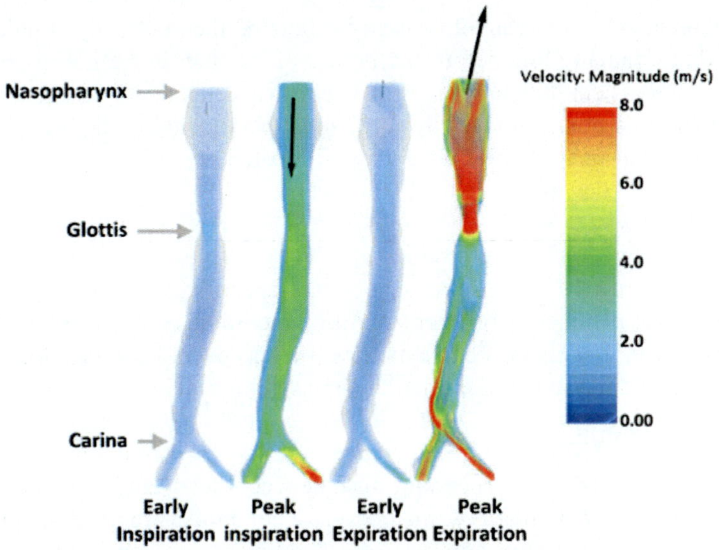

Fig. 7.12 Volume renderings of airflow velocity at four instances in the period of a tidal breath in a neonate. Unsteady flow modelling allows the velocity of air during different periods of the breath to be compared. Black arrows at the top of the models represent the local velocity vectors. This patient has high velocity airflow during peak expiration. Figure courtesy of Chamindu Gunatilaka, Cincinnati Children's Hospital Medical Center

Unsteady CFD simulations can be performed with RANS models, despite their temporal averaging technique. Unsteady RANS (or URANS) simulations still perform temporal averaging, although they separate the period of change due to changing boundary conditions from the period over which turbulence effects occur. Therefore, they require that the period over which they average the Reynolds stresses is much smaller than that of the fluctuations of interest. LES simulations are inherently designed for unsteady simulations.

Unsteady simulations require that the boundary conditions are defined for the entire period of interest. Some boundary conditions may remain constant (for example the exterior air pressure at the nostrils may remain at 0 Pa), but others will change with time, for example the pressure at the distal end of the airway geometry. Initial conditions (i.e. maps of pressure, airflow velocity, and other quantities of interest) are also required, since the flow at each time point depends on what came before it. It is often easiest to start simulations with the flow stationary. The solver will solve the flow equations for the first time step based on the boundary conditions at that time and the initial conditions. Once the desired convergence is obtained, the solver moves on to the next time step, updating the boundary conditions for the new time. The flow at each time point will depend on the flow at a number of previous time steps. The number of previous time-steps that are used to calculate the flow at the current time-step depends on the order of accuracy of the simulation. Figure 7.12 shows the velocity field in the trachea of a neonate at four points within a single tidal breath.

If the initial condition cannot be easily estimated then unsteady simulations can be started to simulate flow before the period of interest. If the initial condition is inaccurate, a transient pressure force can occur to accelerate the flow to the required flow rate and this transient pressure force must be allowed to die away before the period of interest starts. Some studies have simulated several consecutive breaths to ensure the correct flow is calculated in the last breath, by which time transient forces will have died away.

Unsteady flow simulations produce all CFD variables of interest at every time step. Therefore, they can consume a huge amount of storage space and it is vital to decide which parameters of interest will be recorded, and how often, before the simulation is run. Very often coarse metrics, such as overall pressure loss are easier to store at every time step, rather than the pressure in every cell.

Quasi-steady Simulations

Quasi-steady simulations are a compromise between steady and fully unsteady simulations. Like steady simulations, the boundary conditions are fixed in quasi-steady simulations and do not change with time. The boundary conditions are chosen to represent the instant of most interest that occurs during a breath, such as peak inhalation. However, the simulation is still solved as if it was unsteady, allowing temporal dynamics (such as oscillating regions of separated flow, which occur in the nose and trachea, for example) to be calculated.

As the initial flow field is not known, the simulation is often started from a stationary initial flow condition. As in unsteady simulations, a transient pressure force will be required to accelerate the flow (assuming the driving boundary condition is a given flow rate), and this transient pressure force must be allowed to die away over time before analysing the solution. An alternative approach, which may reduce the effect of this transient, is to run a steady simulation first and use the result from the steady simulation as the initial condition for the quasi-steady simulation.

In both approaches, the first few time steps of a quasi-steady simulation are often discarded before meaningful results are recorded. With quasi-steady simulations, temporal averages and fluctuating values may be recorded, as opposed to just temporal averages in the steady approach. The quasi-steady approach also allows computationally inexpensive steady-like simulations to be performed with unsteady turbulence models such as LES.

7.4.3 Temporal Considerations in Respiratory Simulations

The choice of temporal modelling approach between steady, quasi-steady, and unsteady will depend on the quantity of interest for a given simulation. Studies have shown that the main flow patterns within the nose (e.g. regions of fast moving flow, separated regions, etc) develop very quickly and that comparison between quasi-steady and unsteady simulations yield relatively small differences in terms of pressure loss at the same flow rate [4]. Steady and quasi-steady simulations do not compute the pressure force needed to accelerate the flow and only compute the pressure loss due to frictional losses. Therefore, if the quantity of interest is airway resistance, and the flow is relatively steady with respect to time, a computationally inexpensive steady simulation may suffice. However, if the boundary conditions vary throughout the breath—for example if the walls move due to airway motion, then the full breath must be simulated.

Running unsteady simulations provides much richer information regarding the airflow within the breath, but at the additional computational expense of simulating many time steps. It may also be necessary to run several coarse preliminary simulations to determine the required size of time step.

7.5 Practical Respiratory Airflow Modelling

There are many decisions to make when starting a new respiratory CFD project. For example, how many mesh elements does your simulation need, how many prism layers does your mesh require, and which turbulence model should you use? There are few absolute rules to follow in making these decisions, therefore this section aims to provide some general guidance on what to consider when making these decisions.

7.5.1 Simulation Resolution

One important factor to note when choosing simulation parameters is that different quantities of interest will have different simulation requirements. For example, if the quantity of interest is airway resistance at peak inhalation, a low resolution, steady simulation using a RANS turbulence model may match the prediction of a highly resolved DNS calculation over a full breathing cycle to within an acceptable degree of error. However, the deposition of inhaled particles is much more sensitive to local flow patterns, particularly at the wall. Therefore, two simulations that give approximately the same pressure loss may give widely different particle deposition patterns. To determine the required mesh and time-step resolution, simulations are typically repeated at different mesh resolutions and time-step values. The simulation is repeated at ever-increasing mesh-resolutions until the results do not change. The result is then said to be mesh-insensitive (and/or time-step insensitive). However, it is important that the quantities of interest used to determine the required mesh resolution are the same quantities of interest of the main study. Similarly, the geometries used in a mesh resolution study must be similar to those used for the main study as airflow characteristics can differ strongly between different airway anatomies, for example those with pathology. Previous studies have shown that in a healthy trachea, 86% of the losses which occur below the glottal jet are due to friction at the airway wall. However, in pathological tracheas, friction at the wall accounted for only 21% of the losses, with the remainder occurring due to disturbances within the main flow [2].

Therefore, a mesh with high resolution at the wall, but low resolution in the bulk would work well in the healthy case, but would under-resolve losses in pathological cases.

An alternative to the traditional mesh resolution approach of increasing mesh resolution from a low base is to run the highest resolution study that is computationally affordable and then to try to match the results from that high-resolution study with more affordable modelling approaches. A good example of a high-resolution reference simulation would be a full-breathing-cycle LES calculation with of order 10 million cells and sub-millisecond temporal resolution. If high-resolution reference studies cannot be performed, there are several published studies with geometries and results available for comparison [3, 4, 8].

7.5.2 Turbulence Modelling

Most of the decisions that go into choosing simulation parameters are based on the expected level of turbulence in the simulation. When estimating how turbulent the flow will be, one indicator is the Reynolds number which can be estimated prior to performing the simulation. The mean Reynolds number at any cross-section in the geometry can be calculated from the flow rate and geometric analysis. In respiratory CFD, the geometry of the airway anatomy of interest is usually obtained via medical imaging, and the respiratory flow rate can be measured directly in the subject during imaging [5, 6] or standard values can be obtained from previous studies. If a full breathing cycle is to be simulated, the highest Reynolds number will occur at the highest flow rate (assuming static geometry). Using the peak flow rate, or if a steady or quasi-steady simulation is to be performed, the steady flow rate, the spatial mean velocity, $\bar{u}$ at any cross-section in the geometry can be calculated from the equation

$$\bar{u} = \frac{Q}{A} \tag{7.51}$$

where Q is the flow rate and A is the cross-sectional area of the airway lumen at any given location. As we saw in (7.2), the Reynolds number is then given by $Re = \frac{\rho \bar{u} D}{\mu}$. In Sect. 7.2.2, we saw that in idealised long straight pipes, flow is laminar at Reynolds numbers less than 2400 and fully turbulent above 4000. Since the airway contains regions of changes in cross-sectional shape and size, 90° turns in the nasopharynx, motion, and vibration, these values must only be used as extremely rough guidelines for airway turbulence. Better estimates of turbulence can be obtained by considering worse-case scenarios. For example, a jet forms as flow passes through the constriction at the vocal folds and then abruptly opens into the trachea. Therefore, the flow in the upper trachea will consist of a fast moving jet and recirculation regions. The velocity of the jet will be approximately equal to the velocity in the narrow cross-section of the glottis. Using this value of jet velocity may give a better indication of the turbulence in this region than the mean velocity. Similarly, the nasal passages have a very complex geometry. The majority of flow passes along the floor of the nasal cavity or between the middle and inferior turbinates, with little flow in the upper regions of the nose. Therefore, for the same flow rate, local velocities may be higher in the nose than a pipe with a similar cross-sectional area.

Previous studies have shown that flow is generally laminar or mildly transitional in the nose and turbulent in the region of the laryngeal jet [4, 8, 9, 25]. As flow progresses further through the bronchial tree, it will decelerate and viscous forces will start to become more significant.

The choice of turbulence model is generally between no turbulence model, RANS models, and LES models. However, this choice is often confused by the terminology employed by commercial CFD packages, which offer 'laminar' turbulence models which really means no turbulence model is being employed. However, it is still possible to simulate turbulent flow without a turbulence model. If the mesh is of sufficiently high resolution to capture the smallest scales of flow, then the solver will directly calculate all the turbulent structures without the need of turbulence modelling (this is DNS). An alternative approach to achieving fully resolved DNS simulations is to choose an LES turbulence model. If the mesh is sufficiently refined, the filter size, below which the LES performs turbulence modelling, will be smaller than the smallest scale of turbulence (the Kolmogorov length scale) and again, a DNS simulation will have been performed. In practice, DNS simulations are rarely achieved except in cases with very low flow rates or if large computational resources are available.

In most circumstances, LES models provide more accurate results (i.e. better correlation with experimental results) than RANS models, however they also require higher mesh resolution. With current computational availability, this increase in resource demands is often justifiable. In addition to higher mesh resolution requirements, another additional cost of LES models is that they cannot perform steady simulations, however this can be largely mitigated by using a quasi-steady approach. Where LES models require prohibitive resources, k-ω RANS models are often the best 'affordable' alternative for internal flow behaviours found within the airways.

Often, all the requirements of a CFD simulation cannot be determined a priori. For example, it is hard to predict the y^+ values in the first prism layer of cells in a mesh without knowing the wall shear stress values. It is therefore often worth performing a coarse, steady simulation before the main simulation(s) to obtain estimates for these parameters.

7.5.3 Comparison to Experimental Results

The gold standard with which to compare the results of a CFD simulation would be results obtained in vivo. However flow measurements within the airway are extremely difficult to perform due to the difficulty in placing instrumentation within the airway. Placing probes within the airway also causes the local flow of air to change, further disrupting results. One promising approach involves obtaining velocity fields of inhaled gases via phase-contrast MRI. However, this technique results in relatively low spatial-resolution velocity maps [7]. The alternative is to compare CFD results to physical experiments.

Physical experiments involve measuring velocity fields or pressure values within in vitro models. Velocity fields can be measured to a high degree of accuracy via particle image velocimetry (PIV) and the results can be compared to CFD velocity fields. However, it should be remembered that in vitro models are not fully representative of in vivo conditions. For example, they often omit airway motion, or the motion is just simple elastic collapse that lacks neuromuscular control. In vitro models also usually omit considerations such as heating and humification of air, which are often easier to model via CFD rather than recreate physically. Therefore in vitro models should be thought of as an alternative modelling approach to CFD, rather than a gold-standard reference.

7.5.4 Boundary Conditions

To finish this chapter, we should note that a very significant aspect of achieving realistic CFD models of respiratory airflow is the choice of boundary conditions. The process of obtaining airway anatomy from medical imaging, via segmentation and subsequent surface generation, can have a huge effect on subsequent CFD results; perhaps even more effect than the choice of turbulence model. The choice of which physical parameters need capturing in boundary conditions also affects results. The choice of whether or not to model the heating and humidification of inhaled air will depend on the quantities of interest for a particular simulation and may or may not be significant. In addition to comparison of results at different mesh resolutions or with different turbulence models it is often worth comparing choices of boundary conditions to determine how sensitive the results are to a particular boundary condition. In general, to obtain results that are representative of in vivo flow conditions, the boundary conditions, mesh decisions, and flow modelling all need to be considered and optimised, depending on the quantities of interest.

Acknowledgements Alister Bates would like to thank Dr Qiwei Xiao, Dr Neil Stewart, Dr Nara Higano, Deep Gandhi, and Chamindu Gunatilaka for their help in the preparation of this chapter.

References

1. A. Bates, R. Cetto, D. Doorly, R. Schroter, N. Tolley, A. Comerford, The effects of curvature and constriction on airflow and energy loss in pathological tracheas. Respir. Physiol. Neurobiol. **234**, 69–78 (2016)
2. A. Bates, A. Comerford, R. Cetto, R. Schroter, N. Tolley, D. Doorly, Power loss mechanisms in pathological tracheas. J. Biomech. **49**(11), 2187–2192 (2016)

3. A. Bates, A. Comerford, R. Cetto, D. Doorly, R. Schroter, N. Tolley, Computational fluid dynamics benchmark dataset of airflow in tracheas. Data Brief **10**, 101–107 (2017)
4. A.J. Bates, D.J. Doorly, R. Cetto, H. Calmet, A. Gambaruto, N. Tolley, G. Houzeaux, R. Schroter, Dynamics of airflow in a short inhalation. J. R. Soc. Interface **12**(102), 20140880 (2015)
5. A.J. Bates, A. Schuh, K. McConnell, B.M. Williams, J.M. Lanier, M.M. Willmering, J.C. Woods, R.J. Fleck, C.L. Dumoulin, R.S. Amin, A novel method to generate dynamic boundary conditions for airway CFD by mapping upper airway movement with non-rigid registration of dynamic and static MRI. Int. J. Numer. Methods Biomed. Eng. **34**(12), e3144 (2018)
6. A.J. Bates, A. Schuh, G. Amine-Eddine, K. McConnell, W. Loew, R.J. Fleck, J.C. Woods, C.L. Dumoulin, R.S. Amin, Assessing the relationship between movement and airflow in the upper airway using computational fluid dynamics with motion determined from magnetic resonance imaging. Clin. Biomech. **66**, 88–96 (2019)
7. A.J. Bates, M.M. Willmering, R. Thomen, C. Gunatilaka, M.M. Hossain, C. Dumoulin, J. Woods, In vivo validation of upper airway respiratory computational fluid dynamics (CFD) with phase-contrast MRI of hyperpolarized 129xe, p. 4138
8. H. Calmet, A.M. Gambaruto, A.J. Bates, M. Vázquez, G. Houzeaux, D.J. Doorly, Large-scale CFD simulations of the transitional and turbulent regime for the large human airways during rapid inhalation. Comput. Biol. Med. **69**, 166–180 (2016)
9. H. Calmet, G. Houzeaux, M. Vázquez, B. Eguzkitza, A. Gambaruto, A. Bates, D. Doorly, Flow features and micro-particle deposition in a human respiratory system during sniffing. J. Aerosol Sci. **123**, 171–184 (2018)
10. X.B. Chen, H.P. Lee, V.F.H. Chong, D.Y. Wang, Numerical simulation of the effects of inferior turbinate surgery on nasal airway heating capacity. Am. J. Rhinol. Allergy **24**(5), e118–e122 (2010)
11. G.J. Garcia, N. Bailie, D.A. Martins, J.S. Kimbell, Atrophic rhinitis: a CFD study of air conditioning in the nasal cavity. J. Appl. Physiol. **103**(3), 1082–1092 (2007)
12. K. Inthavong, Z. Tian, J. Tu, CFD simulations on the heating capability in a human nasal cavity
13. K. Inthavong, J. Wen, J. Tu, Z. Tian, From CT scans to CFD modelling-fluid and heat transfer in a realistic human nasal cavity. Eng. Appl. Comput. Fluid Mech. **3**(3), 321–335 (2009)
14. D. Joseph, S. Carmi, Stability of Poiseuille flow in pipes, annuli, and channels. Q. Appl. Math. **26**(4), 575–599 (1969)
15. A. Karl, F.S. Henry, A. Tsuda, Low Reynolds number viscous flow in an alveolated duct. J. Biomech. Eng. **126**(4), 420–429 (2004)
16. J. Kimbell, D. Frank, P. Laud, G. Garcia, J. Rhee, Changes in nasal airflow and heat transfer correlate with symptom improvement after surgery for nasal obstruction. J. Biomech. **46**(15), 2634–2643 (2013)
17. P.K. Kundu, I.M. Cohen, *Fluid Mechanics* (Elsevier, Amsterdam, 2001)
18. C.-L. Lin, M.H. Tawhai, G. McLennan, E.A. Hoffman, Characteristics of the turbulent laryngeal jet and its effect on airflow in the human intra-thoracic airways. Respir. Physiol. Neurobiol. **157**(2–3), 295–309 (2007)
19. F. Menter, Zonal two equation kw turbulence models for aerodynamic flows, in *23rd Fluid Dynamics, Plasmadynamics, and Lasers Conference* (1993), p. 2906
20. F.R. Menter, Two-equation eddy-viscosity turbulence models for engineering applications. AIAA J. **32**(8), 1598–1605 (1994)
21. W. Pfenniger, Transition in the inlet length of tubes at high Reynolds numbers, in Boundary Layer and Flow Control, ed. by G. Lachman (1961), pp. 970–980
22. D.C. Wilcox, Reassessment of the scale-determining equation for advanced turbulence models. AIAA J. **26**(11), 1299–1310 (1988)
23. D.C. Wilcox et al., *Turbulence Modeling for CFD*, vol. 2 (DCW Industries La Canada, 1998)
24. A. Willis, J. Peixinho, R. Kerswell, T. Mullin, Experimental and theoretical progress in pipe flow transition. Philos. Trans. R. Soc. A: Math. Phys. Eng. Sci. **366**(1876), 2671–2684 (2008)
25. Q. Xiao, R. Cetto, D.J. Doorly, A.J. Bates, J.N. Rose, C. McIntyre, A. Comerford, G. Madani, N.S. Tolley, R. Schroter, Assessing changes in airflow and energy loss in a progressive tracheal compression before and after surgical correction. Ann. Biomed. Eng. **48**(2), 822–833 (2020)

26. D.A. Yablonskiy, A.L. Sukstanskii, J.C. Leawoods, D.S. Gierada, G.L. Bretthorst, S.S. Lefrak, J.D. Cooper, M.S. Conradi, Quantitative in vivo assessment of lung microstructure at the alveolar level with hyperpolarized 3He diffusion MRI. Proc. Natl. Acad. Sci. **99**(5), 3111–3116 (2002)
27. D.A. Yablonskiy, A.L. Sukstanskii, J.C. Woods, D.S. Gierada, J.D. Quirk, J.C. Hogg, J.D. Cooper, M.S. Conradi, Quantification of lung microstructure with hyperpolarized 3He diffusion MRI. J. Appl. Physiol. **107**(4), 1258–1265 (2009)

Chapter 8
Clinical Implications of Nasal Airflow Simulations

Dennis Onyeka Frank-Ito and Guilherme Garcia

Abstract This chapter provides an up-to-date literature review on the relevance of CFD modeling results in clinical settings and the efforts made in the field to establish normative ranges for some CFD-derived variables. We discuss the significance of CFD-derived variables such as unilateral nasal airflow partitioning, nasal airway resistance, heat flux and wall shear stress on nasal function or symptomatology, and the strong associations established with these variables. Two important issues are discussed: (i) a need to establish meaningful representation involving combinations of different computed variables in order to accurately capture the full dynamical nature of patient-reported symptomatology; (ii) a need to develop a robust database of normative values for CFD-derived variables since typical healthy human sinonasal airway anatomies are characterized by substantial intersubject variability that confound concise description of normal airflow profile.

8.1 Introduction

Although computational fluid dynamics (CFD) modeling is becoming an increasingly important tool in providing important insights into the understanding of nasal airway physiology and pathophysiology, there is still work to be done in understanding the clinical relevance of the many variables that can be computed from CFD simulations. The significance of CFD-derived variables such as unilateral nasal airflow partitioning, nasal airway resistance, heat flux and wall shear stress on nasal function or symptomatology have been explored and strong associations have been established with some of these variables. Nonetheless, one possible explanation why the clinical implications of additional CFD-derived variables has not yet been established is the fact that the complexity and variable nature of nasal and sinus diseases

D. O. Frank-Ito (✉)
Duke University, Durham, NC, USA
e-mail: dennis.frank@duke.edu

G. Garcia
Medical College of Wisconsin, Wauwatosa, WI, USA

K. Inthavong et al. (eds.), *Clinical and Biomedical Engineering in the Human Nose*, Biological and Medical Physics, Biomedical Engineering,
https://doi.org/10.1007/978-981-15-6716-2_8

may not be adequately represented by a single CFD-derived variable, but a mathematical representation involving combinations of different computed variables in order to accurately capture the full dynamical nature of patient-reported symptomatology. Another enormous potential of CFD modeling is in surgical planning for nasal and sinus diseases, particularly for complicated cases. Unnecessary surgical procedures can be minimized by allowing the surgeon to better select surgically treatable patients and to target specific anatomical regions of concern for correction to airflow-related normative ranges without 'guessing' which procedures can provide the most benefit. One current challenge is developing a robust database of normative values for CFD-derived variables since typical healthy human sinonasal airway anatomies are characterized by substantial intersubject variability that confound concise description of normal airflow profile. This chapter aims to provide an up-to-date overview of the literature on the relevance of CFD modeling results in clinical settings and efforts made in the field to establish normative ranges for some CFD-derived variables.

8.1.1 Importance of Nasal Airflow

Nasal respiration is vital for human survival—the nasal airways serve as the primary pathway for modulation and humidification of inspired air from ambient atmosphere into the lungs [46]. While mouth breathing can sustain life, this leads to dryness in the oral cavity as the mouth does not have efficient air-conditioning functionality as the nose [107, 138]. Further, nasal breathing has been reported to produce more ventilation than mouth breathing, which suggests that nasal airflow may have a stimulant effect on respiration [102]. Another importance of nasal airflow is the sense of smell; when exposed to odorants, inhaled nasal airflow delivers odorant-laden air to the olfactory epithelium for smell detection [29, 41, 44, 45]. In spite of the significant role of the nasal airways in human respiration, there remains challenges in proper assessment of what constitutes a normal nasal airway and a normal airflow profile in the nose.

The current clinical standard for objective evaluation of nasal airway patency includes the use of rhinomanometry, acoustic rhinometry, and peak nasal inspiratory flow. Briefly, rhinomanometry provides global information related to nasal airflow and resistance, while acoustic rhinometry gives detailed analysis of the nasal anatomy via information related to nasal airway cross-sectional areas [25, 125, 126]. As the name implies, peak nasal inspiratory flow measures peak nasal airflow during maximal forced inspiration [11, 19, 85]. Nonetheless, these instruments cannot describe the relationship between nasal form and function simultaneously because each instrument assesses different nasal attributes [91, 125]. In addition, the instruments have low correlation with patient-reported quality of life measures [19]. On the contrary, (CFD) modeling provides highly accurate description of local and global nasal airflow field, and the role of temperature and humidity on nasal airflow dynamics [91, 117]. To sum up, the complex relationship between variability in nasal anatomy and differences in nasal airflow pattern can be accurately described using CFD modeling [91].

8.2 Interpreting CFD Simulation Results

CFD simulations of nasal airflow is a highly interdisciplinary field at the confluence of engineering, physiology, and medicine. Thus, it is essential that CFD practitioners speak a common language. It is also crucial that both engineers and clinicians understand the assumptions underlying CFD simulations and to what extent these assumptions affect the precision and clinical relevance of CFD variables. In this section, we discuss the need for physical accuracy, physiological accuracy, and clinical relevance of CFD models.

8.2.1 Physical Accuracy

Modern CFD codes provide numerical solutions that satisfy conservation of mass, momentum, and energy. The challenge to CFD users is to select numerical methods that provide a good balance between physical accuracy and computational time. For example, computational grids with higher mesh density provide more accurate solutions but require additional computational time [70]. Similarly, CFD methods that quantify turbulence at multiple scales, such as direct numerical simulation (DNS) or large eddy simulation (LES), provide more accurate solutions for turbulent flows, but these methods require hardware (computer clusters) that are not universally available. In vitro experiments under controlled laboratory conditions can be used to validate the numerical methods and determine the precision of each numerical scheme. It is important to keep in mind that numerical methods that accurately predict one CFD variable do not necessarily have the same precision for other CFD variables. For example, a laminar flow solver may provide a reasonable estimate of the pressure drop along the nasal cavity for low-turbulence flows, but it will not capture the turbulent velocity profile [93], thus it cannot predict the trajectories of individual particles in turbulent flow. Therefore, experimental validation plays a key role in selecting numerical methods that provide the desired precision, while minimizing computational cost (see Sect. 8.3).

8.2.2 Physiological Accuracy

The physiological accuracy of CFD simulations is perhaps more difficult to establish than the physical accuracy. While in vitro experiments can be repeated multiple times until a desired precision is achieved, in vivo measurements require patient cooperation and can only be repeated a limited number of times. Consequently, in vivo measurements typically have a larger percent error than in vitro experiments performed in well-controlled laboratory conditions. In addition, current CFD models do not include all the complexities of nasal physiology, such as wall compliance and spatial gradients in mucosal temperature.

Another major challenge is that the nasal anatomy is not fixed, but dynamic, thus, each CT scan is only a snapshot of a patient's nasal anatomy. Mucosal engorgement can change in a few minutes due to the nasal cycle or in response to changes in posture, air temperature, and humidity [87]. To control for these factors, some authors advocate for decongesting the nasal mucosa prior to objective assessment [10]. However, this approach has the disadvantage of assessing nasal airflow in a state different from the natural state in which patients live. To minimize these problems, validation of CFD simulations should ideally be based on CT imaging and in vivo measurement of flow variables obtained in rapid succession. In summary, when interpreting CFD simulations of nasal airflow, one must keep in mind the current limitations of CFD technology in reproducing the complexities of nasal physiology.

8.2.3 Clinical Relevance

A major question that remains to be explored is the clinical application of CFD simulations of nasal airflow. It is widely recognized that CFD simulations have potential clinical applications, including virtual surgery planning to select the best surgical procedure for patients with nasal airway obstruction [118, 124, 136]. However, the hypothesis that CFD-informed treatment selection will improve patient outcomes as compared to the current standard of care remains unproven. We anticipate that as the CFD technology evolves, becoming faster and more accessible, studies with larger sample sizes will likely be conducted to test the utility of CFD simulations in a clinical setting.

8.3 Correlating CFD with Objective Clinical Tests

Although CFD simulations of nasal airflow have been validated with in vitro experiments in many studies [38, 83, 93], few studies have compared CFD to objective clinical tests. The subsections below review the current literature comparing CFD to the two most common objective tests of nasal patency, namely rhinomanometry and acoustic rhinometry. Throughout this section, we highlight areas that need further research.

8.3.1 CFD versus Rhinomanometry

Rhinomanometry is a technique to measure the pressure-flow curve of the human nasal cavity. In anterior rhinomanometry, one nostril is sealed with tape or foam insert while airflow in the contralateral cavity is recorded with a mask. A pressure catheter is inserted through the tape (or foam insert) to record the pressure at the nasal choana. In posterior rhinomanometry, the pressure catheter is inserted through the mouth to measure pressure in the oropharynx. The Standardisation Committee on Objective Assessment of the Nasal Airway recommended that rhinomanometry data should be reported for a pressure drop of 150 Pa [24]. However, many patients cannot reach this pressure drop, thus research studies often report nasal resistance and/or airflow at a pressure drop of 75 Pa [48, 148].

The literature is conflicting regarding the correspondence between the pressure-flow curves obtained with CFD versus rhinomanometry. Some authors have reported a perfect match between the two techniques, while others reported a significant discrepancy. For example, Wootton and co-authors found an excellent agreement between CFD and posterior rhinomanometry in one healthy volunteer (Fig. 8.1) [144]. In contrast, in a study of one healthy subject and two NAO patients, Osman and colleagues found that nasal resistance at 150 Pa was roughly 10-fold lower in the CFD simulations as compared to the in vivo measurements [113]. One factor contributing to this discrepancy is that the model in the study by Wootton and co-authors was truncated at the level of the middle turbinate (see panel A in Fig. 8.1), which likely increased the CFD-derived resistance. Another source of discrepancy between CFD and rhinomanometry is that CFD simulations assume rigid walls and thus do not account for nasal valve collapse. Finally, another possibility is that different studies used different radiodensity thresholds when segmenting the nasal airspace from the CT scans. Cherobin and co-authors [20] compared the pressure-flow curve obtained with anterior rhinomanometry with CFD models created with segmentation thresholds of −300, −550, and −800 Hounsfield units (HU). The authors found that unilateral resistance measured at a unilateral flow of 125 mL/s was on average 52% lower in models created with a segmentation threshold of -300 HU as compared to a segmentation threshold of −800 HU (Fig. 8.2) [20]. The authors noted that the segmentation threshold is rarely reported in the CFD literature and recommended that future CFD studies should always report the segmentation threshold.

The first study to compare rhinomanometry and CFD in a large cohort was performed by Zhao and colleagues [148, 149]. A statistically significant correlation (Pearson $r = 0.53$, $p < 0.05$) was found between bilateral nasal resistance measured with anterior rhinomanometry and CFD computed values in a cohort of 22 healthy subjects. Despite the statistically significant correlation, the numerical value obtained with the two techniques was inconsistent. More specifically, the pressure-flow relationship of the nasal cavity is non-linear ($\Delta P \propto Q^{1.75}$ in plastic nasal replicas [51], thus the numerical value of nasal resistance depends on the pressure value at which it is measured. In contrast to this expectation, the average resistance measured with CFD (0.069 ± 0.022 Pa.s/ml) was similar to the resistance measured with rhinomanometry (0.065 ± 0.029 Pa.s/ml) despite a 5-fold difference in the pressure value at which resistance was recorded ($\Delta P = 15$ Pa in the CFD simulations vs. $\Delta P = 75$ Pa in the rhinomanometry measurements) [148]. A similar inconsistency between CFD and anterior rhinomanometry was reported by Radulesco and collaborators [119]. For a pressure drop of 150 Pa, CFD-derived nasal resistance was 2 to 3 times lower

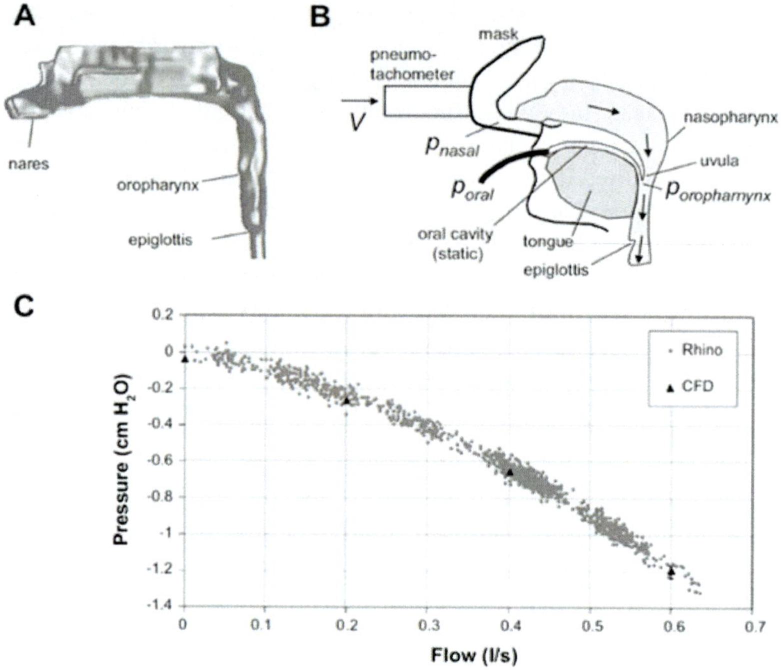

Fig. 8.1 Comparison of the pressure-flow curve calculated with CFD and measured with posterior rhinomanometry in one healthy adult. **A** CFD model. **B** Diagram of posterior rhinomanometry. **C** Pressure-flow curve during inspiration. Figure from Wootton and co-authors [144]

than measured with rhinomanometry (0.8 ± 1.2 Pa.s/mL vs. 1.8 ± 2.2 Pa.s/mL in the most obstructed cavity; 0.23 ± 0.10 Pa.s/mL vs. 0.60 ± 0.37 Pa.s/mL in the least obstructed cavity). This difference was statistically significant, while the correlation between CFD and rhinomanometry was not statistically significant. In contrast to these two studies, excellent agreement was found between CFD and posterior rhinomanometry by Lu and co-authors [95] in a cohort of 10 healthy subjects. For example, the authors reported that the left cavity resistance in the 5 male subjects was 0.52 ± 0.65 Pa.s/ml when calculated with CFD versus 0.55 ± 0.78 Pa.s/ml when measured with rhinomanometry. However, it is unclear if the same pressure drop was used to compare CFD versus rhinomanometry (the authors state that $\Delta P = 150$ Pa for posterior rhinomanometry, but do not specify the pressure drop used to quantify the CFD-derived resistance). The conflicting results reported by these 3 studies highlight the need for additional studies comparing CFD simulations to rhinomanometry.

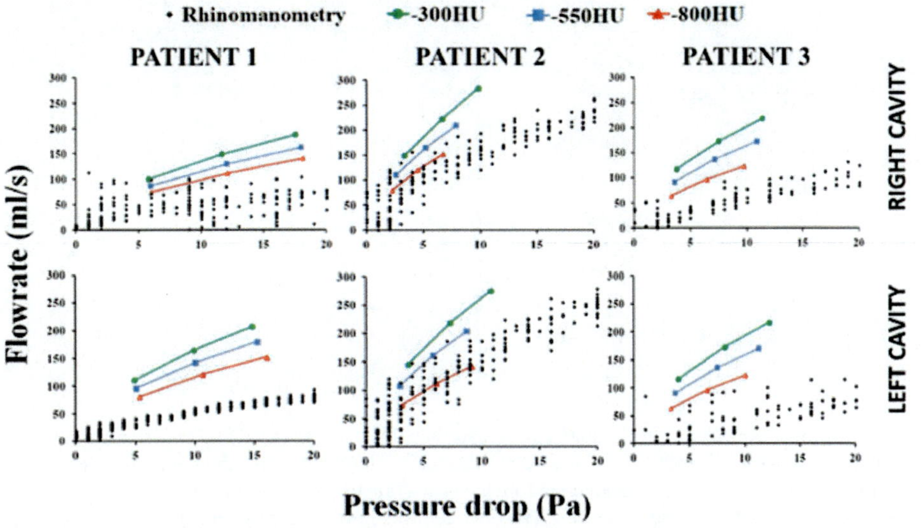

Fig. 8.2 – Pressure-flow curves measured with anterior rhinomanometry and calculated with CFD in three patients with nasal airway obstruction after mucosal decongestion with oxymetazoline. The CFD models were based on 3D reconstructions from CT scans created with segmentation thresholds of −300 HU, −550 HU, or −800 HU. Note the systematic increase in nasal airflow (reduction in nasal resistance) as the segmentation threshold increased from −800 HU to −300 HU. Figure from Cherobin et al. [20]

8.3.2 CFD versus Acoustic Rhinometry

Acoustic rhinometry measures the area-distance curve of the nasal cavity by releasing sound at one nostril and analysing the sound reflected by the nasal cavity [66]. Advantages of this technique include its low cost, non-invasive nature, and ability to provide measurements instantaneously [28]. Like rhinomanometry, acoustic rhinometry has been widely used in rhinologic research. Examples include investigating the nasal cycle [88], age-related changes in nasal anatomy [81], and quantifying surgical outcomes [57]. However, the accuracy of acoustic rhinometry measurements is influenced by many factors, including angle of inclination of the tube and nosepiece sealing quality, thus the operator must be well-trained [24, 47].

To our knowledge, only two studies to date have compared CFD and acoustic rhinometry. Lu and co-authors [95] reported no significant differences between the airspace minimal cross-sectional area (mCSA) obtained with acoustic rhinometry ($0.63 \pm 0.21\,cm^2$ in males and $0.58 \pm 0.19\,cm^2$ in females) and measured in 3D models built from CT scans ($0.64 \pm 0.19\,cm^2$ in males and $0.57 \pm 0.17\,cm^2$ in females). Unfortunately, the methodology used to measure the mCSA in the 3D models was not explained. Garcia and co-authors [52] proposed a CFD-based method to quantify the mCSA. Airspace cross-sectional areas were computed perpendicular to 10 flow streamlines representative of the main flow path and then averaged to produce a single area-distance curve (Fig. 8.3). This technique, dubbed "computational streamline rhi-

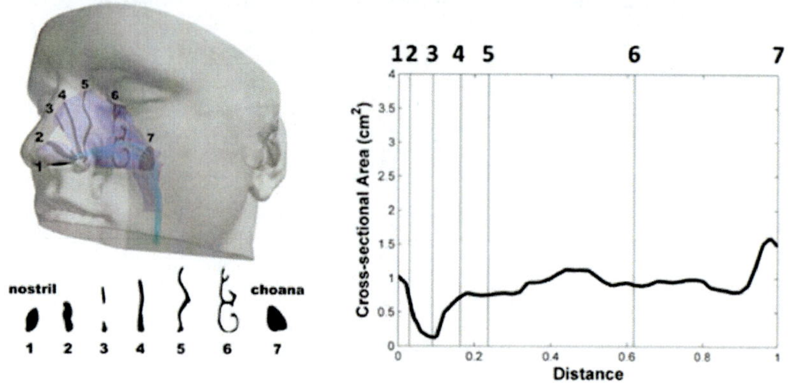

Fig. 8.3 The concept of computational streamline rhinometry. (LEFT) Flow streamlines were calculated using CFD. (RIGHT) Cross-sectional areas, calculated perpendicular to flow streamlines in the anterior nose and perpendicular to the nasal floor in the posterior nose, were plotted as a function of the distance from nostrils. The distance was normalized by the streamline length (i.e., Distance = 0 corresponds to nostril; Distance = 1 corresponds to choana). The shape of seven cross-sections and their locations in the area-distance curve are illustrated. Figure from Garcia et al. [52]

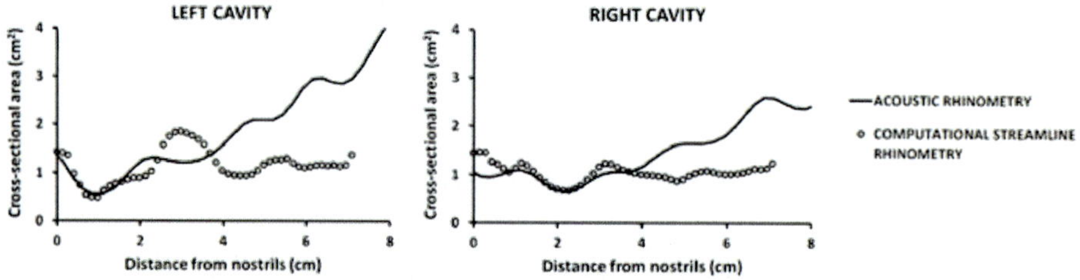

Fig. 8.4 Airspace cross-sectional area versus distance from nostrils in the left and right nasal cavities of one NAO patient obtained with computational streamline rhinometry compared to acoustic rhinometry measurements. Figure from Garcia et al. [52]

nometry", was compared to acoustic rhinometry performed in one patient with NAO. Good agreement was found between the area-distance curve in the anterior nose, including the mCSA, computed with CFD and measured with acoustic rhinometry (Fig. 8.4). In contrast, acoustic rhinometry overestimated the cross-sectional areas in the posterior nose, which is a well-known limitation of acoustic rhinometry [16, 52, 104, 110, 134]. Garcia and co-authors then applied their computational technique to correlate nasal resistance with the aispace mCSA and found that the pressure profile in the nasal cavity is well-described by the Bernoulli Obstruction Theory (orifice flow) [52]. In a subsequent study from the same group, the computational streamline rhinometry technique was applied to quantify the mCSA in 47 healthy subjects [14]. The average mCSA of $0.66 \pm 0.21\,\text{cm}^2$ obtained with this CFD-based technique was consistent with a mCSA around $0.70\,\text{cm}^2$ in healthy Caucasian subjects reported in several acoustic rhinometry studies [14, 52]. Given the importance of airway constrictions in determining nasal resistance to airflow and the availability of acoustic rhinometry for in vivo validation of CFD simulations, future studies should develop more automated methods to quantify the area-distance curve of the nasal cavity.

8.4 Some Open Problems in Comparing CFD to Objective Clinical Tests

One limitation of current CFD models is the assumption of rigid nasal walls. Movement of the lateral nasal wall is known to influence the pressure-flow curve of the human nose [112]. For example, pressure-flow measurements with 4-phase-rhinomanometry often display a loop due to higher flow in the ascending versus the descending parts of inspiration (Fig. 8.5) [137]. This loop is more pronounced in patients with nasal valve collapse, but recent measurements of lateral nasal wall mobility combined with 4-phase-rhinomanometry reveal significant motion of the lateral nasal wall even during quiet breathing in healthy subjects [1]. Another manifestation of lateral nasal wall compliance in rhinomanometric recordings is the greater nasal resistance during inspiration versus expiration. This is the basis of the flow rate inspiratory-expiratory difference (FRIED) test defined as the expiratory flowrate minus the inspiratory flowrate at the maximum absolute pressure common to both expiration and inspiration [100]. In a study of 55 patients with NAO [100], the FRIED test was found to distinguish patients with dynamic nasal valve collapse from patients whose feeling of nasal congestion did not improve with passive abduction of the lateral cartilage with a sensitivity of 82% and specificity of 59%. Therefore, the assumption of rigid nasal walls is expected to contribute to the inconsistent correlation between CFD and rhinomanometry reported in the literature (see Sect. 8.3.2).

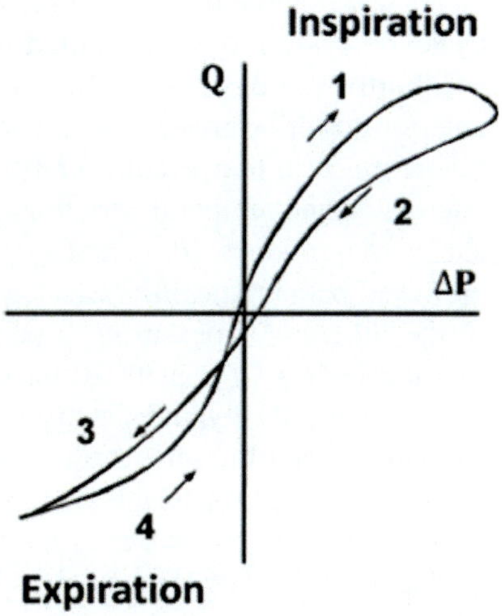

Fig. 8.5 General shape of pressure-flow curve obtained with 4-phase rhinomanometry. Note the higher flow in the ascending part of inspiration (phase 1) as compared to the descending part (phase 2). Figure modified from Vogt and co-authors [137]

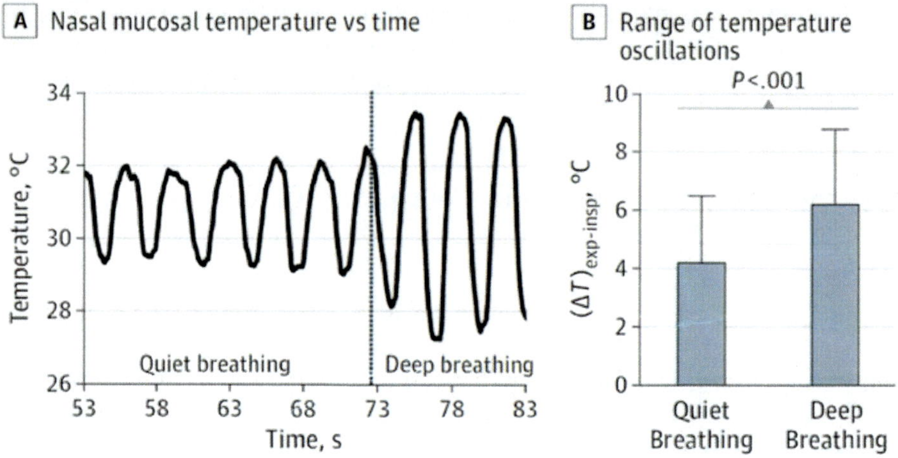

Fig. 8.6 **A** Nasal mucosal temperature versus time recorded at the nasal septum across from the inferior turbinate in one healthy volunteer, showing typical temperature variation during the respiratory cycle for quiet breathing and deep breathing. **B** The range of temperature oscillations during the breathing cycle (i.e., the difference $\Delta T = T_{exp} - T_{insp}$, where T_{exp} is the end-expiratory temperature and T_{insp} is the end-inspiratory temperature) was greater during deep breathing than during quiet breathing in a cohort of 22 healthy volunteers. Figure from Bailey and co-authors [7])

Future studies may improve this correlation using fluid-structure interaction (FSI) techniques that account for wall compliance [89].

Another area in which CFD methods need further validation is the quantification of mucosal cooling. As discussed in Sect. 8.5.4, mucosal cooling plays a central role in the perception of nasal airflow. To date, most CFD studies of nasal air conditioning (heating and humidification of inspired air) assumed a constant, uniform nasal wall temperature [53, 70, 82, 119, 132, 148, 149]. In reality, in vivo measurements have shown that nasal mucosal temperature oscillates during the respiratory cycle, decreasing during inspiration and increasing during expiration (Fig. 8.6) [7, 94]. Furthermore, similar to air temperature (Fig. 8.7), mucosal temperature has an anterior-posterior gradient with a lower mucosal temperature in the nasal valve and higher mucosal temperature in the nasopharynx [94]. Numerical methods to simulate spatial and temporal variations in mucosal temperature have been proposed and shown to reproduce the spatial variations in mucosal temperature qualitatively [78]. However, patient-specific validation is still lacking. Progress in this area is hindered by the difficulty of measuring nasal mucosal temperature in vivo accurately [7]. Also, time-dependent CFD methods have not yet been implemented in studies with large cohorts to test the hypothesis that the correlation between CFD variables and subjective perception of nasal airflow can be improved with more accurate CFD measures of mucosal heat loss. Given the importance of mucosal cooling in sensation of nasal airflow, advances in this area may improve the ability of CFD models to predict patient-specific symptoms in virtual surgery planning platforms [124, 136].

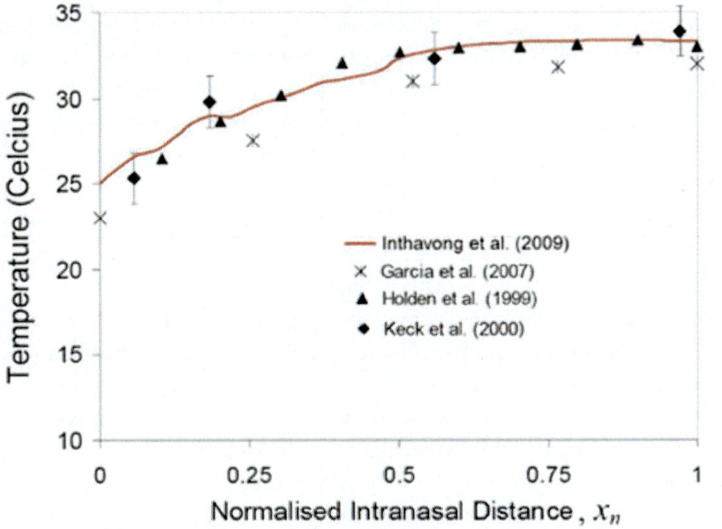

Fig. 8.7 Air temperature in the human nasal cavity during inspiration. The distance from nostrils was normalized by a maximum distance of 96.5 mm. CFD simulations performed by Inthavong et al. (red line) [70] and Garcia et al. (x symbols) [53]—are compared to time-averaged in vivo measurements performed by Holden et al. (black triangles) [67] and Keck et al. (black diamonds) [72]. Figure modified from Inthavong and co-authors [70]

8.5 Correlating CFD with Subjective Measures of Nasal Airflow

Few publications to date had samples sizes large enough to investigate the correlation between CFD variables and subjective measures of nasal airflow. The expectation is that more studies in this area will emerge as CFD technology evolves (i.e., automated segmentation of the nasal passages from CT scans and faster CFD solvers will allow larger sample sizes to be studied.) Nevertheless, the studies available suggest that several CFD variables correlate with subjective nasal patency. The subsections below summarize the key CFD variables that have been shown to correlate with subjective nasal patency.

8.5.1 Subjective Measures of Nasal Airflow

Before we dive into the correlation between CFD variables and subjective perception of nasal airflow, let us briefly review the major subjective measures of nasal airflow. The Nasal Obstruction Symptom Evaluation (NOSE) is a disease-specific quality-of-life instrument for nasal airway obstruction (NAO). Patients rate symptoms of nasal congestion, nasal blockage, trouble breathing through nose, trouble sleeping, and air hunger sensation during exercise on a scale of 0 (not a problem) to 4 (severe problem). Patients are asked to rate these symptoms during the one-month period preceding the survey. The scores are added and multiplied by 5 to give a total score from 0 to 100. The visual analog scale (VAS) provides an instantaneous measure of subjective nasal patency. Patients rate their perception of nasal airflow in a scale from 0 (completely clear) to 10 (completely blocked). This scale may be used to quantify bilateral or unilateral flow. When used for unilateral flow, patients are asked to cover one nostril with their finger and take several breaths through the uncovered nostril before rating their sensation of nasal patency.

A systematic literature review suggested that both bilateral VAS and NOSE scores can distinguish NAO patients from healthy subjects [122]. The mean ($\pm$ standard deviation) bilateral VAS score was 6.9 ± 2.3 in NAO patients and 2.1 ± 1.6 in asymptomatic individuals. The mean NOSE score was 65 ± 22 in NAO patients and 15 ± 17 in asymptomatic individuals. It is interesting to note that the mean NOSE score in each group was roughly 10 x the VAS score, despite the fact that the NOSE score measures NAO symptoms during the one-month period preceding the survey while the VAS score is an instantaneous measure. Readers should note that while the most common VAS scale is from 0 to 10, some studies have used non-standard scales. For example, Radulesco and co-authors quantified subjective perception of unilateral airflow using a VAS scale from 0 to 4 [119].

8.5.2 *Nasal Resistance*

Nasal resistance is a measure of the effort to breathe through your nose. It is defined as the pressure drop across the nasal cavity (ΔP, in Pa) divided by the flowrate (Q, in ml/s), namely

$$R_T = \frac{\Delta P}{Q_T},\ R_L = \frac{\Delta P}{Q_L},\ R_R = \frac{\Delta P}{Q_R} \tag{8.1}$$

where R_T is the total resistance (bilateral resistance), R_L and R_R are the unilateral resistances in the left and right cavities, and Q_T is the total flow given by the sum of flows in the left and right cavities, namely $Q_T = Q_L + Q_R$.

In a study of 10 patients with nasal airway obstruction (NAO) who displayed no signs of nasal cycle in the pre-surgery and post-surgery CT scans, Kimbell and colleagues [82] found that subjective scores of nasal patency have a higher correlation with unilateral than with bilateral nasal resistance (Table 8.1). Zhao and co-authors [149] also found no correlation between bilateral nasal resistance and subjective perception of nasal airflow. In contrast, all four CFD studies that examined the correlation between unilateral nasal resistance and subjective scores of nasal patency found a statistically significant correlation (Table 8.1). These results are consistent with a systematic literature review which found that the correlation of subjective nasal patency with rhinomanometry and acoustic rhinometry is stronger for unilateral measurements and when obstructive symptoms are present [3].

Table 8.1 Summary of studies that reported correlation coefficients (r) and p-values for the correlation between CFD-derived nasal resistance and subjective scores of nasal patency

Subjective score	|r|	p	Sample size	Statistical test	Reference
Bilateral Nasal Resistance (Pa.s/ml)					
Unilateral VAS	N.S.	N.S.	10 NAO patients	Pearson	Kimbell et al. [82]
NOSE	0.33	<0.05	10 NAO patients	Pearson	Kimbell et al. [82]
Bilateral VAS	N.S.	N.S.	22 healthy subjects	Pearson	Zhao et al. [149]
Unilateral Nasal Resistance (Pa.s/ml)					
Unilateral VAS	0.42	<0.05	10 NAO patients	Pearson	Kimbell et al. [82]
NOSE	0.48	<0.05	10 NAO patients	Pearson	Kimbell et al. [82]
Unilateral VAS	0.58	0.003	12 NAO patients	Spearman	Gaberino et al. [49]
NOSE	0.55	0.005	12 NAO patients	Spearman	Gaberino et al. [49]
Unilateral VAS	0.41	<0.001	15 NAO + 15 healthy	Kendall's tau	Casey et al. [18]
NOSE	0.40	<0.001	15 NAO + 15 healthy	Kendall's tau	Casey et al. [18]
Unilateral VAS	0.6	<0.001	22 NAO patients	Spearman	Radulesco et al. [119]

Notes (1) Data from Gaberino et al. [49] are after correction for the nasal cycle ; (2) The sample in Casey et al. [18] was composed of 15 NAO patients and 15 healthy subjects; (3) Unilateral nasal resistance was measured in the most obstructed side (nasal cavity with less flow); (4) Abbreviations: N.S. = not significant

8.5.3 Airflow Partitioning

Airflow partitioning quantifies the symmetry of flow distribution between the left and right nostrils. It is defined as the ratio of unilateral airflow to total airflow:

$$(AP)_L = \frac{Q_L}{Q_T} \tag{8.2}$$

$$(AP)_R = \frac{Q_R}{Q_T} \tag{8.3}$$

where $(AP)_L$ and $(AP)_R$ are the airflow partitioning in the left and right cavities, respectively. By definition, $(AP)_L$+$(AP)_R$=1. Airflow partitioning is determined by the ratio of unilateral resistances in the left and right cavities, namely[1]

$$(AP)_L = \frac{1}{1 + \frac{R_L}{R_R}} \tag{8.4}$$

For example, when the resistances of the left and right cavities are equal ($R_L = R_R$), inhaled air is split equally between the two nostrils, i.e. $(AP)_L = 0.50$. In contrast, if resistance of one cavity is 9-fold higher than the resistance of the contralateral side ($R_L = 9R_R$), only 10% of airflow passes through the obstructed side.

In a study of 10 NAO patients, Kimbell and colleagues [82] found that airflow partitioning in the most obstructed side correlated with both NOSE scores (Pearson r = −0.71) and unilateral VAS (Pearson r = 0.58). However, airflow partitioning cannot be used as an isolated measure. As the discussion in the preceding paragraph illustrates, a patient with bilateral nasal obstruction may have normal airflow partitioning if the resistances of the left and right cavities are equal.

Although the goal of NAO surgery is to achieve symmetric distribution of nasal airflow between the left and right nostrils, asymmetric airflow partitioning should not always be interpreted as pathological. A CFD analysis of 47 healthy subjects found that airflow partitioning had a normal range of (23.8 ± 7.8)–(76.2 ± 7.8)% [14]. This is in good agreement with rhinomanometry measurements performed by Roblin and Eccles [123] in 100 healthy subjects who reported a normal range of 18.6–81.4% [14, 123]. Using (8.4), one finds that a 20–80% partitioning between the left and right nostrils corresponds to unilateral resistance being 4-fold higher in one cavity compared to the contralateral side. Fluctuations of 5-fold or higher in unilateral nasal resistance are routinely observed during the nasal cycle due to spontaneous changes in nasal mucosa engorgement [63]. In summary, a symmetric airflow partitioning averaged over the nasal cycle is likely important for normal sensation of nasal airflow. However, airflow partitioning must be interpreted with caution since neither symmetric partitioning necessarily correspond to normal physiology, nor asymmetric partitioning necessarily correspond to pathology.

8.5.4 Mucosal Cooling

Evidence suggests that the primary mechanism of nasal airflow sensation is stimulation of menthol-sensitive TRPM8 cold receptors [7, 129, 132, 148]. The central role of these cold receptors is illustrated by improved airflow sensation during exposure to menthol vapor in the absence of changes in nasal airflow [15, 42, 43]. A key advantage of CFD compared to traditional objective techniques (i.e., rhinomanom-

[1]This relation is obtained using the definitions of airflow partitioning, nasal resistance, and the fact that the pressure drop is the same in the left and right cavities.

etry, acoustic rhinometry, and peak nasal inspiratory flow) is the ability to quantify mucosal cooling. Several CFD studies have reported that CFD-derived mucosal cooling correlates with subjective perception of nasal airflow (Table 8.2).

One open question that requires further study is determining the optimal CFD variable to quantify mucosal cooling because a myriad of CFD variables have been proposed for this purpose (Table 8.2). In a study applying CFD, rhinomanometry, and acoustic rhinometry to evaluate nasal airflow in 22 healthy subjects, Zhao and colleagues [148] found that "peak heat loss posterior to the nasal vestibule" was the only objective variable that correlated with bilateral VAS scores. Unfortunately, the authors did not provide a clear definition of peak heat loss ("values reported were obtained after application of a 2 mm moving average filter") [148]. In a subsequent study, Sullivan and co-authors [132] defined peak heat flux as the heat flux value above which only 1 cm^2 of mucosa was exposed to. This definition provided a statistically significant correlation with NOSE scores in a cohort of 10 NAO patients studied pre- and post-surgery, but peak heat fluxes did not correlate with unilateral VAS [132] (Table 8.2). These authors explored various definitions of mucosal cooling and found that the best correlation with subjective scores was obtained with the surface area where heat flux exceeds 50 W/m^2 (SAHF50) measured unilaterally in the most obstructed cavity (Table 8.2). In a recent study of 22 NAO patients with septal deviation, Radulesco and collaborators [119] measured the average heat flux in a circular region of the nasal septum (area 1 cm^2) centered 1 cm posterior to the maximum septal deviation and 1.5 cm from the nasal floor [119]. This heat flux measure had the strongest correlation with unilateral VAS scores among all CFD variables studied (pressure, temperature, velocity, and wall shear stress).

The potential benefit of mucosal cooling as a measure of nasal airflow is illustrated by a CFD study of empty nose syndrome. Dayal and co-authors [34] performed a virtual total inferior turbinectomy in 10 patients with inferior turbinate hypertrophy [34]. Complete resection of the inferior turbinate reduced unilateral nasal resistance, increased unilateral airflow, and reduced mucosal cooling (measured by SAHF50). This reduction in mucosal cooling concomitant with an increase in flowrate is consistent with the paradoxical feeling of nasal obstruction reported by patients with empty nose syndrome [34]. Altogether, these studies support the hypothesis that mucosal cooling plays a central role in perception of nasal airflow.

8.5.5 *Intranasal Airflow Distribution*

Another CFD variable that has been shown to correlate with subjective nasal patency is intranasal airflow distribution. In a study of 15 NAO patients and 15 healthy subjects, Casey and colleagues [18] divided coronal sections of the nasal cavity into three regions, namely inferior, middle, and superior thirds. The left and right cavities were defined as narrow or non-narrow sides based on which cavity had the highest flow. Patients with NAO had a deficit in middle airflow in the narrow side (31 ± 18 mL/s in NAO patients vs. 68 ± 10 mL/s in healthy subjects; bilateral flow was

Table 8.2 Summary of studies that reported correlation coefficients (r) and p-values for the correlation between CFD-derived mucosal cooling and subjective scores of nasal patency

Subjective score	|r|	p	Sample size	Statistical test	Reference
Bilateral Heat Flux(W/m^2)					
Unilateral VAS	N.S.	N.S.	10 NAO patients	Pearson	Kimbell et al. [82]
NOSE	0.45	<0.05	10 NAO patients	Pearson	Kimbell et al. [82]
Unilateral Heat Flux(W/m^2)					
Unilateral VAS	0.46	<0.05	10 NAO patients	Pearson	Kimbell et al. [82]
NOSE	0.65	<0.05	10 NAO patients	Pearson	Kimbell et al. [82]
Unilateral VAS	0.51	0.01	12 NAO patients	Spearman	Gaberino et al. [49]
NOSE	0.51	0.01	12 NAO patients	Spearman	Gaberino et al. [49]
Unilateral VAS	0.43	0.02	15 NAO + 15 healthy	Pearson	Casey et al. [18]
NOSE	0.48	0.008	15 NAO + 15 healthy	Pearson	Casey et al. [18]
Peak Heat Flux (W/m^2)					
Bilateral VAS	0.46	<0.01	22 healthy subjects	Pearson	Zhao et al. [149]
Unilateral VAS	N.S.	N.S.	10 NAO patients	Pearson	Sullivan et al. [132]
NOSE	0.44	0.05	10 NAO patients	Pearson	Sullivan et al. [132]
SAHF50 (cm^2)					
Unilateral VAS	0.63	0.002	10 NAO patients	Pearson	Sullivan et al. [132]
NOSE	0.76	<0.001	10 NAO patients	Pearson	Sullivan et al. [132]
Unilateral VAS	N.S.	N.S.	12 NAO patients	Spearman	Gaberino et al. [49]
NOSE	0.43	0.03	12 NAO patients	Spearman	Gaberino et al. [49]
Unilateral VAS	0.51	0.004	15 NAO + 15 healthy	Pearson	Casey et al. [18]
NOSE	0.55	0.002	15 NAO + 15 healthy	Pearson	Casey et al. [18]
Heat Transfer Rate Across Nasal Cavity (W) or Total Heat Loss (W)					
Bilateral VAS	N.S.	N.S.	22 healthy subjects	Pearson	Zhao et al. [149]
Unilateral VAS	0.50	0.02	10 NAO patients	Pearson	Sullivan et al. [132]
NOSE	0.70	<0.001	10 NAO patients	Pearson	Sullivan et al. [132]
Heat Flux Behind a Septal Deviation (W/m^2)					
Unilateral VAS	0.86	<0.001	22 NAO patients	Spearman	Radulesco et al. [119]

Notes (1) Data from Gaberino et al. [49] are after correction for the nasal cycle; (2) Zhao et al. [149] did not clearly define "total heat loss", thus "total heat loss" is listed under the heading Heat Transfer Rate; (3) sample in Casey et al. [18] was composed of 15 NAO patients and 15 healthy subjects; (4) Unilateral nasal resistance was measured in the most obstructed side (nasal cavity with less flow); (5) Abbreviations: N.S. = not significant

250 mL/s in all subjects), while no difference was found between the two groups for inferior or superior airflow. Middle airflow had a significant correlation with NOSE scores (Pearson $r = -0.76$, $p < 0.0001$) and VAS scores (Pearson $r = -0.64$, $p < 0.002$). Importantly, middle airflow had a stronger correlation with subjective scores of nasal patency than other CFD variables (unilateral nasal resistance, unilateral heat flux, and SAHF50). This observation led the authors to speculate that mucosal cooling in the region surrounding the middle turbinate might be especially important for the sensation of nasal airflow.

8.5.6 Other CFD Variables

In our opinion, the primary CFD variables related to nasal airflow sensation are mucosal cooling and nasal resistance (or nasal airflow). Any variable that influences these primary variables, including airspace cross-sectional areas, airflow partitioning, and intranasal airflow distribution, are also expected to play a central role. In addition to these variables, some authors have suggested that other CFD variables are important. For example, some authors emphasize surgical changes in velocity magnitude [64] and wall shear stress has been shown to correlate with subjective nasal patency [82, 119]. However, given the central role of TRPM8 cold-receptors in sensation of nasal airflow, it is unclear if wall shear stresses and air velocity independently affect nasal airflow sensation or if these variables correlate with nasal patency scores simply because they are tightly linked to flowrate. Future studies are needed to firmly determine the primary CFD variables that determine perception of nasal airflow.

8.6 CFD Results in the Normal Nose

8.6.1 Normal Nasal Anatomical Variability

The typical human nasal cavity is characterized by normal intra- and inter-individual morphological variability that can confound concise characterization of a normal nasal airflow profile. Notable variations include nasal cycle, nasal vestibule indentation phenotype, and nasal index (race/ethnicity) [6, 14, 17, 23, 40, 50, 58, 73, 92, 111, 115, 116, 120, 151]. The nasal cycle is the most common intra-individual variability that the nose experiences, albeit a temporal variation (Fig. 8.8 TOP). The classical nasal cycle is a spontaneous, reciprocal fluctuation of nasal patency due to phases of congestion and decongestion of the mucosa in both nasal cavities. Typical duration of a cycle ranges from 30 min to 6 h [59]. Studies have observed various forms of the nasal cycle; some of these forms comprise a combination of reciprocal and nonreciprocal relationship, regular and irregular, and in- and out-of-phase

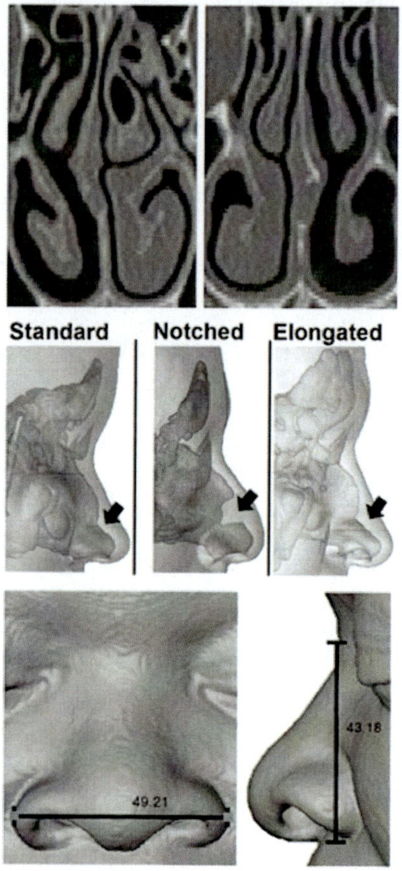

Fig. 8.8 [TOP] Reciprocal fluctuation in nasal mucosa engorgement due to the nasal cycle. Reproduced with permission from Patel et al. [115]. [MIDDLE] Transparent side view of the face showing morphological phenotypes of the nasal vestibule. Reproduced with permission from Ramprasad and Frank-Ito [120]. [BOTTOM] Frontal and side views of the exterior nose showing how the nasal width and height were measured to compute the nasal index. Reproduced with permission from [120]

changes [105]. The classical nasal cycle has been described as a change in nasal patency between both nostrils with a $180/circo$ out-of-phase, reciprocal relationship in which one nasal cavity's mucosa gets congested as the other side becomes decongested. Nonetheless, this instantaneous asymmetry between the left and right sides does not influence bilateral nasal airflow, even though airflow partitioning switches between the left and right cavities [4, 41, 62, 105]. There is no consensus on the number of individuals who experience nasal cycle, earlier studies postulated that about 80% of adults demonstrated some form of cycling [4, 59, 105]; which is contrary to the 22 and 21% of individuals with significant periodic switches in unilateral airflow pattern that were reported by Gilbert [55] and Flanagan and Eccles [48], respectively.

Another main feature of normal variations seen in both intra- and inter-individual nasal anatomy are the distinct nasal vestibule morphological phenotypes described by Ramprasad and Frank-Ito [120], namely Notched, Standard and Elongated (Fig. 8.8-MIDDLE). The Notched phenotype has a prominent notch at the junction of the ala and sidewall; the Standard phenotype is considered the "typical" formation; and the Elongated phenotype is slightly elongated at the level of the nasal ala [120]. Intra-individual differences in morphology exist when the nasal vestibule phenotype on one side of the cavity differs from the phenotype on the contralateral side. Of the 16 healthy individuals studied, four individuals had Standard nasal vestibule phenotype on both sides of their nasal cavities, three subjects had both sides Notched, another three students had Elongated phenotype on both sides, and the phenotype on one side differed from the contralateral side in six subjects. These distinct morphological phenotypes do not affect global airflow dynamics and resistance in the nasal cavity [69, 120]. However, local airflow resistance and air conditioning performance were significantly influenced by nasal vestibule morphological phenotypes [99, 120]. This phenomenon could be attributed to the fact that the human nose naturally compensates for localized constrictions by adjusting physiological response on other regions of the airway so as to modulate nasal airflow, and maintain natural ventilation levels [9, 120, 142]. With respect to transport of particulate matter, nanoparticle transport patterns in the cavity were influenced by nasal vestibule phenotypes [37].

Evolutionary adaptation of humans to diverse climatic environments is another contributing factor to inter-individual variability in nasal morphology [109, 139, 146]. The variabilities that arise from adaptation to varying geographic regions are particularly evident in the ratio of nasal width to nasal height, termed the nasal index (Fig. 8.8 BOTTOM) [30, 32, 65, 135, 139, 143]. The classic Caucasian nose is classified as "leptorrhine", with a nasal index <70; the nasal index for typical Asian and Latin American noses lie between 70 and 85, and belong to the "mesorrhine" category; and a representative nose of individuals from African descent is known as "platyrrhine", with nasal index >85 [31, 90, 92, 111]. In their study, Leong and Eccles [92] stated that nasal physiology is not significantly different across varying nasal indices. Similarly, the studies by Doddi and Eccles [36], and Patki and Frank-Ito [116] also reported a low correlation between nasal index and nasal resistance. With regards to inter-individual variability due to race/ethnicity, Zhu et al. [151] reported increased airflow passing through the middle passageway in Caucasian nasal airways, and Indian airways had increased airflow volume passing through inferior region. In a related study, Keeler et al. [73] compared nasal resistance across individuals with normal nasal cavities from four different races/ethnicities; Blacks, Caucasians, Asians, and Latin Americans. Their findings suggest that on average, Blacks have the lowest total nasal resistance (0.034 Pa.s/mL), followed by Latin Americans (0.037 Pa.s/mL), Asians (0.042 Pa.s/mL), and Caucasians (0.046 Pa.s/mL). Other studies have reported no significant difference in nasal resistance between Blacks and Caucasians [6, 17, 111]. Altogether, these studies suggest the existence of racial variability in nasal anatomy, they had small sample sizes, thus this will remain an open question until much larger samples sizes can be studied for proper statistical significance.

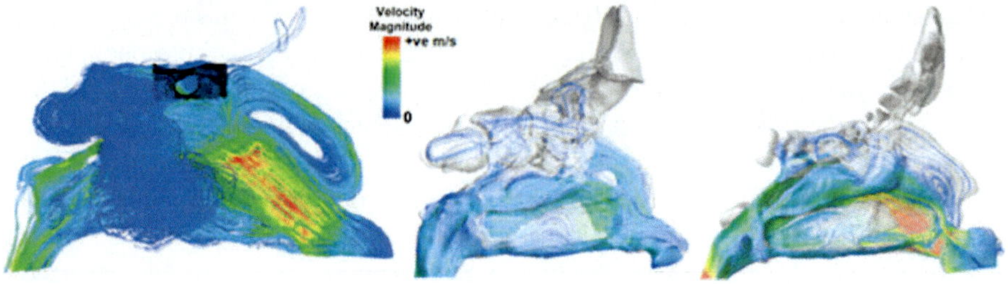

Fig. 8.9 Nasal flow streamline patterns showing velocity magnitude in Notched and Elongated nasal vestibule phenotypes. Reproduced with permission from Zhao and Jiang [148] & Ramprasad and Frank-Ito [120]

8.6.2 Normal Nasal Airflow Description

The nasal vestibule is the portal of entry where inhaled ambient air enters the respiratory tract. Inhaled air flowing in the notched nasal vestibule twist and swirl through to the nasal valve, while lateral flow tend to be smoother, less contorted, and have higher velocities [69]. The velocity of inhaled air entering the narrow nasal valve increases and a jet flow pattern is formed as air exits the nasal valve, where it then changes direction towards the turbinate region of the nasal cavity [91, 133, 151]. Additionally, airflow velocities during inspiration have been reported to be highest in the vicinity of the nasal valve as air enters the turbinate region (Fig. 8.9) [35, 127, 133]. There is a noticeable recirculating zone at the anterior superior region in some nasal passages; in particular, individuals with a notched or elongated nasal vestibule phenotypes tend to display a prominent vortex formation in their airways (Fig. 8.9) [77, 120, 133, 141, 148]. In certain scenarios, depending on nasal anatomical variation, the vortex formed can extend posteriorly to the anterior end of the middle turbinate and beyond; this recirculating zone increases residence time of inhaled air in the anterior superior region [77, 120, 131, 133, 141, 148].

As inhaled air flows into the turbinate region, Keyhani, Scherer, and Mozell [76] noted that about 30% of flow volume passes through the inferior turbinate and 10% through the olfactory recess. Xiong et al. [145] and Yu et al. [147] reported that the bulk of air flowing in the nasal airway pass through the common nasal meatus. In another study involving 4 subjects, Segal, Kepler, and Kimbell [127] reported 3–18% (median = 6%) of flow volume passing through the superior region of the nasal cavity, with 38–59% (median = 53%) and 36–44% (median = 42%) of inhaled flow passing through the middle and inferior regions of the nasal cavity, respectively. In a recent study involving 47 healthy adults, Borojeni and colleagues [14] computed unilateral airflow allocation across 5 coronal cross-sections, average unilateral airflow through the superior, middle and inferior regions were 12% ± 8%, 51% ± 12%, and 37% ± 17%, respectively; while Zhao and Jiang [148] reported approximately 16% ± 10%, 48 %± 12%, and 36% ± 12%, respectively, in corresponding regions

Table 8.3 Airflow distribution through different regions (olfactory cleft, superior, middle, and inferior regions) in the nasal cavity of 22 healthy subjects. Bilateral nasal resistance values obtained from CFD simulations at a transnasal pressure drop of ΔP = 15 Pa and measured with Rhinomanometry at ΔP = 75 Pa. Reproduced with permission from Zhao and Jiang [148]

	Subject ID	**Regional Airflow Allocation (%)**				**Nasal Resistance (Pa.s/ml)**	
		Olfact.	**Superior**	**Middle**	**Infer.**	**Simulated**	**Measured**
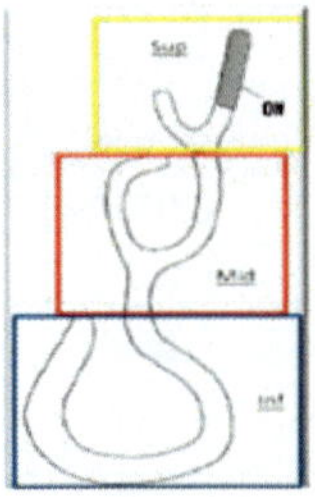	**2**	4.7	14.1	49.4	36.5	0.072	0.074
	3	2.1	5.4	58.9	35.7	0.053	0.062
	4	0.9	3.6	46.2	50.2	0.046	0.036
	5	14.4	32.2	29.9	38	0.05	0.06
	6	4.4	20.7	41.2	38	0.057	0.043
	7	16.1	38.2	26.1	35.7	0.102	0.047
	8	1.2	2.8	44.4	52.8	0.08	0.089
	9	1.1	2.6	70.9	26.5	0.118	0.074
	10	0.4	1	38	61	0.091	-
	11	5.6	15.2	60	24.7	0.113	0.17
	12	2.9	15.9	46.8	37.3	0.08	0.073
	13	2.2	7.6	61.7	30.7	0.058	0.044
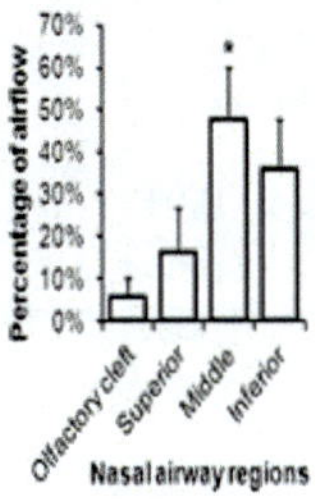	**14**	7.4	25	34.5	40.6	0.071	0.06
	15	4.7	18.3	64.8	17	0.054	0.056
	16	6.1	18.2	52.9	29	0.063	0.078
	18	6	13.7	55.4	30.9	0.06	0.074
	19	5.3	14.1	55.4	30.6	0.079	0.088
	20	5.5	11.4	26.1	62.5	0.071	0.057
	21	7.6	19.4	50.4	30.1	0.046	0.058
	22	11	28.9	43.6	27.6	0.059	0.041
	23	11.3	30.3	50.3	19.4	0.053	0.028
	24	7.9	19.7	46.8	33.5	0.04	0.047
	Mean	5.8	16.3	47.9	35.8	0.069	0.065
	SD	4.3	10.3	12.1	11.8	0.022	0.029

among 22 healthy subjects, including about 6%±4% in the olfactory recess (Table 8.3).

Borojeni and colleagues [14] also reported that airflow partition between the left and right sides of a normal nasal cavity ranged from 23.8 to 76.2% in healthy subjects. The 90% confidence interval lower and upper normative ranges for unilateral nasal airflow resistance were 0.029 ± 0.01 Pa.s/ml and 0.32 ± 1.32 Pa.s/ml, respectively; and normative bilateral airflow resistance ranges were 0.017 ± 0.004 Pa.s/ml (lower) and 0.12 ± 0.13 Pa.s/ml (upper) (Table 8.4) [14]. According to Zhao and Jiang [148] computed average bilateral airflow resistance was 0.069 Pa.s/ml (standard deviation = 0.022 Pa.s/ml), and measured bilateral resistance from the same cohort of subjects during normal breathing via anterior rhinomanometry at a reference pressure of 75 Pa was 0.065 Pa.s/ml (standard deviation = 0.029 Pa.s/ml). In general, intranasal pressure drop during inhalation is highest at the anterior region of the nasal cavity

Table 8.4 Computed airflow and nasal resistance in 47 healthy subjects at a bilateral flow of 15 L/min. Reproduced with permission from the supplementary material in Borojeni et al. [14]

Patient ID	Airflow (ml/s)		Airflow Partitioning (%)		Nasal Resistance (Pa.s/ml)		Bilateral Nasal Resistance (Pa.s/ml)
	LEFT	RIGHT	LEFT	RIGHT	LEFT	RIGHT	
1	145.62	104.77	58.16	41.84	0.0324	0.045	0.0188
3	121.35	128.88	48.5	51.5	0.0653	0.0615	0.0317
4	117.08	133.11	46.8	53.2	0.0411	0.0361	0.0192
5	116.9	134.04	46.58	53.42	0.1369	0.1194	0.0638
6	78.03	172.74	31.12	68.88	0.1005	0.0454	0.0313
7	157.51	92.48	63.01	36.99	0.0679	0.1157	0.0428
8	110.55	140.3	44.07	55.93	0.1381	0.1088	0.0608
9	75.33	174.99	30.09	69.91	0.0739	0.0318	0.0222
10	79.43	170.59	31.77	68.23	0.132	0.0615	0.0419
11	150.94	99.6	60.25	39.75	0.0989	0.1499	0.0596
12	145.04	106.08	57.76	42.24	0.1157	0.1581	0.0668
13	93.58	156.55	37.41	62.59	0.072	0.043	0.0269
14	125.26	124.92	50.07	49.93	0.1045	0.1048	0.0523
15	100.76	149.39	40.28	59.72	0.0864	0.0583	0.0348
16	108.38	142.67	43.17	56.83	0.0887	0.0674	0.0383
17	69.2	180.58	27.7	72.3	0.3399	0.1303	0.0942
18	163.91	86.12	65.56	34.44	0.1558	0.2965	0.1021
19	64.65	185.51	25.84	74.16	0.1714	0.0597	0.0443
20	181.27	68.3	72.63	27.37	0.0523	0.1387	0.038
21	114.7	135.62	45.82	54.18	0.0578	0.0489	0.0265
22	123.59	126.48	49.42	50.58	0.0796	0.0778	0.0394
23	102.65	148.42	40.89	59.11	0.1358	0.0939	0.0555
24	77.98	172.2	31.17	68.83	0.1542	0.0698	0.0481
25	191.48	58.55	76.58	23.42	0.0259	0.0848	0.0199
26	105.88	144.18	42.34	57.66	0.0455	0.0334	0.0193
27	78	172.11	31.18	68.82	0.0798	0.0362	0.0249
28	94.63	155.56	37.82	62.18	0.0483	0.0294	0.0183
29	138.9	111.23	55.53	44.47	0.1698	0.212	0.0943
31	180.99	69.13	72.36	27.64	0.0289	0.0756	0.0209
32	115.04	135.32	45.95	54.05	0.0369	0.0314	0.017
33	153.33	96.77	61.31	38.69	0.0697	0.1104	0.0427
35	189.34	61.55	75.47	24.53	0.0672	0.2067	0.0507
36	121.02	129.07	48.39	51.61	0.0561	0.0526	0.0271
37	128.02	121.19	51.37	48.63	0.054	0.057	0.0277

(continued)

Table 8.4 (continued)

Patient ID	Airflow (ml/s)		Airflow Partitioning (%)		Nasal Resistance (Pa.s/ml)		Bilateral Nasal Resistance (Pa.s/ml)
38	129.21	121.19	51.6	48.4	0.0666	0.071	0.0344
40	167.54	83.67	66.69	33.31	0.0893	0.1788	0.0595
41	113.93	136.28	45.53	54.47	0.0371	0.031	0.0169
42	118.02	132.26	47.15	52.85	0.0628	0.056	0.0296
46	178.51	72.34	71.16	28.84	0.0729	0.1799	0.0519
47	114.24	135.85	45.68	54.32	0.2668	0.2243	0.1219
48	124.48	126.14	49.67	50.33	0.0346	0.0341	0.0172
49	151.73	99.63	60.36	39.64	0.0815	0.124	0.0492
51	140.74	109.02	56.35	43.65	0.1693	0.2186	0.0954
52	139.29	111.63	55.51	44.49	0.0799	0.0997	0.0443
53	56.39	193.99	22.52	77.48	0.1928	0.056	0.0434
54	122.6	128.01	48.92	51.08	0.0724	0.0693	0.0354
55	176.56	73.53	70.6	29.4	0.1581	0.3796	0.1116
Mean $\pm$ SD	125 ± 35		50 ± 14		0.10 ± 0.07		0.045 ± 0.027

(nostrils to anterior inferior turbinate), accounting for more than 50% of the total pressure drop in the airway, with pressure drop decreasing gradually as air flows posteriorly to the nasopharynx [14, 69, 79, 133, 148].

Nasal flow rate up to 15 l/min is considered resting breathing [131]. The flow at resting inspiration is generally assumed to be laminar, while transitional flow is between 15 and 25 l/min and turbulent at flow rate in excess of 25 l/min [60, 76, 77, 84, 127, 128, 141]. In their review article Leong et al. [91], noted that airflow in the nasal vestibule is relatively laminar, with evidence of turbulent flow profile from the nasal valve region and into the anterior turbinate region, where flow becomes less turbulent at it passes through the turbinate. On the other hand, Zhao and Jiang [148] reported the presence of turbulence in nasal airflow at the anterior nasal cavity and through the nasal valve region during resting inspiration, which quickly dwindled to zero as airflow passes the nasal valve and into the turbinate region. Even so, as acknowledged by Doorly et al. [38] there is ambiguity as to the exact anatomical region(s) where laminar nasal flow transitions to turbulent flow and if indeed nasal flow can become fully turbulent.

8.6.3 CFD Results in the Normal Sinuses

8.6.3.1 The Paranasal Sinuses

The paranasal sinuses are air-filled cavities located in the frontal, maxillary, ethmoid and sphenoid bones, and are named after the bones where they are located. These sinuses, which surround the nasal cavity are connected to it via narrow openings referred to as ostia. The primary functions of the paranasal sinuses have long been deliberated [12, 13, 74, 75]. For nearly two millennia, many theories pertaining to anatomical and/or functional significance of the paranasal sinuses have been proposed and debunked [12, 13, 74, 75].

Some of the hypothetical roles that have been proposed over the years include (Fig. 8.10): (1) resonance to influence voice quality; (2) ventilation, heating and humidification of inspired air; (3) mucus secretion; (4) reduce weight of the skull; (5) facial growth and architecture; (6) evolutionary remnants; (7) increase olfactory area; and (8) immune defense and nitric oxide secretion. As Kellman and Schmidt [75], and previous authors have noted, most of these hypothetical roles have been either scientifically discredited or lack empirical evidence [12, 13, 74, 75]. According to Keir [74], the theory that the paranasal sinuses play a critical role in immune defense of the nasal cavity and secretion of nitric oxide is perhaps the only hypothesis that has been proven to be valid.

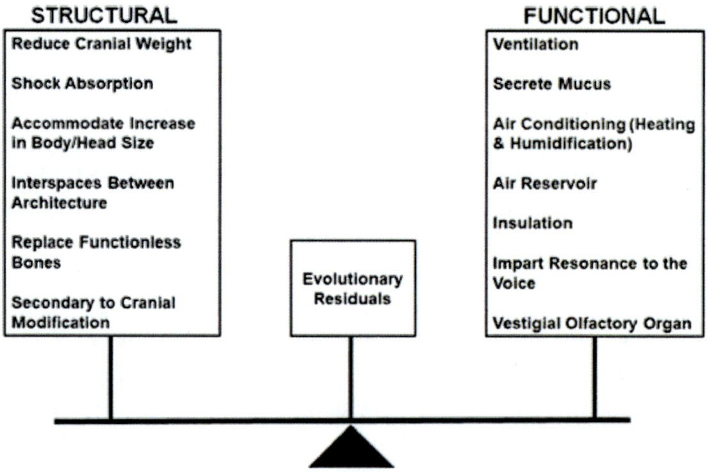

Fig. 8.10 Proposed roles of the human paranasal sinuses reproduced with permission from Blaney [12]

8.6.3.2 Ventilation in Sinuses

The paranasal sinuses are not significantly involved during normal respiration; there is minimal exchange of airflow between the sinuses and the main nasal cavity, and the sizes and locations of the natural sinus ostia may help prevent the sinus mucosa from excessive ventilation [26, 91, 145]. In their paper, Xiong et al. [145] reported the natural sinus ostia diameters on the left side of a 40-year-old female subject as 2.1, 1.2, and 1.4 mm for the maxillary ostium, sphenoid ostium, and frontal ostium, respectively. Similarly, Ge et al. [54] reported maxillary ostium diameters of 4.6 and 3.8 mm from a healthy sinonasal airways of a 51-year-old Asian male. The maxillary sinus ostium cross-sectional area was computed for each of the 16 healthy subjects studied in Keeler et al. [73], Patki and Frank-Ito [116], and Ramprasad and Frank-Ito [120]; the minimum maxillary ostium area in this cohort was 2.6 mm^2, and maximum area was 33.4 mm^2, the mean and median areas were 12.3 mm^2 and 11.8 mm^2, respectively. In another study by Kumar et al. [86], the natural maxillary ostia area for the healthy subject in this study were 62 mm^2 (left side) and 65 mm^2 (right side), and the combined bilateral frontal ostia surface area was 166 mm^2.

In a cohort of 20 (11 females and 9 males) healthy Chinese subjects with normal sinonasal cavities ranging in age from 25 to 35 years, Tan and colleagues [133] found that the peak airflow velocity at the maxillary ostium was 0.18 ± 0.14 m/s (left cavity) and 0.19 ± 0.13 m/s (right cavity) during inspiration, and 0.15 ± 0.13 m/s (left cavity) and 0.15 ± 0.15 m/s (right cavity) during expiration. Findings reported by Xiong et al. [145] depicted a particularly slow airflow velocity at each paranasal sinus ostium, less than 0.01 m/s during inspiration. In work conducted by Kumar and colleagues [86], average airflow velocities (during inspiration or expiration) at the maxillary ostium were approximately 0.155m/s on the left ostium, and 0.405 m/s on the right ostium; maximum velocities were 0.45 (left ostium) and 0.96 m/s (right ostium). Ge et al. [54] and Kumar et al. [86] observed airflow recirculation in the maxillary and frontal sinuses, forming complex vortices, notwithstanding that the volume of air flowing from the main nasal cavity through the natural maxillary ostium was very small; about 0.006 l/min for inflow and outflow [86]. Further, Ge et al.[54] noted that the flow rate through the left maxillary ostium was 0.0006 l/min and 0.0003 l/min on the right maxillary ostium, which accounted for less than 0.006% of total inspiratory flow rate in the sinonasal cavity.

8.6.3.3 Maxillary Accessory Ostium

The presence of an accessory ostium in the maxillary sinus may be due to either a natural occurrence, disease or surgical antrostomy (Fig. 8.11 TOP) [108, 121]. In the case of a natural occurrence, Davis [33] described the maxillary accessory ostium as the result of a progressive thinning of the medial wall between the mucosa linings of the sinus and that of the middle meatus, with no embryologic significance in its development. Davis [33] also noted that the youngest subject with evidence of an accessory ostium was from a 4-year-old child. Reported incidence of accessory

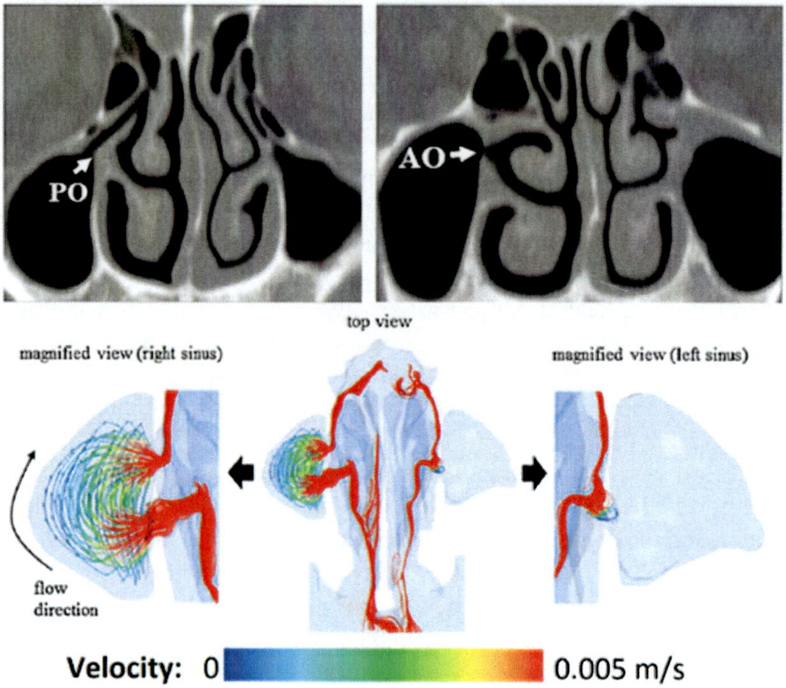

Fig. 8.11 [TOP] Radiographic images of the sinonasal cavity showing primary natural ostium (PO; Left Panel) and accessory ostium (AO; Right Panel). Reproduced with permission from Na et al. [108]. [BOTTOM] Flow streamline patterns in the sinonasal cavity showing flow behaviors on right side with an accessory ostium and left side with only natural ostium. Reproduced with permission from Na et al. [108]

ostium from literature ranged widely, 4–50% with a lower prevalence rate among healthy individuals [21, 22, 33, 39, 71, 106, 114]. Radiographic examination based on computed tomographic images found evidence of an accessory ostium in 13.75% of individuals, of which 6% had bilateral maxillary accessory ostia [22, 39]. The study by Ozel et al. [114] reported the presence of unilateral accessory ostium in a little over 24% of their study cohort and bilateral accessory ostia in 12.6%.

In idealized models used to investigate maxillary sinus ventilation, Hood et al. [68] and Rennie et al. [121] indicated that the presence of an accessory ostium increases ventilation in the maxillary sinus by several orders of magnitude. In a computational sinonasal airway model of a 43-year-old female subject with one maxillary accessory ostium on the right sinus and two accessory ostia on the left side, Zhu et al. [150] indicated that at 15 l/min respiration rate, airflow volume on the right maxillary sinus with one accessory ostium was 0.54, and 0.46 l/min on the left maxillary sinus with two accessory ostia. Additionally, their findings suggest that sinus ventilation increased by more than an order of magnitude with accessory ostium present compared to maxillary sinuses with only natural ostia [150], which is in agreement with Hood et al. [68] and Rennie et al. [121]. In another study, Na et al. [108] reported that the effects of an accessory ostium on velocity field created only localized alteration of airflow pattern, and that the volume of airflow by-pass due to an accessory ostium during inspiration was roughly 0.0216 l/min (Fig. 8.11 BOTTOM).

8.6.4 Nitric Oxide and Sinuses

The paranasal sinuses are known to be the main producer of nasal nitric oxide [96, 98]. However, the physiological functions of nasal nitric oxide is yet to be completely understood though it is thought to play significant role in the inhibition of viral and bacterial growth, regulation of ciliary motility, and act as an airborne transport of a biological messenger to the lungs [2, 27, 61, 96, 98]. Menzel et al. [103] noted that individuals with lower concentration of nitric oxide in the paranasal sinuses may be more likely to experience sinus and nasal airway infections, which may be attributed to their inability to produce enough nitric oxide [97]. A number of studies have showed that during humming the concentration of nitric oxide in nasally exhaled air increases [56, 101, 103, 140]. In addition, normative ranges for nasal nitric oxide have been extensively reported in the literature; among children, these normative values have varied widely due to differences in measurement techniques and ages [27]. In a study involving six healthy term newborns between 2 and 24 d old, Stehling et al. [130] reported measured nitric oxide levels using tidal breathing to range from 100ppb to 232ppb. In another study, Baraldi et al. [8] reported nitric oxide levels using tidal breathing in five healthy children (aged: 1–7) to be between 225 and 379 ppb. A meta-analysis conducted by Corbelli and Hammer [27] revealed average normative nasal nitric oxide between 101 and 239 ppb (measurement technique: tidal breathing) among 4–18 years old, and between 223.7 and 1024 ppb (measurement technique: Breath hold) among 4.5–24 years old. In healthy adults involving 141 subjects with mean age ($\pm$ standard deviation) of 25.6 ± 9.8 years, Antosova et al. [5] reported mean nasal nitric oxide ($\pm$standard deviation) on the right sinonasal cavity as 379.6 ± 170.4 and 401.6 ± 207.8 ppb on the left sinonasal cavity (measurement technique: tidal breathing).

CFD modeling techniques have demonstrated some relevance in analysis of nasal nitric oxide concentrations in the sinonasal cavity. In a 2009 publication, Hood and colleagues [68] estimated the transport rate of nasal nitric oxide based on Fick's law for diffusion and CFD convective transport values. Their findings suggest that if the sinus produces only enough nitric oxide to compensate for estimated convective and diffusive losses, then the rate of nitric oxide production per unit area of the sinus mucosa will be $4 \times 10^{-8} - 1 \times 10^{-6}$ mol.m^{-2}.s^{-1}. In addition, the estimated nitric oxide transport passing through the ostia from the sinuses may not supply all the nitric oxide concentration in nasally exhaled air. Hood et al. [68] and Na et al. [108] adopted the volume of fluid method to investigate concentration of nitric oxide in the maxillary sinuses. During respiration, Na and colleagues [108] noted that the concentration of nitric oxide in the maxillary sinus was relatively stable with a low

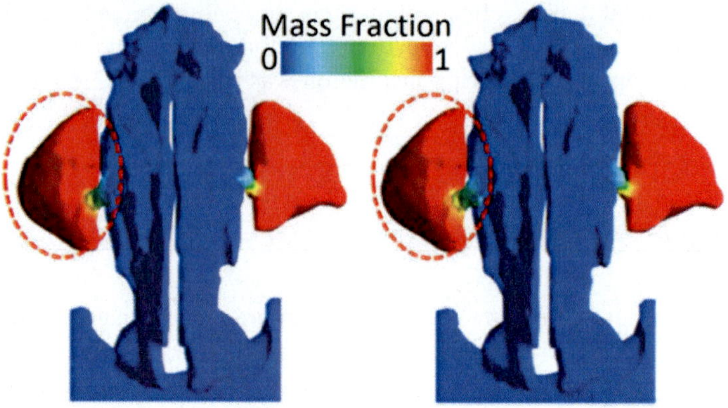

Fig. 8.12 Concentration of nitric oxide in the sinonasal cavity. Initial concentration level was assumed to be 10% at the start of respiration. Left Panel—Maxillary sinus concentration was 99.6% at the time of maximum flow rate during inspiration. Right Panel—Maxillary sinus concentration was 99.1% at the time of maximum flow rate during expiration. Reproduced with permission from Kim et al. [80]

rate of diffusion, in contrast with the scenario when an accessory ostium was present. From time 0.11 to 0.65 s, average volume fraction of nitric oxide in the maxillary sinus decreased from 0.9799 to 0.8913 at a rate of 0.1641 per second during inspiration (0.981–0.8933 at a rate of 0.1624 per second during expiration) compared to when an accessory ostium was present that went from 0.9437 to 0.7409 at a rate of 0.3756 per second during inspiration (0.9457–0.7421 at a rate of 0.3770 per second during expiration) [108] Similarly, Rennie et al. [121] suggested that increased ventilation in the sinuses due to an accessory ostium will likely result in reduction of nitric oxide concentration in the sinuses. As depicted in Fig. 8.11, Kim et al. [80] reported that the concentration of nitric oxide exchange between maxillary sinus and nasal cavity was not significant (Fig. 8.12).

8.7 Summary

In this chapter, we have described the challenges in computational modeling of upper airway in deciding on the numerical scheme that strike a good balance between physical accuracy and computational efficiency. In addition, we highlighted the conflicting literature reports among the relatively few studies that have compared pressure-flow curves and nasal resistance between CFD generated results and those measured using rhinomanometry. Comparisons between CFD and acoustic rhinometry were also discussed, as well as a CFD -based method to quantify the minimal cross-sectional area in the nasal cavity, named "computational streamline rhinometry". Similarly, we reviewed a number of studies demonstrating varying degrees of correlation between CFD variables and patient-reported outcomes for nasal and sinus diseases.

It is noteworthy that the typical human nasal passage is characterized by several normal intra- and inter-individual temporal and morphological variabilities, such as nasal cycle, nasal vestibule indentation phenotype, and nasal index (race/ethnicity). In general, these variations confound the characterization of normal nasal airflow dynamics to varying degrees; however, studies have reported that variations due to nasal vestibule indentation phenotype and nasal index do not significantly alter inter-subject nasal physiology. Next, it is important to note that despite all the debate over the functional significance of the paranasal sinuses, there is an overwhelming consensus that the sinuses play a critical role in immune defense of the nasal cavity and secretion of nitric oxide, and there is minimal exchange of airflow between the healthy normal sinuses and the main nasal cavity.

In conclusion, we noted a number of open questions in the field that will benefit from further research. First, the assumption of rigid nasal walls while performing CFD airflow simulations. Second, in-depth analysis of the role of mucosal cooling in the perception of nasal airflow and a robust validation (in vivo or in vitro) of the this role. Third, clarity on the degree of agreement between CFD generated results and objective clinical metrics. Forth, a strong evidence of how well different CFD-derived variables correlate with patient-reported quality of life measures for varying nasal and sinus diseases. Lastly, development of a robust database of normative ranges for different CFD-derived variables that have demonstrated strong evidence of clinical associations with nasal function or symptomatology.

References

1. L. Akmenkalne, M. Prill, K. Vogt, Nasal valve elastography: quantitative determination of the mobility of the nasal valve. Rhinol. Online **2**, 81–86 (2019)
2. J. Alberty, W. Stoll, C. Rudack, The effect of endogenous nitric oxide on mechanical ciliostimulation of human nasal mucosa. Clin. Exp. Allergy **36**(10), 1254–1259 (2006)
3. R.F. Andre, H.D. Vuyk, A. Ahmed, K. Graamans, G.J. Nolst Trenite, Correlation between subjective and objective evaluation of the nasal airway. a systematic review of the highest level of evidence. Clin .Otolaryngol. **34**(6), 518–25 (2009)
4. W.T. Anselmo-Lima, V.J. Lund, The effects of endoscopic sinus surgery on the nasal cycle as assessed by acoustic rhinometry. Am. J. Rhinol. **15**(3), 165–8 (2001)
5. M. Antosova, D. Mokra, I. Tonhajzerova, P. Mikolka, P. Kosutova, M. Mestanik, L. Pepucha, J. Plevkova, T. Buday, V. Calkovsky, Nasal nitric oxide in healthy adults-reference values and affecting factors. Physiolog. Res. **66**, S247 (2017)
6. F.D. Babatola, Nasal resistance values in the adult negroid nigerian. Rhinology **28**(4), 269–73 (1990)
7. R.S. Bailey, K.P. Casey, S.S. Pawar, G.J. Garcia, Correlation of nasal mucosal temperature with subjective nasal patency in healthy individuals. JAMA Facial Plast. Surg. **19**(1), 46–52 (2017)
8. E. Baraldi, M. Pasquale, A. Cangiotti, S. Zanconato, F. Zacchello, Nasal nitric oxide is low early in life: case study of two infants with primary ciliary dyskinesia. Eur. Respir. J. **24**(5), 881–883 (2004)
9. G. Berger, I. Hammel, R. Berger, S. Avraham, D. Ophir, Histopathology of the inferior turbinate with compensatory hypertrophy in patients with deviated nasal septum. The laryngoscope **110**(12), 2100–2105 (2000)

10. E.R. Berkinshaw, P.M. Spalding, P.S. Vig, The effect of methodology on the determination of nasal resistance. Am. J. Orthod. Dentofac. Orthop. (1987)
11. C. Bermüller, H. Kirsche, G. Rettinger, H. Riechelmann, Diagnostic accuracy of peak nasal inspiratory flow and rhinomanometry in functional rhinosurgery. The Laryngoscope **118**(4), 605–610 (2008)
12. S. Blaney, Why paranasal sinuses? J. Laryngol. Otol. **104**(9), 690–693 (1990)
13. P.L. Blanton, N.L. Biggs, Eighteen hundred years of controversy: the paranasal sinuses. Am. J. Anat. **124**(2), 135–147 (1969)
14. A.A. Borojeni, G.J. Garcia, M.G. Moghaddam, D.O. Frank-Ito, J.S. Kimbell, P.W. Laud, L.J. Koenig, J.S. Rhee, Normative ranges of nasal airflow variables in healthy adults. International Journal of Computer Assisted Radiology and Surgery 1–12 (2019)
15. A. Burrow, R. Eccles, A.S. Jones, The effects of camphor, eucalyptus and menthol vapour on nasal resistance to airflow and nasal sensation. Acta Otolaryngol. **96**(1–2), 157–61 (1983)
16. O. Cakmak, E. Tarhan, M. Coskun, M. Cankurtaran, H. Celik, Acoustic rhinometry: accuracy and ability to detect changes in passage area at different locations in the nasal cavity. Ann. Otol. Rhinol. Laryngol. **114**(12), 949–57 (2005)
17. K.H. Calhoun, W. House, J.A. Hokanson, F.B. Quinn, Normal nasal airway-resistance in noses of different sizes and shapes. Otolaryngol. Head Neck Surg. **103**(4), 605–609 (1990)
18. K.P. Casey, A.A. Borojeni, L.J. Koenig, J.S. Rhee, G.J. Garcia, Correlation between subjective nasal patency and intranasal airflow distribution. Otolaryngol. Head Neck Surg. (United States) (2017)
19. R.K. Chandra, M.O. Patadia, J. Raviv, Diagnosis of nasal airway obstruction. Otolaryngol. Clin. N. Am. **42**(2), 207–225 (2009)
20. G.B. Cherobin, R.L. Voegels, E. Gebrim, G.J.M. Garcia, Sensitivity of nasal airflow variables computed via computational fluid dynamics to the computed tomography segmentation threshold. PLoS One **13**(11), e0207178 (2018)
21. S. Chung, H. Dhong, D. Na, Mucus circulation between accessory ostium and natural ostium of maxillary sinus. J. Laryngol. Otol. **113**(9), 865–867 (1999)
22. S.-K. Chung, D.Y. Cho, H.J. Dhong, Computed tomogram findings of mucous recirculation between the natural and accessory ostia of the maxillary sinus. Am. J. Rhinol. **16**(5), 265–268 (2002)
23. S.E. Churchill, L.L. Shackelford, J.N. Georgi, M.T. Black, Morphological variation and airflow dynamics in the human nose. Am. J. Hum. Biol.: Off. J. Hum. Biol. Assoc. **16**(6), 625–638 (2004)
24. P.A. Clement, F. Gordts, Standardisation Committee on Objective Assessment of the Nasal Airway, I. R. S., Ers. Consensus report on acoustic rhinometry and rhinomanometry. Rhinology **43**(3), 169–79 (2005)
25. P. Clements, F. Gortds, Standardisation committee on objective assessment of the nasal airway, irs, and ers consensus report on acoustic rhinometry and rhinomanometry. Rhynology **43**(3), 169–179 (2005)
26. P. Cole, Physiology of the nose and paranasal sinuses. Clin. Rev. Allergy Immunol. **16**(1–2), 25–54 (1998)
27. R. Corbelli, J. Hammer, Measurement of nasal nitric oxide. Paediatr. Respir. Rev. **8**(3), 269–272 (2007)
28. J.P. Corey, Acoustic rhinometry: should we be using it? Curr. Opin. Otolaryngol. Head Neck Surg. **14**(1), 29–34 (2006)
29. B.A. Craven, T. Neuberger, E.G. Paterson, A.G. Webb, E.M. Josephson, E.E. Morrison, G.S. Settles, Reconstruction and morphometric analysis of the nasal airway of the dog (canis familiaris) and implications regarding olfactory airflow. Anat. Rec. (Hoboken) **290**(11), 1325–40 (2007)
30. E. Crognier, Climate and anthropometric variations in Europe and the mediterranean area. Ann. Hum. Biol. **8**(2), 99–107 (1981)
31. R.K. Daniel, Hispanic rhinoplasty in the united states, with emphasis on the Mexican American nose. Plast. Reconstr. Surg. **112**(1), 244–56; discussion 257–8 (2003)

32. A. Davies, A re-survey of the morphology of the nose in relation to climate. J. R. Anthropol. Inst. G. B. Irel. **62**, 337–359 (1932)
33. W.B. Davis, Lv. anatomy of the nasal accessory sinuses in infancy and childhood. Ann. Otol. Rhinol. Laryngol. **27**(3), 940–967 (1918)
34. A. Dayal, J.S. Rhee, G.J. Garcia, Impact of middle versus inferior total turbinectomy on nasal aerodynamics. Otolaryngol. Head Neck Surg. **155**(3), 518–25 (2016)
35. H.P.L. De Yun Wang, B.R. Gordon, Impacts of fluid dynamics simulation in study of nasal airflow physiology and pathophysiology in realistic human three-dimensional nose models. Clin. Exp. Otorhinolaryngol. **5**(4), 181 (2012)
36. N.M. Doddi, R. Eccles, The relationship between nasal index and nasal airway resistance, and response to a topical decongestant. Rhinology **49**(5), 583–6 (2011)
37. J. Dong, J. Ma, Y. Shang, K. Inthavong, D. Qiu, J. Tu, D. Frank-Ito, Detailed nanoparticle exposure analysis among human nasal cavities with distinct vestibule phenotypes. J. Aerosol Sci. **121**, 54–65 (2018)
38. D.J. Doorly, D.J. Taylor, R.C. Schroter, Mechanics of airflow in the human nasal airways. Respir. Physiol. Neurobiol. **163**(1–3), 100–10 (2008)
39. J. Earwaker, Anatomic variants in sinonasal ct. Radiographics **13**(2), 381–415 (1993)
40. R. Eccles, A role for the nasal cycle in respiratory defence. Eur. Respir. J. **9**(2), 371–376 (1996)
41. R. Eccles, Nasal airflow in health and disease. Acta Otolaryngol. **120**(5), 580–595 (2000)
42. R. Eccles, D.H. Griffiths, C.G. Newton, N.S. Tolley, The effects of menthol isomers on nasal sensation of airflow. Clin. Otolaryngol. Allied Sci. **13**(1), 25–9 (1988)
43. R. Eccles, A.S. Jones, The effect of menthol on nasal resistance to air flow. J. Laryngol. Otol. **97**(8), 705–9 (1983)
44. T.P. Eiting, J.B. Perot, E.R. Dumont, How much does nasal cavity morphology matter? patterns and rates of olfactory airflow in phyllostomid bats. Proc. Biol. Sci. **282**(1800), 20142161 (2015)
45. T.P. Eiting, T.D. Smith, J.B. Perot, E.R. Dumont, The role of the olfactory recess in olfactory airflow. J. Exp. Biol. **217**(Pt 10), 1799–803 (2014)
46. D. Elad, R. Liebenthal, B.L. Wenig, S. Einav, Analysis of air flow patterns in the human nose. Med. Biol. Eng. Comput. **31**(6), 585–92 (1993)
47. E.W. Fisher, D.P. Morris, J.M. Biemans, C.R. Palmer, V.J. Lund, Practical aspects of acoustic rhinometry: problems and solutions. Rhinology **33**(4), 219–23 (1995)
48. P. Flanagan, R. Eccles, Spontaneous changes of unilateral nasal airflow in man. a re-examination of the 'nasal cycle'. Acta Otolaryngol. **117**(4), 590–595 (1997)
49. C. Gaberino, J.S. Rhee, G.J. Garcia, Estimates of nasal airflow at the nasal cycle mid-point improve the correlation between objective and subjective measures of nasal patency. Respir. Physiol. Neurobiol. 238:23–32 (2017). https://doi.org/10.1016/j.resp.2017.01.004
50. A.M. Gambaruto, D.J. Taylor, D.J. Doorly, Decomposition and description of the nasal cavity form. Ann. Biomed. Eng. **40**(5), 1142–59 (2012)
51. G.J. Garcia, E.W. Tewksbury, B.A. Wong, J.S. Kimbell, Interindividual variability in nasal filtration as a function of nasal cavity geometry. J. Aerosol. Med. Pulm. Drug Deliv. **22**(2), 139–55 (2009)
52. G.J.M. Garcia, B.M. Hariri, R.G. Patel, J.S. Rhee, The relationship between nasal resistance to airflow and the airspace minimal cross-sectional area. J. Biomech. **49**(9), 1670–1678 (2016)
53. G. Garcia, G. Mitchell, B. N. T. D. W. J. K. J. Visualization of nasal airflow patterns in a patient affected with atrophic rhinitis using particle image velocimetry. J. Phys.: Conf. Ser. **85** (2007)
54. Q.J. Ge, K. Inthavong, J.Y. Tu, Local deposition fractions of ultrafine particles in a human nasal-sinus cavity cfd model. Inhal. Toxicol. **24**(8), 492–505 (2012)
55. A.N. Gilbert, Reciprocity versus rhythmicity in spontaneous alternations of nasal airflow. Chronobiol. Int. **6**(3), 251–257 (1989)
56. S. Granqvist, J. Sundberg, J.O. Lundberg, E. Weitzberg, Paranasal sinus ventilation by humming. J. Acoust. Soc. Am. **119**(5), 2611–2617 (2006)

57. L.F. Grymer, P. Illum, O. Hilberg, Septoplasty and compensatory inferior turbinate hypertrophy: a randomized study evaluated by acoustic rhinometry. J. Laryngol. Otol. **107**(5), 413–7 (1993)
58. R.A. Guilmette, Y.S. Cheng, W.C. Griffith, Characterising the variability in adult human nasal airway dimensions. Ann. Occup. Hyg. **41**(1), 491–496 (1997)
59. A. Gungor, R. Moinuddin, R.H. Nelson, J.P. Corey, Detection of the nasal cycle with acoustic rhinometry: techniques and applications. Otolaryngol. Head Neck Surg. **120**(2), 238–47 (1999)
60. I. Hahn, P. Scherer, M. Mozell, Velocity profiles measured for air-flow through a large-scale model of the human nasal cavity. J. Appl. Physiol. **75**(5), 2273–2287 (1993)
61. J.S. Haight, P. Djupesland, W. Qjan, J. Chatkin, H. Furlott, J. Irish, I. Witterick, P. McClean, R. Fenton, V. Hoffstein, Does nasal nitric oxide come from the sinuses? J. Otolaryngol. **28**(4), 197–204 (1999)
62. M. Hasegawa, E. Kern, Variations in nasal resistance in man: a rhinomanometric study of the nasal cycle in 50 human subjects. Rhinology **16**(1), 19–29 (1978)
63. M. Hasegawa, E.B. Kern, Variations in nasal resistance in man: a rhinomanometric study of the nasal cycle in 50 human subjects. Rhinology **16**(1), 19–29 (1978)
64. K. Hemtiwakorn, V. Mahasitthiwat, S. Tungjitkusolmun, K. Hamamoto, C. Pintavirooj, Patient-specific aided surgery approach of deviated nasal septum using computational fluid dynamics. IEEJ Trans. Electr. Electron. Eng. **10**(3), 274–286 (2015)
65. J. Hiernaux, A. Froment, Correlations between anthropo-biological and climatic variables in sub-saharan Africa - revised estimates. Hum. Biol. **48**(4), 757–767 (1976)
66. O. Hilberg, A.C. Jackson, D.L. Swift, O.F. Pedersen, Acoustic rhinometry: evaluation of nasal cavity geometry by acoustic reflection. J. Appl. Physiol. (1985) **66**(1), 295–303 (1989)
67. W.E. Holden, J.P. Wilkins, M. Harris, H.A. Milczuk, G.D. Giraud, Temperature conditioning of nasal air: effects of vasoactive agents and involvement of nitric oxide. J. Appl. Physiol. (1985) **87**(4), 1260–5 (1999)
68. C. Hood, R. Schroter, D. Doorly, E. Blenke, N. Tolley, Computational modeling of flow and gas exchange in models of the human maxillary sinus. J. Appl. Physiol. **107**, 1195–1203 (2009)
69. K. Inthavong, J. Ma, Y. Shang, J. Dong, A.S. Chetty, J. Tu, D. Frank-Ito, Geometry and airflow dynamics analysis in the nasal cavity during inhalation. Clinical Biomechanics (2017)
70. K. Inthavong, J. Wen, J.Y. Tu, Z.F. Tian, From ct scans to cfd modelling - fluid and heat transfer in a realistic human nasal cavity. Eng. Appl. Comput. Fluid Mech. **3**(3), 321–335 (2009)
71. M. Jog, G. McGarry, How frequent are accessory sinus ostia? J. Laryngol. Otol. **117**(4), 270–272 (2003)
72. T. Keck, R. Leiacker, A. Heinrich, S. Kuhnemann, G. Rettinger, Humidity and temperature profile in the nasal cavity. Rhinology **38**(4), 167–71 (2000)
73. J.A. Keeler, A. Patki, C.R. Woodard, D.O. Frank-Ito, A computational study of nasal spray deposition pattern in four ethnic groups. J. Aerosol. Med. Pulm. Drug Deliv. **29**(2), 153–66 (2016)
74. J. Keir, Why do we have paranasal sinuses? J. Laryngol. Otol. **123**(1), 4–8 (2009)
75. R.M. Kellman, C. Schmidt, The paranasal sinuses as a protective crumple zone for the orbit. The Laryngoscope **119**(9), 1682–1690 (2009)
76. K. Keyhani, P.W. Scherer, M.M. Mozell, Numerical simulation of airflow in the human nasal cavity. J. Biomech. Eng. **117**(4), 429–41 (1995)
77. K. Keyhani, P.W. Scherer, M.M. Mozell, A numerical model of nasal odorant transport for the analysis of human olfaction. J. Theor. Biol. **186**(3), 279–301 (1997)
78. D.W. Kim, S.K. Chung, Y. Na, Numerical study on the air conditioning characteristics of the human nasal cavity. Comput. Biol. Med. **86**, 18–30 (2017)
79. S.K. Kim, G.E. Heo, A. Seo, Y. Na, S.K. Chung, Correlation between nasal airflow characteristics and clinical relevance of nasal septal deviation to nasal airway obstruction. Respir. Physiol. Neurobiol. **192**, 95–101 (2014)

80. S.K. Kim, Y. Na, J.-I. Kim, S.-K. Chung, Patient specific cfd models of nasal airflow: overview of methods and challenges. J. Biomech. **46**(2), 299–306 (2013)
81. S.W. Kim, J.H. Mo, J.W. Kim, D.Y. Kim, C.S. Rhee, C.H. Lee, Y.G. Min, Change of nasal function with aging in Korean. Acta Otolaryngol. Suppl. **558**, 90–4 (2007)
82. J. Kimbell, D. Frank, P. Laud, G. Garcia, J. Rhee, Changes in nasal airflow and heat transfer correlate with symptom improvement after surgery for nasal obstruction. J. Biomech. **46**(15), 2634–2643 (2013)
83. J.S. Kimbell, S. Basu, G.J. Garcia, D.O. Frank-Ito, F. Lazarow, E. Su, D. Protsenko, Z. Chen, J.S. Rhee, B.J. Wong, Upper airway reconstruction using long-range optical coherence tomography: effects of airway curvature on airflow resistance. Lasers Surg. Med. (2019)
84. J.S. Kimbell, R.A. Segal, B. Asgharian, B.A. Wong, J.D. Schroeter, J.P. Southall, C.J. Dickens, G. Brace, F.J. Miller, Characterization of deposition from nasal spray devices using a computational fluid dynamics model of the human nasal passages. J. Aerosol Med. **20**(1), 59–74 (2007)
85. T. KjÆrgaard, M. Cvancarova, S.K. SteinsvÅg, Relation of nasal air flow to nasal cavity dimensions. Arch. Otolaryngol. Head Neck Surg. **135**(6), 565–570 (2009)
86. H. Kumar, R. Jain, R.G. Douglas, M.H. Tawhai, Airflow in the human nasal passage and sinuses of chronic rhinosinusitis subjects. PloS one **11**(6), e0156379 (2016)
87. D. Lal, M.L. Gorges, G. Ungkhara, P.M. Reidy, J.P. Corey, Physiological change in nasal patency in response to changes in posture, temperature, and humidity measured by acoustic rhinometry. Am. J. Rhinol. **20**(5), 456–62 (2006)
88. C. Lang, S. Grutzenmacher, B. Mlynski, S. Plontke, G. Mlynski, Investigating the nasal cycle using endoscopy, rhinoresistometry, and acoustic rhinometry. Laryngoscope **113**(2), 284–9 (2003)
89. T.B. Le, M.G. Moghaddam, B.T. Woodson, G.J.M. Garcia, Airflow limitation in a collapsible model of the human pharynx: physical mechanisms studied with fluid-structure interaction simulations and experiments. Physiol. Rep. **7**(10), e14099 (2019)
90. J. Leach, Aesthetics and the hispanic rhinoplasty. Laryngoscope **112**(11), 1903–16 (2002)
91. S.C. Leong, X.B. Chen, H.P. Lee, D.Y. Wang, A review of the implications of computational fluid dynamic studies on nasal airflow and physiology. Rhinology **48**(2), 139–45 (2010)
92. S.C. Leong, R. Eccles, A systematic review of the nasal index and the significance of the shape and size of the nose in rhinology. Clin. Otolaryngol. **34**(3), 191–198 (2009)
93. C. Li, J. Jiang, H. Dong, K. Zhao, Computational modeling and validation of human nasal airflow under various breathing conditions. J. Biomech. (2017)
94. J. Lindemann, R. Leiacker, G. Rettinger, T. Keck, Nasal mucosal temperature during respiration. Clin. Otolaryngol. Allied Sci. **27**(3), 135–9 (2002)
95. J. Lu, D. Han, L. Zhang, Accuracy evaluation of a numerical simulation model of nasal airflow. Acta Otolaryngol. **134**(5), 513–9 (2014)
96. J. Lundberg, G. Settergren, S. Gelinder, J. Lundberg, K. Alving, E. Weitzberg, Inhalation of nasally derived nitric oxide modulates pulmonary function in humans. Acta Physiol. Scandinavica **158**(4), 343–347 (1996)
97. J. Lundberg, E. Weitzberg, S. Nordvall, R. Kuylenstierna, J. Lundberg, K. Alving, Primarily nasal origin of exhaled nitric oxide and absence in kartagener's syndrome. Eur. Respir. J. **7**(8), 1501–1504 (1994)
98. J.O. Lundberg, Nitric oxide and the paranasal sinuses. Anat. Rec.: Adv. Integr. Anat. Evol. Biol.: Adv. Integr. Anat. Evol.y Biol. **291**(11), 1479–1484 (2008)
99. J. Ma, J. Dong, Y. Shang, K. Inthavong, J. Tu, D.O. Frank-Ito, Air conditioning analysis among human nasal passages with anterior anatomical variations. Med. Eng. Phys. **57**, 19–28 (2018)
100. Maalouf, R., Bequignon, E., Devars du Mayne, M., Zerah- Lancner, F., Isabey, D., Coste, A., Louis, B., and Papon, J. F. A functional tool to differentiate nasal valve collapse from other causes of nasal obstruction: the fried test. *J Appl Physiol (1985) 121*, 1 (2016), 343–7

101. M. Maniscalco, E. Weitzberg, J. Sundberg, M. Sofia, J. Lundberg, Assessment of nasal and sinus nitric oxide output using single-breath humming exhalations. Eur. Respir. J. **22**(2), 323–329 (2003)
102. W.T. Mcnicholas, M. Coffey, T. Boyle, Effects of nasal airflow on breathing during sleep in normal human. Am. Rev. Respir. Dis. **147**, 620–620 (1993)
103. L. Menzel, A. Hess, W. Bloch, O. Michel, K.-D. Schuster, R. Gabler, W. Urban, Temporal nitric oxide dynamics in the paranasal sinuses during humming. J. Appl. Physiol. **98**(6), 2064–2071 (2005)
104. Y.G. Min, Y.J. Jang, Measurements of cross-sectional area of the nasal cavity by acoustic rhinometry and ct scanning. Laryngoscope **105**(7 Pt 1), 757–9 (1995)
105. R. Moinuddin, B. Mamikoglu, S. Barkatullah, J.P. Corey, Detection of the nasal cycle. Am. J. Rhinol. **15**(1), 35–9 (2001)
106. M. Myerson, Natural orifice of the maxillary sinus: Ii. clinical studies. Arch. Otolaryngol. **15**(5), 716–733 (1932)
107. N. Mygind, R. Dahl, Anatomy, physiology and function of the nasal cavities in health and disease. Adv. Drug Deliv. Rev. **29**(1–2), 3–12 (1998)
108. Y. Na, K. Kim, S.K. Kim, S.-K. Chung, The quantitative effect of an accessory ostium on ventilation of the maxillary sinus. Respir. Physiol. Neurobiol. **181**(1), 62–73 (2012)
109. M.L. Noback, K. Harvati, F. Spoor, Climate-related variation of the human nasal cavity. Am. J. Phys. Anthropol. **145**(4), 599–614 (2011)
110. J. Numminen, P. Dastidar, T. Heinonen, T. Karhuketo, M. Rautiainen, Reliability of acoustic rhinometry. Respir. Med. **97**(4), 421–7 (2003)
111. M. Ohki, K. Naito, P. Cole, Dimensions and resistances of the human nose - racial-differences. Laryngoscope **101**(3), 276–278 (1991)
112. G. O'Neill, N.S. Tolley, The complexities of nasal airflow - theory and practice. J. Appl. Physiol. (1985) (2019)
113. J. Osman, F. Großmann, K. Brosien, U. Kertzscher, L. Goubergrits, T. Hildebrandt, Assessment of nasal resistance using computational fluid dynamics. Curr. Dir. Biomed. Eng. **2**(1), 617–621 (2016)
114. H.E. Ozel, F. Ozdogan, E. Esen, M.G. Genc, S. Genc, A. Selcuk, The association between septal deviation and the presence of a maxillary accessory ostium. In: International Forum of Allergy & Rhinology, vol. 5, pp. 1177–1180. Wiley Online Library
115. R.G. Patel, G.J. Garcia, D.O. Frank-Ito, J.S. Kimbell, J.S. Rhee, Simulating the nasal cycle with computational fluid dynamics. Otolaryngol. Head Neck Surg. (2014)
116. A. Patki, D.O. Frank-Ito, Characterizing human nasal airflow physiologic variables by nasal index. Respir. Physiol. Neurobiol. **232**, 66–74 (2016)
117. M. Quadrio, C. Pipolo, S. Corti, R. Lenzi, F. Messina, C. Pesci, G. Felisati, Review of computational fluid dynamics in the assessment of nasal air flow and analysis of its limitations. Eur. Arch. Oto-Rhino-Laryngol. **271**(9), 2349–2354 (2014)
118. C.W. Quammen, R.M. Taylor, P. Krajcevski, S. Mitran, A. Enquobahrie, R. Superfine, B. Davis, S. Davis, C. Zdanski, The virtual pediatric airways workbench. In: Studies in Health Technology and Informatics (2016)
119. T. Radulesco, L. Meister, G. Bouchet, A. Varoquaux, J. Giordano, J. Mancini, P. Dessi, P. Perrier, J. Michel, Correlations between computational fluid dynamics and clinical evaluation of nasal airway obstruction due to septal deviation: An observational study. Clin. Otolaryngol. **44**(4), 603–611 (2019)
120. V.H. Ramprasad, D.O. Frank-Ito, A computational analysis of nasal vestibule morphologic variabilities on nasal function. J. Biomech. **49**(3), 450–7 (2016)
121. C.E. Rennie, C.M. Hood, E.J. Blenke, R.S. Schroter, D.J. Doorly, H. Jones, D. Towey, N.S. Tolley, Physical and computational modeling of ventilation of the maxillary sinus. Otolaryngol. Head Neck Surg. **145**(1), 165–170 (2011)
122. J.S. Rhee, C.D. Sullivan, D.O. Frank, J.S. Kimbell, G.J. Garcia, A systematic review of patient-reported nasal obstruction scores: defining normative and symptomatic ranges in surgical patients. JAMA Facial Plast. Surg. **16**(3), 219–25; quiz 232 (2014)

123. D.G. Roblin, R. Eccles, Normal range for nasal partitioning of airflow determined by nasal spirometry in 100 healthy subjects. Am. J. Rhinol. **17**(4), 179–83 (2003)
124. E. Sanmiguel-Rojas, M.A. Burgos, F. Esteban-Ortega, Nasal surgery handled by cfd tools. Int. J. Numer. Method Biomed. Eng. **34**(10), e3126 (2018)
125. M.J. Schumacher, Nasal congestion and airway obstruction: The validity of available objective and subjective measures. Curr. Allergy Asthma Rep. **2**(3), 245–51 (2002)
126. M.J. Schumacher, Nasal dyspnea: the place of rhinomanometry in its objective assessment. Am. J. Rhinol. **18**(1), 41–46 (2004)
127. R.A. Segal, G.M. Kepler, J.S. Kimbell, Effects of differences in nasal anatomy on airflow distribution: a comparison of four individuals at rest. Ann. Biomed. Eng. **36**(11), 1870–82 (2008)
128. K.T. Shanley, P. Zamankhan, G. Ahmadi, P.K. Hopke, Y.S. Cheng, Numerical simulations investigating the regional and overall deposition efficiency of the human nasal cavity. Inhal. Toxicol. **20**(12), 1093–100 (2008)
129. J. Sozansky, S.M. Houser, The physiological mechanism for sensing nasal airflow: a literature review. Int. Forum Allergy Rhinol. **4**(10), 834–8 (2014)
130. F. Stehling, C. Roll, F. Ratjen, H. Grasemann, Nasal nitric oxide to diagnose primary ciliary dyskinesia in newborns. Arch. Dis. Childhood-Fetal Neonatal Ed. **91**(3), F233–F233 (2006)
131. R. Subramaniam, R. Richardson, K. Morgan, R. Guilmette, J. Kimbell, Computational fluid dynamics simulations of inspiratory airflow in the human nose and nasopharynx. Inhal. Toxicol. **10**, 91–120 (1998)
132. C.D. Sullivan, G.J. Garcia, D.O. Frank-Ito, J.S. Kimbell, J.S. Rhee, Perception of better nasal patency correlates with increased mucosal cooling after surgery for nasal obstruction. Otolaryngol. Head Neck Surg. **150**(1), 139–47 (2014)
133. J. Tan, D. Han, J. Wang, T. Liu, T. Wang, H. Zang, Y. Li, X. Wang, Numerical simulation of normal nasal cavity airflow in Chinese adult: a computational flow dynamics model. Eur. Arch. Otorhinolaryngol. **269**(3), 881–9 (2012)
134. E. Tarhan, M. Coskun, O. Cakmak, H. Celik, M. Cankurtaran, Acoustic rhinometry in humans: accuracy of nasal passage area estimates, and ability to quantify paranasal sinus volume and ostium size. J. Appl. Physiol. (1985) **99**(2), 616–23 (2005)
135. A. Thomson, L.H.D. Buxton, Man's nasal index in relation to certain climatic conditions. J. R. Anthropol. Inst. G. B. Ireland **53**, 92–122 (1923)
136. D.L. Vanhille, G.J.M. Garcia, O. Asan, A.A.T. Borojeni, D.O. Frank-Ito, J.S. Kimbell, S.S. Pawar, J.S. Rhee, Virtual surgery for the nasal airway: a preliminary report on decision support and technology acceptance. JAMA Facial Plast. Surg. **20**(1), 63–69 (2018)
137. K. Vogt, A.A. Jalowayski, W. Althaus, C. Cao, D. Han, W. Hasse, H. Hoffrichter, R. Mosges, J. Pallanch, K. Shah-Hosseini, K. Peksis, K.D. Wernecke, L. Zhang, P. Zaporoshenko, 4-phase-rhinomanometry (4pr)-basics and practice 2010. Rhinol. Suppl. **21**, 1–50 (2010)
138. D.W. Warren, A.F. Drake, J.U. Davis, Nasal airway in breathing and speech. Cleft Palate-Craniofacial J. **29**(6), 511–519 (1992)
139. J.S. Weiner, Nose shape and climate. Am. J. Phys. Anthropol. New Ser. **12**(4), 615–618 (1954)
140. E. Weitzberg, J.O. Lundberg, Humming greatly increases nasal nitric oxide. Am. J. Respir. Critical Care Med. **166**(2), 144–145 (2002)
141. J. Wen, K. Inthavong, J. Tu, S.M. Wang, Numerical simulations for detailed airflow dynamics in a human nasal cavity. Respir. Physiol. Neurobiol. **161**(2), 125–135 (2008)
142. D. Willatt, The evidence for reducing inferior turbinates. Rhinology **47**(3), 227–236 (2009)
143. M.H. Wolpoff, Climatic influence on the skeletal nasal aperture. Am. J. Phys. Anthropol. **29**(3), 405–23 (1968)
144. D.M. Wootton, H. Luo, S.C. Persak, S. Sin, J.M. McDonough, C.R. Isasi, R. Arens, Computational fluid dynamics endpoints to characterize obstructive sleep apnea syndrome in children. J. Appl. Physiol. (1985) **116**(1), 104–12 (2014)
145. G.-X. Xiong, J.-M. Zhan, H.-Y. Jiang, J.-F. Li, L.-W. Rong, G. Xu, Computational fluid dynamics simulation of airflow in the normal nasal cavity and paranasal sinuses. Am. J. Rhinol. **22**, 477–482 (2008)

146. T.R. Yokley, Ecogeographic variation in human nasal passages. Am. J. Phys. Anthropol. **138**(1), 11–22 (2009)
147. S. Yu, Y. Liu, X. Sun, S. Li, Influence of nasal structure on the distribution of airflow in nasal cavity. Rhinology **46**(2), 137–143 (2008)
148. K. Zhao, J. Jiang, What is normal nasal airflow? a computational study of 22 healthy adults. Int. Forum Allergy Rhinol. **4**(6), 435–46 (2014)
149. K. Zhao, J. Jiang, K. Blacker, B. Lyman, P. Dalton, B.J. Cowart, E.A. Pribitkin, Regional peak mucosal cooling predicts the perception of nasal patency. Laryngoscope **124**(3), 589–95 (2014)
150. J.H. Zhu, H.P. Lee, K.M. Lim, B.R. Gordon, D.Y. Wang, Effect of accessory ostia on maxillary sinus ventilation: a computational fluid dynamics (cfd) study. Respir. Physiol. Neurobiol. **183**(2), 91–99 (2012)
151. J.H. Zhu, H.P. Lee, K.M. Lim, S.J. Lee, D.Y. Wang, Evaluation and comparison of nasal airway flow patterns among three subjects from caucasian, chinese and indian ethnic groups using computational fluid dynamics simulation. Respir. Physiol. Neurobiol. **175**(1), 62–69 (2011)

Chapter 9
Clinical CFD Applications 1

Chengyu Li, Kai Zhao, Dennis Shusterman, Hadrien Calmet, Alister J. Bates, Joey Siu, and Richard Douglas

Abstract This chapter is the first of two chapters demonstrating the wide variety of CFD studies in clinical applications presented from leading researchers in their respective fields. This chapter covers the latest research techniques and outcomes in airflow and conditioning in the nasal cavity; fluid and particle dynamics from sniffing; nasal obstruction and empty nose syndrome; nasal nitric oxide (nNO) dynamics and the ostiomeatal complex.

9.1 Airflow and Conditioning in the Nasal Cavity

Joey Siu and Richard Douglas

9.1.1 Introduction

There are many studies that have investigated normal human nasal airflow [11, 31, 35, 53, 109, 117, 122, 126]. These studies have produced varied results, primarily due to differences in nasal geometry between individuals and populations and the

C. Li (✉)
Villanova University, Villanova, USA
e-mail: chengyu.li@villanova.edu

K. Zhao
Ohio State University, Columbus, USA

D. Shusterman
University of California, San Francisco, USA

H. Calmet
Barcelona Supercomputing Center, Barcelona, Spain

A. J. Bates
Cincinnati Children's Hospital, Cincinnati, USA

J. Siu · R. Douglas
University of Auckland, Auckland, New Zealand

K. Inthavong et al. (eds.), *Clinical and Biomedical Engineering in the Human Nose*, Biological and Medical Physics, Biomedical Engineering,
https://doi.org/10.1007/978-981-15-6716-2_9

effect of the physiological nasal cycle. However, a consensus of studies of healthy subjects largely supports the presence of predominantly laminar flow during restful breathing, with turbulent flow confined to the anterior nasal cavity, where vortices form posterior to the internal nasal valve [109, 126]. Major flow paths occur through the middle meatus, and the nasal airway medial to the inferior and middle turbinates. Only a small fraction of flow goes through the superior meatus and olfactory clefts [109, 117, 126].

The smooth and even flow of air through most of regions of the nasal cavity assists respiration, as well as enhancing the perception of nasal patency and the diffusion of odor molecules to the olfactory region [109, 117, 126]. Slow flow in the olfactory region is probably a defence mechanism that prevents airborne toxins such as ammonia, hairdressing chemicals, gasoline, and heavy metals being deposited onto the olfactory mucosa, where they could potentially induce chemosensory dysfunction [102]. The presence of vortices, turbulence and turbinates collectively act to increase the residence time of airflow in the nasal cavity, allowing air to be humidified and warmed [109].

The efficiency of the turbinates in humidifying and warming air likely relies very little on the size of surface area contacted and more on how they influence flow [100]. In a virtual CFD study, evaluating the aerodynamic impact of a total inferior turbinectomy resection and inferior turbinoplasty by removing the inferior bulk of the turbinate in eight patients with inferior turbinate hypertrophy (enlargement), a more altered airflow distribution, reduced air-mucosa contact and prominent jet-like course of the main flow stream was observed following total resection [100]. Representative streamline plots for one patient and a healthy subject comparison are shown in Fig. 9.1.

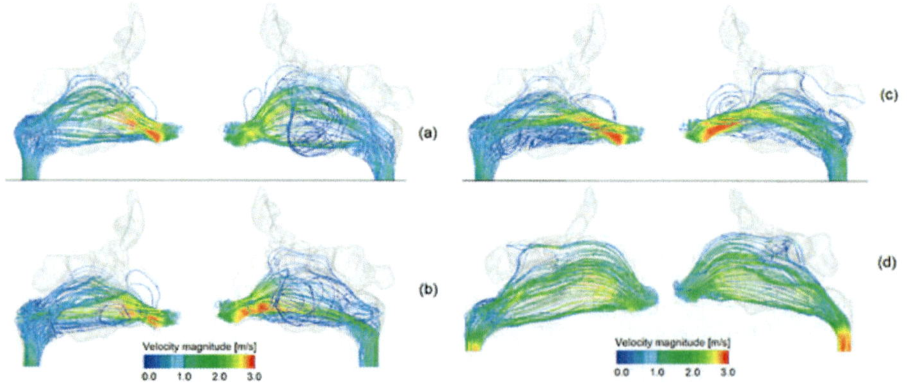

Fig. 9.1 Streamline plots for a patient with nasal airway obstruction (NAO) secondary to inferior turbinate hypertrophy **a** NAO patient following virtual turbinoplasty **b** NAO patient following virtual turbinectomy **d** healthy subject comparison; right sinonasal cavity (left), left sinonasal cavity (right)

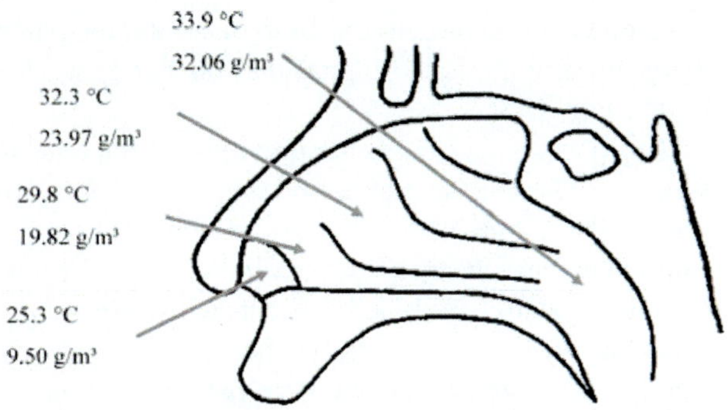

Fig. 9.2 Increase of air temperature and absolute humidity during inspiration of air (25 °C; 8.06 g Wasser/m^3), image from Keck and Lindemann [49]

Despite the large mean difference in remnant mucosa surface area in the surgery groups, the differences in aerodynamic outcomes were small and are more likely only to influence physiological function in extremely cold or dry environments.

In vivo measurements of intranasal air temperature and humidity (RH) under ambient conditions (temperature 25 °C; RH 35%) during inspiration show a synchronous rise from the nostril to the nasopharynx, with the highest increase in both variables occurring anterior to the nasal valve and head of the inferior turbinate as shown in Fig. 9.2 [48, 49]. Once inhaled air reaches the nasopharynx, its temperature is close to 34 °C and its RH is greater than 90% at the end of expiration [48]. Webb [115] showed that when the ambient air temperature of 7 °C, nasopharyngeal temperature is approximately 29 °C. The contact time of inspired air with the surrounding mucosa of the anterior nasal cavity is sufficient to warm the inspired air despite the high velocity of flow in this region of the nose. Nasal mucosal temperature also varies with distance from the nostril and phase of inspiration due to heat exchange between air and the mucosa. Recent studies suggest that a change in mucosal temperature (mucosal cooling) correlates better than transnasal resistance with nasal patency [10, 15, 30, 82, 104, 105].

Without humidification or warming occurring in the nasal passage mucociliary clearance and nasal particle deposition may be impaired. This is because mucociliary activity depends on airway temperature and viscoelasticity of the mucosal surface [81]. When heat and water fluxes in the nasal cavity are diminished alveolar gas exchange can also be adversely impacted, especially in extreme environments [22, 84].

9.1.2 Airflow and Conditioning in the Paranasal Sinuses

The paranasal sinuses are air-filled cavities located above, between and below the orbits, each of which drains into the nasal passages through small openings (known as ostia). Their physiological function is uncertain and a number of theories have been generated [43]. The most plausible is that their presence thins the medial and inferior orbital walls, so that blunt orbital trauma can prolapse these walls, protecting the globe from severe injury. More recently, the paranasal sinuses have been shown to behave as a modulator of nasal airflow and reduce air turbulence generated in the nasal cavity [35].

Airflow into the sinuses has previously been considered to be negligible in healthy subjects [17] and the paranasal sinuses have frequently been excluded from nasal cavity models. However, in the past decade, computational fluid dynamics (CFD) techniques have been used to evaluate sinus airflow in both healthy subjects and chronic rhinosinusitis (CRS) patients who have not undergone functional endoscopic sinus surgery [11, 31, 53, 122]. These show that despite the airflow exchanges occurring between the nasal cavity and paranasal sinuses being very small during calm breathing, a functional interdependence exists as they are connected by open passages or the ostia. An in vivo manometric study in 50 healthy subjects demonstrated that during inhalation, negative pressure is created within the nasal cavity [35]. This results in a Venturi effect whereby air flows from the sinuses and mixes with inhaled air resulting in a reduction of the velocity of inhaled air and laminar flow even with deep inhalation. The temperature of inhaled air is likely to be regulated by a similar mechanism.

The ostia act to reduce local pressure distribution to the paranasal sinuses [11]. A small amount of airflow through the ostia helps to maintain nitric oxide (NO) and humidity exchanges [11, 35, 41]. NO is produced in the upper airways mainly by the mucosa of the nose and paranasal sinuses [85]. A stable concentration of NO contributes to the human host defence by its bacteriostatic and antiviral effects and its impact on mucociliary clearance [19, 75, 93]. NO is also likely to have a role in the local immune responses, although this has not been well defined [85]. On the other hand, patients with CRS have reduced levels of exhaled NO and elevated levels of NO and NO metabolites which may contribute to the pathogenesis of CRS [85].

Airflow and local circulations in the normal sinuses during the respiratory cycle have been modelled using CFD techniques [109, 122]. The direction of flow changes according to the phases of the respiratory cycle. Swirling patterns of circulation have been identified in the maxillary sinuses [100, 126]. The velocity distributions in unoperated sinuses are stable during the inspiratory, expiratory and quiet phases [11, 53]. On the contrary, patients who have undergone functional endoscopic sinus surgery (FESS) have different vortices and regions of maximal velocity which fluctuate between inspiration and expiration [11, 100]. Sinus aeration is enhanced following standard FESS, particularly in the maxillary and ethmoid sinuses, but not in the frontal sinuses [11, 31, 53, 101, 122]. These changes are mainly attributed to the larger size of the respective openings (ostia) after surgery. Velocity distributions at selected openings for a comprehensive FESS model are displayed in Fig. 9.3. When more extensive surgery is performed to increase the patency of the frontal ostia, such as in a modified Lothrop procedure, greater airflow also occurs to the frontal sinuses [53, 100]. The relationship between ostium size and corresponding sinus inflow appears to be strongest for the maxillary and sphenoid sinuses [100]. In addition, the rate of inhalation airflow impacts the intensity of airflow more greatly following FESS, and this remains true even during quiet phases where airflow is continuous and velocity magnitudes are six to eight times higher due to inertia effects [11, 100]. This

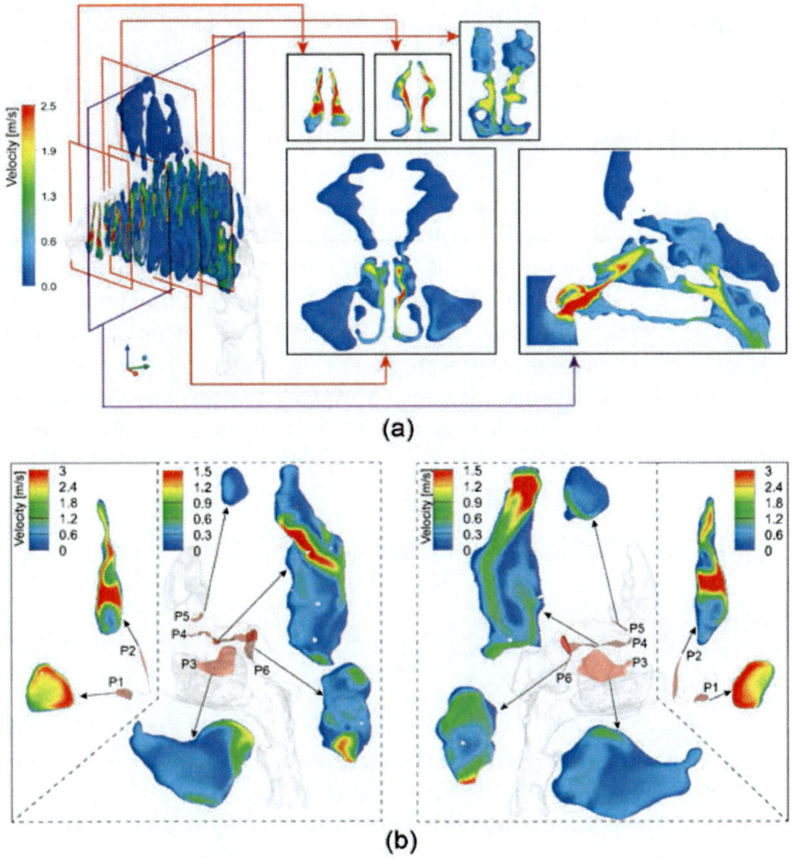

Fig. 9.3 Velocity magnitude contours for comprehensive FESS model; inhalation flow rate = 15 L/min **a** Velocity distributions in selected coronal and sagittal sections of the nasal cavity in coronal and sagittal planes **b** Velocity distributions at selected openings for comprehensive FESS model displayed from left lateral aspect (pictured left) and from right lateral aspect (picture right); inhalation flow rate = 15 L/min P1 = nostril; P2 = nasal valve; P3 = opening of maxillary sinus; P4 = opening of ethmoid sinus in axial plane; P5 = opening of frontal sinus; P6 = opening of sphenoid sinus

suggests that with increasing extensive surgery, the sinus and nasal cavity become more interconnected and functionally interdependent. The changes in aerodynamic patterns in post-FESS patients are likely to impact not only topical drug deposition in the sinuses [101, 119] and also nasal physiological functions including thermal and humidity regulation [35, 41]. Gas exchange as measured by NO redistributions are influenced by airflow and circulation in the sinuses following FESS [41, 85].

9.1.3 *Summary*

Predominantly laminar flow occurs in the nasal cavity during restful breathing. Major flow paths occur through the middle meatus, and the nasal airway medial to the inferior and middle turbinates. The smooth and even flow of air through most regions of the nasal cavity assists respiration, heat exchange between air and the mucosa, the perception of nasal patency and the diffusion of odour molecules to the olfactory region. Despite the airflow exchanges occurring between the nasal cavity and paranasal sinuses being very small during calm breathing, a functional interdependence exists as they are connected by the ostia. With increasing extensive surgery, the sinus and nasal cavity this relationship is enhanced. The changes in aerodynamic patterns in post-FESS patients are likely to impact topical drug deposition in the sinuses, bacterial and allergen deposition and nasal physiological function.

9.2 Fluid and Particle Dynamics from Sniffing

Hadrien Calmet and Alister Bates

9.2.1 *What Is Sniffing?*

Sniffing involves a shorter breath that is both rapid and with high acceleration. It can be quantified by measuring intra-nasal pressure or flow or air and is often related to increase odour perception for humans [54]. There have been a number of studies of sniffing focussing on odour reception [125], but it also has influence on inhalation exposure to particles and drug delivery. This section describes some recent work related to the consequences of sniffing on both airflow and particle deposition.

Inhaled air traverses a series of bends and changes in passage area as it passes via the nose to the early generations of the bronchi (Fig. 9.4). The speed of the air and the width of the passages vary greatly, and in most parts, the Reynolds number of the flow is at least in the several hundreds, so that a wide range of airflow patterns occur, influencing local physiological function.

Many studies have adopted a steady flow based on the assumption of a quasi-steady flow phenomena because the time scale of a breath is sufficiently long during quiet, restful breathing and therefore flow inertia does not excessively dominate viscous forces. This is quantified through the Womersley parameter, α (defined as $\alpha = R(\omega/\nu)^{1/2}$, where R is a length scale, ω is the angular frequency and ν is the kinematic viscosity). With increased frequency ventilation or higher flow rates, differences between quasi-steady and unsteady simulations have been observed. Horschler et al. [44] found significant hysteresis in the pressure—flow relationship in accordance with the low-frequency flow variation during the breathing cycle, while Doorly [27] found that even at low flow rates, a disturbed laminar flow behaviour was apparent. Rapid acceleration of the flow during inhalation increases the frequency content

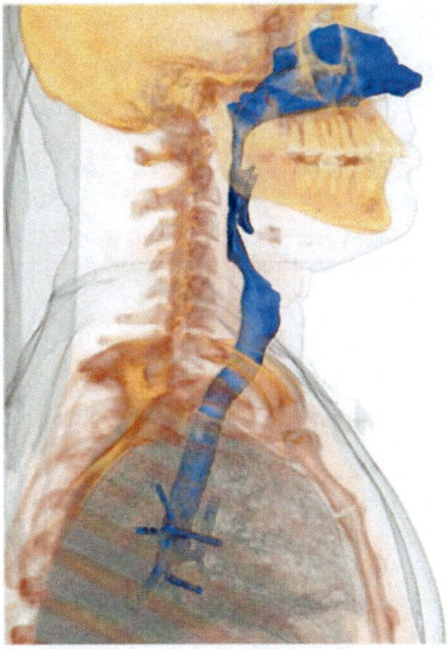

Fig. 9.4 Volume rendering of a neck and chest CT scan with the segmented airway surface highlighted showing the upper respiratory airway relative to the body

of the flow waveform and thus the value of α. Air flow in a rapid inhalation is of physiological interest to understand sniffing for olfaction or during manoeuvres commonly recommended when taking aerosol medication. It also reveals the time scales over which flow responds.

9.2.2 Modelling

The computational model was reconstructed from a CT scan of a 48-year-old male subject. The scan was performed in the supine position and the resulting data set comprised of 912 images in the axial plane, with 1 mm slice thickness and in-plane 0.65×0.65 mm pixel size. An unstructured mesh was generated using an octree-based method for the surface, while the Delaunay method used to fill the volume with tetrahedral cells with a smooth cell transition ratio of approximately 1.2. a multi-layer prism mesh was created to resolve the high velocity gradients at the wall, with a few pyramids needed to ensure the transition between the prism layer and tetrahedral mesh in the core of the computational domain. The flow field was based on the incompressible Navier–Stokes equation, and solved using an in-house code Alya. The numerical model was based on a stabilized finite element method using the Variational MultiScale method which is considered as an implicit Large Eddy Simulation method.

The computational domain (in order of inspired flow direction) comprised a hemisphere of the subject's face, followed by the bilateral nasal passages, through to the pharynx, larynx, trachea and down to the third branch generation in the lungs (Fig. 9.5.

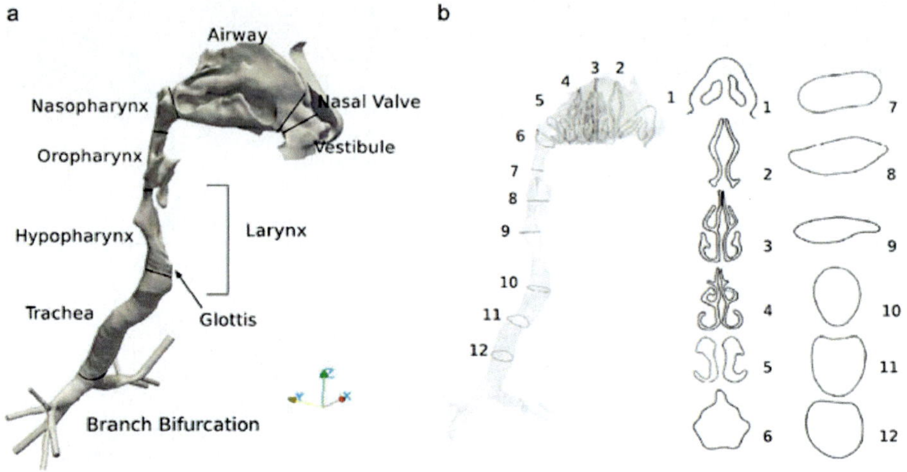

Fig. 9.5 **a** Nomenclature of the approximate regions of the upper airways. **b** Location of different slices along the airways. **c** Locations of sections and points used as measurement locations for the results. Positions of planes within the airways are: plane A: coronal through nasal cavity, plane B: sagittal through right nasal cavity, plane C: sagittal through descending airways and plane D: coronal through supra-glottic region

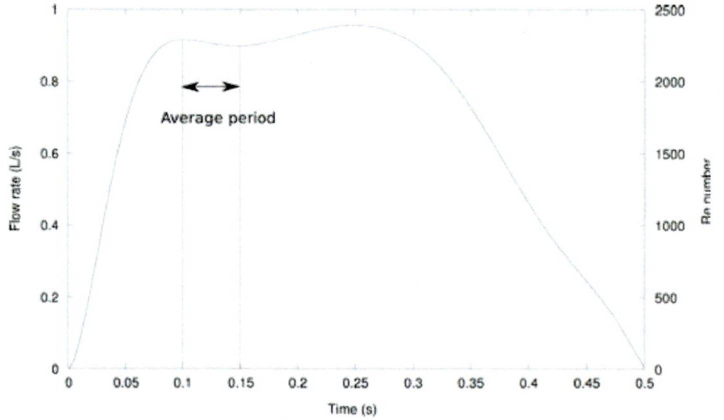

Fig. 9.6 Flow rate profile with the local Reynolds number and average period

The inflow velocity was imposed as a Dirichlet condition on the hemisphere dome. It was prescribed as a time varying uniform velocity with direction normal to the hemisphere. The radius of the hemisphere was 0.5 m and was chosen to be large enough to avoid the inlet boundary conditions affecting the flow field within the airways on expiration. To model the sniff, a polynomial function of order 10 was used (Fig. 9.6) derived from the experimental work detailed in Rennie et al. [90].

9.2.3 *Sniffing Airflow Behaviour*

A sniff is a short, transient and unsteady inspiration, which can be divided into three phases: acceleration [0–0.1 s], steady [0.1–0.4 s] and deceleration [0.4–0.5 s]. Each phase results in a distinct flow field, largely due to the resulting forces driving the inspiration. Velocity contours on four different slices (described in Fig. 9.2) with the time average velocity and the instantaneous velocity u. It is apparent that the mean velocity does not differ significantly from the time instant one within the nasal passages, however in the descending tract a noticeable difference is evident (Fig. 9.7).

The larynx is a highly complex valve involved in breathing, producing sound and protecting the entrance to the bronchial tree. It exhibits a tight constriction and expansion geometry which leads to disturbed flow and regions of turbulence. This can be seen through the location of vortical structures in the pharyngolaryngeal region in Fig. 9.8. There is a transition of the vortical structures' direction, initially tangentially aligned with the flow direction in the nasopharynx region, while further downstream in the larynx, the flow turbulence has undergone a cascade from larger to smaller structures which become aligned transversally to the flow direction. At the nasopharynx there is a severe stenosis that forms a jet with longitudinal vortical structures, which enters the oropharynx where there is a sudden expansion in the cross-sectional area of the passage. The fluid mechanics behaviour of free jets involves Kelvin–Helmholtz instability, which occurs when there is a velocity difference across the interface between the higher velocity jet and the surrounding air into which it discharges. These instabilities promote a cascade from large vortex structures to smaller ones, resulting in a turbulent flow.

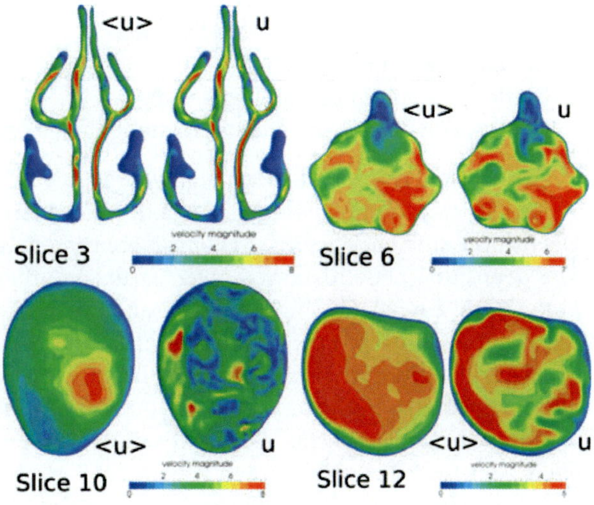

Fig. 9.7 Different cross-sectional slices of the airways, referenced in Fig. 9.2, the time average velocity (computed for the time period [0.1–0.15 s]), and u the instantaneous velocity (0.15 s)

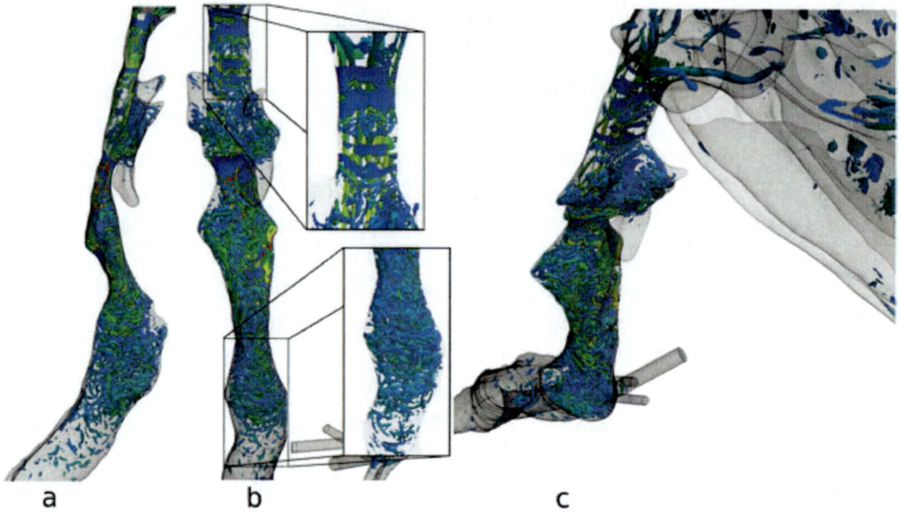

Fig. 9.8 Iso-surface of the Q-criterion (value=), colored by the angular velocity, in the pharyngo-laryngeal region at. **a** Lateral view. **b** Front view. **c** Isometric view

9.2.4 Particle Deposition

The transport of particles were simulated in a Lagrangian frame of reference with the following assumptions:

- Particles are assumed sufficiently small to neglect their effect on the air: therefore, a one way coupling is considered;
- Particles do not interact with each other;
- Particle rotation is neglected;
- Thermophoretic forces are neglected;
- The forces considered are drag gravitational and buoyancy.

Figure 9.9 shows the deposition pattern for 1, 5, 10, 20 μm particles during sniffing. During the *acceleration phase*, the inlet flow rate increased from zero to 1 L/s in 0.1 s. Deposition efficiency increased with increasing flow rate and particle diameters. The deposition pattern showed lager particles (green and blue spheres) deposited in the nasal cavity. The higher inertia and deviation from the flow streamlines of larger particles during acceleration significantly affected the deposition. Deposition of smaller particles (red and orange spheres) were scattered along the respiratory tract, particularly in the laryngeal region. During the *plateau phase*, the flow rate was approximately constant during for 0.2 s. The deposition fractions were almost equivalent for lager particles and were located exclusively in the nasal cavity. The deposition fraction for larger particles were found in the nasal cavity, while for

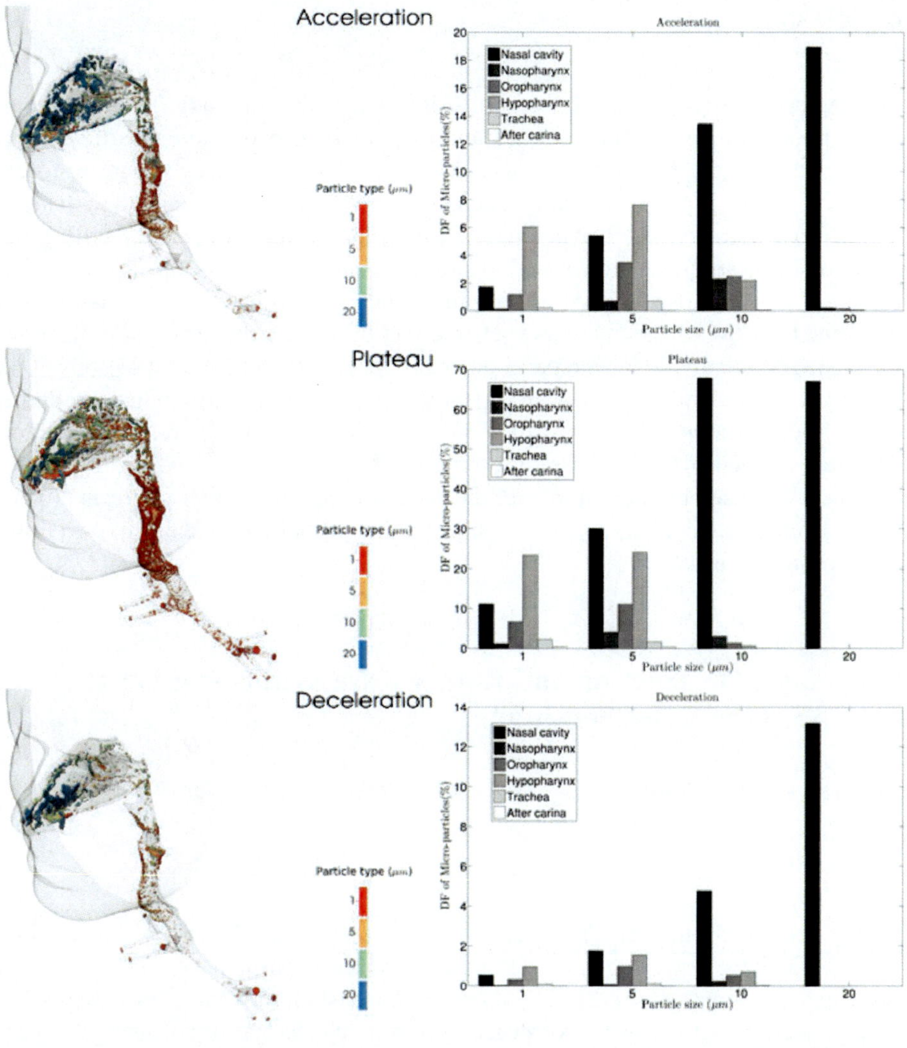

Fig. 9.9 Deposition patterns and regional deposition fraction for the three phases of the sniff: acceleration, plateau and deceleration

smaller particles were scattered along the respiratory tract with a higher value in the hypopharynx. The regional deposition fraction for the plateau phase is proportional for the entire duration of the sniff. During the deceleration phase particle deposition was the lowest of the three phases.

9.2.5 *Summary*

Sniffing produced different flow features of in particular unsteady flow that occurs in the upper and descending airways and as a result affects micron particle deposition. An inspiratory waveform was selected as an extreme case of strong forced breathing, leading to complex fluid mechanics, which is often associated with methods of drug delivery or other behaviours, for example short repeated sniffing is a common inspiration waveform used to smell objects. Analysis of the airflow within the nasal cavity showed flow partitioning, which was affected by the nasal valve directing the flow far upstream as a jet, and the turbinate structures that project into the passages and divide the airspace, diverting the airflow to different regions of the cavity. The turbulent flow in the laryngeal region illustrated the complex configuration vortical structures. This was the dominant site of turbulence production, more precisely located in the supra-glottic region. Micron particle deposition was predominantly caused by inertial impaction. As a consequence larger particles (10 and 20 μm) deposited in the nasal cavity while smaller particles (1 and 5 μm) penetrated further into the airway.

9.3 Nasal Obstruction and Empty Nose Syndrome: What Are Our Noses Sensing?

Chengyu Li and Kai Zhao

9.3.1 *Introduction*

Empty nose syndrome (ENS) is a rare but debilitating clinical syndrome that affects people who have undergone endoscopic sinus surgery, particularly those who have undergone turbinectomy [95]. The most common symptom for ENS patients is a paradoxical sensation of nasal obstruction despite a clear nasal passages [42]. ENS patients also suffer from varying degrees of nasal crusting, dryness, nasal discharge, and facial pain. In some severe cases, ENS patients complain about sleeping difficulty, disrupted concentration, elevated anxiety, and a constant feeling of suffocation [57]. Although ENS is a rare rhinologic disorder, it can severely affect the patients' quality of life, with many patients feeling suicidal due to its horrific symptoms.

9.3.1.1 Clinical Evaluations

So far, there is still a lack of reliable objective measures to quantify these symptoms. The diagnosis of ENS in clinical practice mainly relies on subjectively reported symptomology. The Empty Nose Syndrome 6-item Questionnaire (ENS6Q) has been validated to identify patients who are suspected as developing ENS [112]. Participants can rate their perceived nasal symptoms from "0" (not applicable) to 5 (extremely severe) for six different symptoms, including "dryness", "sense of diminished nasal airflow (cannot feel air flowing through your nose)", "suffocation", "nose feels too open", "nasal crusting", and "nasal burning". A score of 11 out of a possible total of 30 was determined as the cut-off criterion to predict ENS. In addition, the cotton test has also been used to aid in the diagnosis of ENS [42, 110]. This test involves placement of a dry cotton plug into the region of the inferior lateral nasal wall to restore turbinate-associated tissue bulk. The ENS patients report improvement in their nasal breathing within minutes in the office, while patients with non-ENS sinonasal conditions typically present insignificant difference between pre-cotton and post-cotton.

9.3.2 Image-Based Computational Fluid Dynamics (CFD) Simulations

The precise pathogenesis of ENS is poorly understood. Altered nasal aerodynamics is often suspected as a major contributing factor. However, experimental measurements of nasal airflow can be challenging and impractical during clinical visit. Instead, image-based computational fluid dynamics (CFD) simulation is a nonintrusive alternative to achieve quantitative evaluation of nasal aerodynamics for the individual patient (as the example shown in Fig. 9.10). The simulated data, such as local airflow velocity, breathing flow rate, wall shear stress along the mucus surface, can be synthesized to generate "virtual data" that can be used improve our understanding of ENS.

To obtain accurate nasal airflow patterns, anatomically correct models of the nasal cavity need to be reconstructed based on MRI or CT scans [127, 128]. It allows defining the boundaries of fluid domain to solve the governing equation of fluid motions—the Navier–Stokes equations. Even if we have the governing equations to resolve the nasal airflow, the numerical simulation is not a plug-and-chug work. Depending on the local Reynolds numbers (Re) in nasal cavity, nasal airflow patterns may undergo different flow regimes from laminar flow to turbulent flow. At a moderate flow rate, the entrance flow from nares is generally considered as laminar flow, which accelerates in the nasal vestibule to become transitional flow, gradually developing into turbulent flow due to the reduction of cross-sectional area [27]. A transition to turbulent flow usually takes place at a local Re of approximately 2000. Depending on the complexity of the nasal airway, the transition to turbulence may occur at a Re lower than this critical value. Thus, an accurate prediction of these airflow properties requires careful selection of CFD models. The CFD models that have been commonly adopted for simulating nasal airflow include laminar model [18, 50, 113] and Reynolds-averaged Navier–Stokes (RANS) models (including $k-\varepsilon$ model [109, 125], standard $k-\omega$ model [123], Shear Stress Transport (SST) $k-\omega$ model [83]. With the advancement of high performance computing, in recent

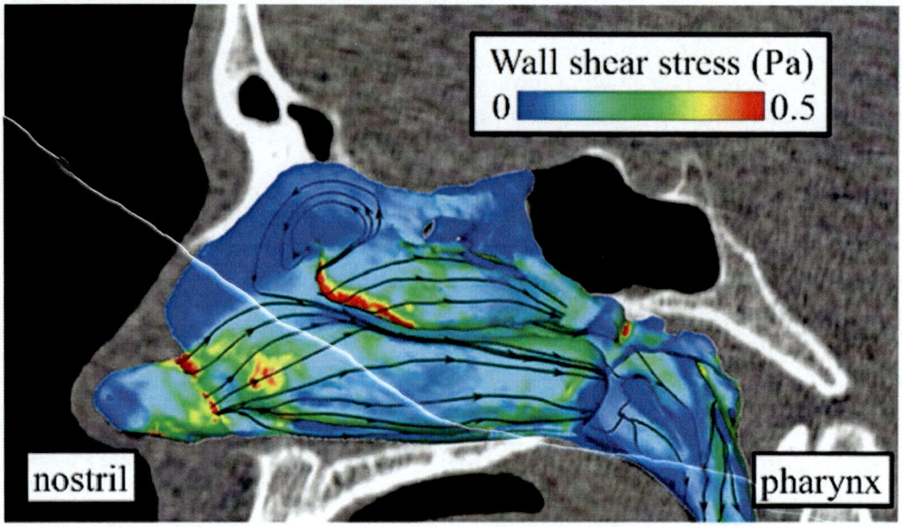

Fig. 9.10 CT-based CFD modeling of nasal airflow. Sagittal view of right nasal airway color-coded by wall shear stress and overlapping with airflow streamlines

years, many groups have considered large eddy simulation (LES) as an alternative approach for modeling airflow in human respiratory systems [65, 121]. Unlike the RANS approach, LES can directly solve the equations that describe the evolution of a large range of turbulence scales. Only the smallest scales are modeled using sub-grid-scale (SGS) models. Direct numerical simulation (DNS) does not require any turbulence models to resolve turbulent fluctuations if the time and length scales used are small enough and is often considered as the golden standard in numerical simulations [62, 114]. In general, LES and DNS can provide accurate predictions of the nasal airflow under various flow conditions but with high computational cost. The laminar model can achieve decent predictions for restful breathing condition, but not for high breathing flow rates. Among the commonly used RANS models, the $k - \omega$ model represents the best overall performance at high and moderate breathing flow rates in terms of accuracy and computational cost [62].

9.3.3 Menthol Lateralization Detection (LDT) Thresholds

As a byproduct of nasal surgery, the impaired nasal sensory function has been suspected as another underlying pathogenesis of ENS. Measurement of menthol lateralization detection (LDT) is a common assessment of nasal trigeminal sensitivity. Similar to other trigeminal agonists, menthol produces both an odor (via olfaction) and a cooling sensation (via the TRPM8 trigeminal pathway). The mechanism of this test is based on the fact that non-irritating odorants cannot be lateralized, and thus the lateralization threshold is a measure of nasal trigeminal sensitivity to volatile compounds that are independent of the chemical's effect on the olfactory system [59, 60].

In the LDT test, menthol is diluted into mineral oil in a 20-step binary dilution series with each step diluting the previous solution to half. In each trial, the subject simultaneously sniffs from a pair of bottles, one in each nostril. One bottle has Menthol and the other bottle is blank with only mineral oil. The subject is required to identify which side contains Menthol. The trials are conducted based on a force-choice ascending method of limits [3]. Each stimulus concentration is presented once at most, in ascending concentrations. Testing ends when the subject: (1) has responded correctly in four consecutive trials and (2) is confident that the last response is correct (equal to or above 7 on a confidence rating of 0–10). The latter criterion represents an attempt to reduce the chances of false positive, as four consecutive "lucky guesses" is not a rare event (6.25%). The dilution bottles are numbered 1–20 accordingly. The threshold can be determined as the first of the four consecutive correct trials and reported in dilution step or bottle number.

9.3.4 Clinical Studies of Empty Nose Syndrome Patients

Historically, ENS has been a poorly understood iatrogenic disorder, and its diagnosis heavily relied on clinical suspicion and physical examination. The underlying pathogenesis has been attributed to possible radical surgery and poor mucosal wound healing [12]. It has been reported that a smaller inferior turbinate volume of ENS patients was significantly associated with their syndrome scores [40]. It makes a sort of sense at the first glance, since the inferior turbinate plays a critical role in providing resistance to airflow and regulating airflow distribution [58]. However, if the degree of turbinate reduction could be treated as a good predictor of ENS symptoms, then the clinical surgeons and scientists would not have been stumped by this medical mystery for decades. Most post-turbinate surgery or even post-turbinectomy patients will not continue to develop ENS [79]. Millions of turbinate surgeries are performed world-wide with a vast majority of patients having desirable outcomes—why would a very small subset of patients unfortunately develop ENS?

Among ENS patients' data in literature [59, 60, 73, 74, 79], the one common nasal aerodynamic distortion was the congregated nasal airflow forming into a narrow band in the middle meatus region, leaving the airway surrounding the inferior turbinate with little or no airflow (see Figs. 9.11 and 9.12). The distortion in aerodynamics among ENS patients seems paradoxical, as inferior turbinate reduction should open up the airway surrounding the inferior turbinate, thus allowing more flow through that region. On the contrary, among ENS patients, the flow rate and intensity surrounding the inferior turbinate significantly decreased after reduction surgery. This is opposite to the common assumption concerning the therapeutic mechanism of inferior turbinate reduction. We know that airflow is driven not only by the path of least resistance, but also by its momentum. The post-surgery ENS patient may be

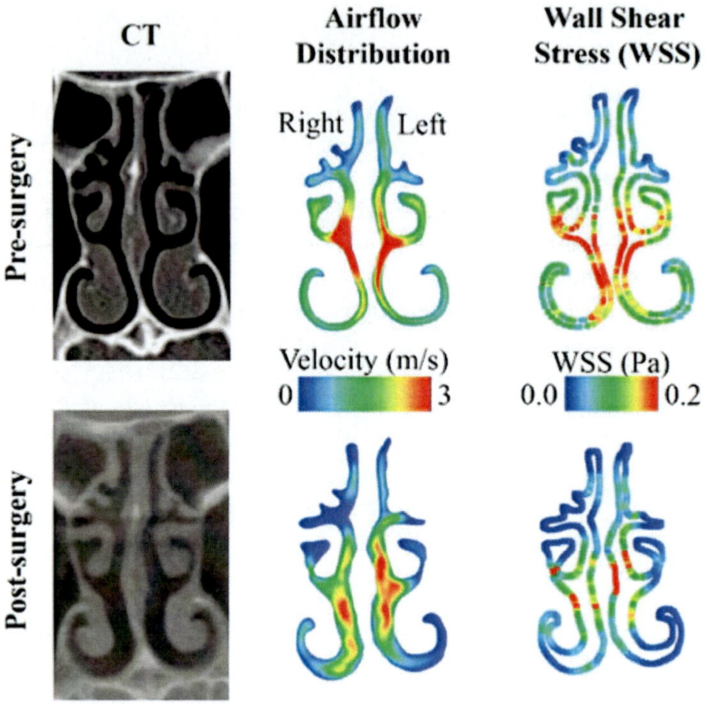

Fig. 9.11 Pre-surgery and post-surgery data of an ENS patient. Post-surgery CT shows a bilat-eral turbinate reduction. CFD analysis shows the post-surgery nasal airflow (2nd column) congregates towards the middle meatus and leaves some inferior and lateral airway region with no airflow (blue color = 0 air velocity). In the 3rd column, wall shear stress (WSS), the shear force of airflow exerted onto the nasal mucosa, is much lower surrounding the infe-rior turbinate post-surgery (blue color = 0 WSS) as compared to that of pre-surgery. Figure is adapted from [60]

akin to a large room in which a jet stream of air is passed. Resistance is low in all directions and so the airstream maintains its original path and directly shoots towards the middle nasal airway. On the other, studies have shown patients underwent similar degree of aggressive surgery or skull base surgery that removes turbinates and septum who do not report ENS symptoms do not exhibit this paradoxical "jet flow" pattern [73, 79]. Among these patients, the nasal airflow increases more uniformly, including the inferior turbinate region, as one would expect. This provides further evidence that the paradoxically airflow distortions among ENS patients are unique to their symptomatology, beyond degree of surgery.

In general, healthy nasal turbinates serve to partition the nasal airflow uniformly throughout the nasal cavity to facilitate the normal nasal physiology. For instance, airflow peaks of healthy population would occur equally at all three nasal meatuses: inferior, middle, and superior meatuses (see Fig. 9.12). Although regions of higher or lower flow may exist, it is very rare to find regions with zero velocity (dark blue) as in the post-surgery ENS cases. A uniformly distributed nasal airflow theoretically

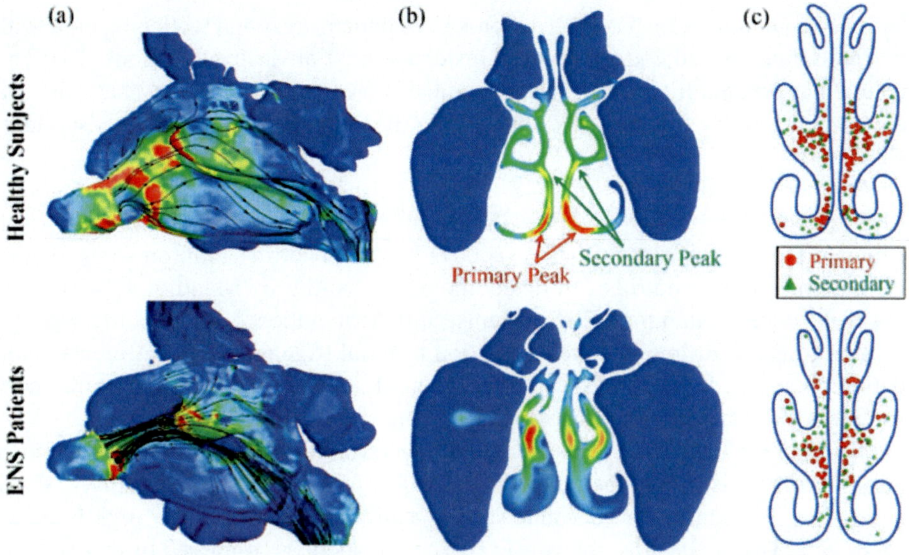

Fig. 9.12 Examples of airflow distribution for healthy subject and ENS patient. **a** Nasal cavity color coded by wall shear stress and overlapping with airflow streamlines. **b** Airflow distribution peaks from coronal view. **c** The location of all primary and secondary peaks of healthy subjects (n = 42) and ENS patients (n = 27) are color coded onto a generic cross-sectional plane: red circles for the primary airflow peaks and green triangles for secondary peaks, respectively. For the healthy subjects, the airflow peaks were widely spread out in the inferior meatus, middle meatus, and lower/upper common meatus (between the turbinates and septum). For the ENS patients, the airflow peaks shifted upward, congregated above the inferior turbinate. Figure is adapted from Li et al. [59]

improves efficiency, resulting in an enhanced capability to warm, humidify and filter the inspired airflow, while preventing unwanted desiccation of the mucosa. Many ENS patients complain excessive nasal dryness or crusting, potentially due to their over-congregated nasal airflow that over-stresses a small patch of the nasal mucosa.

In addition, the more widely distributed airflow could also improve the capacity of the nasal mucosal to detect airflow. It has been reported mucosal detection sensitivity among a normal population has a significant regional variability [14, 120], and this is also applicable to ENS patients. Theoretically, assuming normal sensitivity of at least some portion of nasal mucosa, a widely-distributed nasal airflow pattern would lead to the perception of normal airflow via receptors in that region. Thus, patients with a normal widely spread airflow pattern may have better "opportunity" to sense available flow. Conversely, patients in whom the airflow pattern is directed to a small subsite of the nasal cavity, or in whom a significant portion of the nasal mucosa is either removed or insensate (due to scar, poor healing or desiccation), have potentially less "opportunity" to sense airflow, even in the absence of significant nasal resistance. The paradoxical sensation of nasal obstruction in ENS patients may be related to a combination of loss of neural sensitivity and reduction in the surface area stimulated by mucosal cooling [23, 124, 126]. Some ENS patients respond well to "cotton test" when a saline-soaked cotton is placed inside the nose at various locations [74]. The cotton may potentially re-distribute the nasal airflow to mucosa regions that still have the normal sensory function, and thus enable the patient to regain their sensation to nasal airflow.

Another key contributor of ENS is the nasal neurosensory impairment among ENS patients. Study showed that ENS patients had significant lower Menthol LDT compared to healthy controls ($p < 0.0001$) [59]. Converting dilution steps to concentration, healthy controls can detect the nasal cooling by Menthol at ≈50 times lower in concentration than ENS patients. Sufficient evidence suggests that the perception of nasal airflow involves activation of nasal trigeminal cool afferents when cool ambient air is inspired [29, 124]. Menthol activates TRPM8 trigeminal cool receptors [6], thus impaired Menthol LDT may be indicative of patients' lack of sensation of airflow. The consistency and low variability in Menthol LDTs among healthy control is a demonstration of the validity of this test. Surprisingly, study has shown that aggressive turbinate surgery patients w/o ENS had a slightly better Menthol LDT than healthy controls [73]. These evidences combined further suggest that the paradoxical sensation of nasal obstruction in ENS patients may be related to a combination of nasal airflow distortion and loss of neural sensitivity.

9.3.5 Conclusion Remarks and Future Directions

To advance our understanding of ENS, we need to adopt nasal aerodynamics and neurosensory measures to facilitate our diagnostic assessment in addition to conventional clinical examination. The integrative study combining CFD simulations and trigeminal sensory examinations is an effective approach to explore the etiology of ENS. The findings in literature suggest that the development of ENS is possibly related to multiple factors, when aligned, result in this devastating disease. Such factors include: surgical impact, distorted nasal aerodynamics, as well as the overlapping neurosensory losses.

The exact causes of the sensory loss remain unknown. However, the distorted aerodynamics in ENS patients may have an impact on mucosal sensory recovery, with either too little simulation or too high stresses. Evidence suggested that nasal trigeminal senses undergo adaptive processes during acute or prolonged exposure and disease processes [7, 21, 97]. Future research could further explore how to promote mucosal neurosensory recovery to find alternative treatment options for ENS.

Measurement of Menthol LDT does have one pitfall, which is, it cannot probe regional mucosal sensitivity, as the whole nasal cavity is exposed to the inhaled Menthol. Mapping of mucosal sensitivity to menthol [82] or to temperature probes might find mucosal regions still with good neurosensory functions among ENS patients. Future treatment can then direct nasal airflow to these regions to restore patient's sensation of airflow.

In future, CFD simulations combined with LDT-like sensory testing could be used to further investigate the surgical impact on nasal aerodynamics and neurosensory functions, and the collective knowledge would potentially lead to optimized surgical outcome, thus preventing ENS.

9.4 Nasal Nitric Oxide (nNO) Dynamics and the Ostiomeatal Complex: Fertile Ground for CFD?

Dennis Shusterman

9.4.1 Background

Nitric oxide (NO) is an endogenously produced compound in the human body, with widespread distribution both intra- and extra-cellularly. Within the respiratory system, it is found in gaseous form intraluminally in both the upper and lower airway. Lower airway NO is measured as 'fractional excreted NO' or 'FeNO' [4]. FeNO has found clinical utility in monitoring bronchial inflammation in asthma (in which it is generally elevated) and in primary ciliary dyskinesia and cystic fibrosis (in which it is depressed) [63, 67]. Since inspired air reaching the lower airway is enriched in NO concentration as it passes through the nasal cavity, NO has been termed an 'aerocrine messenger', with potential downstream physiological effects as enumerated below [67, 72].

In the upper airway, the paranasal sinuses serve as a reservoir of NO, and nasal NO ('nNO') levels (which are generally higher than FeNO) are affected by the patency of communication between the sinuses and nasal cavities [69, 70]. Because the maxillary sinuses have the largest volume (hence, make the largest potential contribution to NO flux), a major focus of nNO studies is the patency of the ostiomeatal complex ('OMC'), which connects the maxillary sinuses to the nasal airways via the middle meatus. Because the OMC can be obstructed in inflammatory sinus disease, there is a complex relationship between upper airway inflammation (i.e., rhinitis and sinusitis) and nNO concentrations, in contrast to asthmatic inflammation and FeNO in the lower airway. This relationship is reviewed below.

The objective of this brief review is to survey the potential utility of nNO as an index of the severity of upper airway inflammatory disease, as well as the potential value of CFD in modeling NO flux in the upper airway.

9.4.2 Normal NO Physiology

Physiologically, NO exerts varied effects in the airway, including bacteriostasis, vasodilation, bronchodilation, and stimulation of mucociliary function [28, 34, 67,

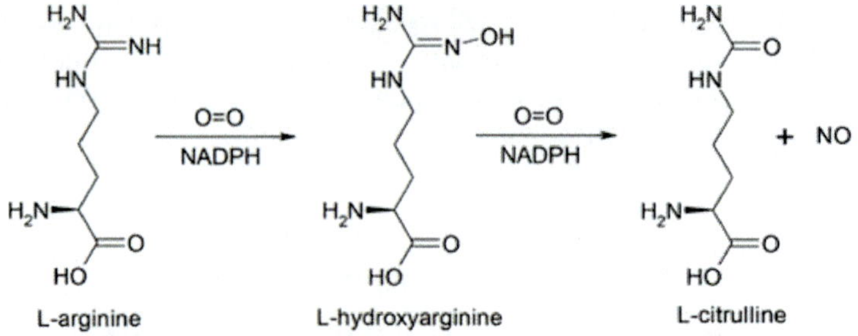

Fig. 9.13 Metabolic pathway for NO generation via NOS

94]. NO is generated, metabolically, by the action of nitric oxide synthase (NOS) on the amino acid, L-arginine (Fig. 9.13)

NOS occurs as three major isoenzymes: the constitutively expressed eNOS (endothelial) and nNOS (neural), and the inducible isoenzyme, iNOS [52]. iNOS supports higher NO output than the constitutive isoenzymes and is most often found in the presence of inflammation. However, the discovery that iNOS is expressed in healthy (non-inflamed) sinus tissue supports the concept that high sinus NO concentrations may play an essential role in normal respiratory physiology [66, 67, 71].

Given the antimicrobial and ciliostimulatory effects of NO, it has been speculated that elevated NO concentrations may help control pathogen growth in healthy sinuses [92, 94]. In the lower airway, the fact that NO delivery in inspired air is augmented during nasal breathing suggests that antimicrobial defenses might also benefit in the tracheobronchial tree and deep lung. Regarding NO's potential vasodilatory effect in the deep lung, inhaled NO has been successfully administered therapeutically in some pulmonary conditions involving vasoconstriction, although at concentrations higher than are found physiologically [37, 103].

9.4.3 *nNO: Clinical Studies*

As alluded to above, the relationship between upper airway inflammatory conditions (rhinitis and sinusitis) and nNO is more complex than is the relationship between bronchial asthma and FeNO. Since the inflammatory processes in rhinosinusitis and asthma are closely related, this "disconnect" in the upper airway merits investigation. In exploring this issue, we will first consider currently documented relationships between upper airway inflammatory conditions and nNO.

(1) Allergic Rhinitis and nNO:
Table 9.1 summarizes published studies of nNO levels sampled in individuals with allergic rhinitis (compared with normal controls). A majority of studies (9 vs. 3) show findings consistent with nNO as an inflammatory barometer in isolated allergic

Table 9.1 Clinical studies of nNO in relationship to allergic rhinitis

Consistent	Inconsistent
↑ nNO in rhinitis	→ nNO in rhinitis
Arnal [1]	Henriksen [39]
Kharitono [51]	Palm [88]
Baraldi [5]*	Williamson [118]
Maniscalco [76]	
Djupesland [26]	
Lee [56]	
Suojalehto [106]	
Nesic [86]	
Asano [2]	

*also nNO with steroid Tx

rhinitis. Of note, one of these studies [5] also shows a decrease in nNO after medical therapy for allergic rhinitis.

Despite a predominance of 'positive' studies, it is unclear whether nNO can be used a reliable index of nasal inflammation in uncomplicated allergic rhinitis using existing sampling techniques, as NO flux from the maxillary sinuses can overshadow (and statistically confound) inflammatory changes on the smaller flux deriving directly from the nasal mucosa. Two approaches have been suggested to neutralize this effect: (a) Recent work on regional nNO sampling within the nasal cavity suggests that there may be significant local differences in nNO. Specifically, sampling near the middle turbinate reportedly yields higher nNO levels that do not vary by rhinitis status, whereas the lower levels found near the inferior turbinate do vary by allergic rhinitis status [107, 108]. This regional difference may derive from the proximity of the middle turbinate sampling site to the middle meatus (i.e., sinus ostium). (b) Deliberate depletion of maxillary sinus NO (through repeated humming maneuvers) has been shown to lower sampled nNO levels, and to decrease variability in their measurement. It has been suggested that this may unmask rhinitis-related changes in NO flux from the nasal mucosa which might otherwise be overshadowed by flux from the sinuses [76].

(2) Rhinosinusitis and nNO:
Virtually all studies of nNO in chronic rhinosinusitis (CRS)—with or without accompanying polyposis—show paradoxically low NO levels compared to normal controls, despite the presence of extensive upper airway inflammation (Table 9.2). Similarly, intervention—both medical and surgical—in CRS patients produces a paradoxical increase in nNO (Table 9.3).

The consensus interpretation of the above studies is that chronic inflammation—in particular with ostial occlusion results—in lower sampled nNO due to impaired NO flux between the maxillary sinuses and the nasal cavity. Reversal of occlusion, either by surgical or medical measures, restores this flux and paradoxically raises nNO.

Table 9.2 Paradoxically low nNO in rhinosinusitis

With polyposis	With or without polyposis
Arnal [1]	Dabholkar [20]
Colantonio [16]	Liu [64]
Bommarito [8]	
McKinlay [80]	
Jeong [46]	
Frendo [33]	

Table 9.3 Paradoxical increases in nNO after sinusitis therapy

Medical Tx	Surgical Tx
Colantonio [16]	Colantonio [16]
Ragab [89]	Ragab [89]
Delclaux [24]	Deroee [25]
Dabholkar [20]	Lee [55]

9.4.4 Studies of NO Dynamics in the Upper Airway

(1) Cannulation Studies
Several studies have directly documented elevated NO levels in the maxillary sinuses. In the earliest cannulation (sinus puncture) study, sinus NO levels ranged between approximately 900–23,000 ppb, compared to nasal NO levels in the 100–200 ppb range. Subsequent displacement of sinus contents by injecting clean air through the punctum produced a transient ≈5-fold increase in nNO sampled from the nasal cavity [70]. In a follow-up study, Lundberg and colleagues [68] replicated their initial finding and further documented a profound suppression of sinus NO levels after sinus lavage using a solution containing L-NAME (a NOS inhibitor) versus saline control, thereby confirming the enzymatic origin of sinus NO. More recently, real-time aspiration of intra-sinus NO with non-occlusive OMC cannulation showed breathing-related cyclic variation in the 200–700 ppb range, consistent with a sinus reservoir model (as diluted by makeup air entering around the cannula; [38]).

(2) Acoustic Stimulation Studies
For nearly two decades it has been recognized that acoustic stimulation of the upper airway (e.g., by humming) transiently increases nNO levels [69, 116]. During humming, real-time nNO sampling typically demonstrates a 'spike' (Fig. 9.2). The underlying mechanism is thought to be increased gas mixing between the sinus and nasal airways via a 'Helmholtz resonator' phenomenon [76]. Repeated humming maneuvers produce NO 'washout', blunting the nNO spike absent a recovery period [78]. Acoustic stimulation of upper airway NO flux has been modeled mechanically, with

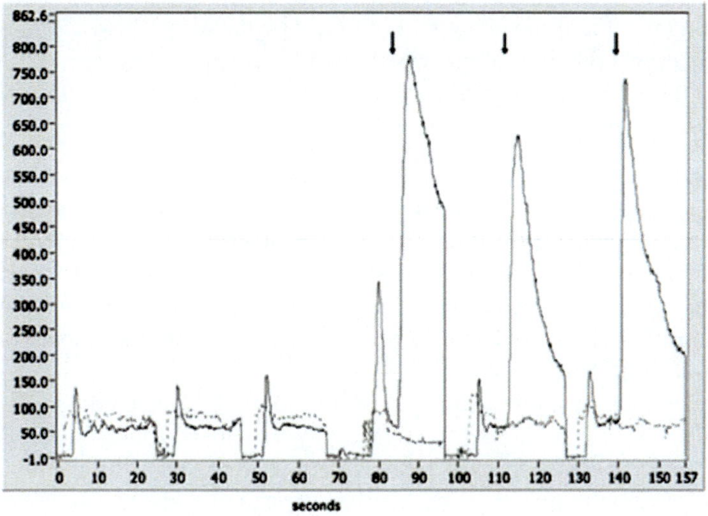

Fig. 9.14 nNO during steady nasal exhalation under quiet (no arrows) and acoustically stimulated (arrows) conditions. Flow (target, 50 mL/s); NO concentration (parts-per-billion) [98]

Table 9.4 Studies comparing quiet and acoustically stimulated nNO

Study	CT or endoscopy	Quiet nNO versus obstruction	Humming nNO versus obstruction
Maniscalco, [77]	Endoscopy	N.S.	Significant
Vaidyanathan, [111]	Endoscopy	N.S.	Significant
Shusterman, [99]	CT	Significant	N.S.
Oliver, [87]	CT	Significant	N.S.

blunting of the effect occurring with decreasing aperture ('ostium') diameter, albeit on a non-linear basis [78] (Fig. 9.14).

A total of four published studies have compared baseline (quiet) and acoustically stimulated (humming) nNO levels in human subjects who have had measures of OMC patency (CT or endoscopy). Results have been mixed and selective, with both CT studies linking decreased OMC obstruction with baseline (quiet) nNO, and both endoscopy studies linking OMC obstruction with blunting of acoustically stimulated nNO (Table 9.4):

An early goal of the acoustic stimulation line of investigation was to develop a non-invasive measure of OMC obstruction, potentially serving as an alternative to more invasive procedures during initial screening for CRS [71]. So far, this objective has only been partially realized, likely due to: (1) limitations in the sampling or computational techniques employed, or (2) limitations in the underlying biological rationale. From a sampling standpoint, inter-individual variability in humming frequency and intensity has been addressed by using an external source of acoustic stimulation (as pilot tested in a group of normal volunteers; [96]). From a computational standpoint, although simple OMC cross-sectional area models (i.e., predicting diffusion) have been employed to-date, potential utilization of CFD models could account for convective mixing as well.

(3) CFD Studies

CFD applies knowledge of boundary layer phenomena, shear forces and turbulence to model mass flow, heat flux, and mixing in geometrically defined fluid systems. It has found application in rhinology in the modeling of airflow resistance, localized cooling and perceived nasal patency, elastic airway changes, and vapor and aerosol distribution [9, 45, 61]. It has also been used to simulate the impact of potential surgical interventions [32, 36].

To date, published studies have applied the tools of CFD to modeling gaseous (and in particular, NO) flux through the OMC. Two of these studies utilized idealized anatomic models, and two used models derived from real-life anatomy (i.e., CT scans). All dealt with modeled—rather than measured—nNO levels.

Hood and colleagues [41] reviewed earlier studies of sinus ventilation; based on these, they constructed a simple geometric model of gas exchange, with variation in the ostium diameter, length and the presence or absence of an auxiliary ostium. They concluded that diffusion from the maxillary sinuses alone could not explain observed nNO levels, and that the potential role of convection in sinus-nasal admixture (as modeled by CFD) increased with increasing ostium diameter.

In a subsequent study by the above group, Rennie et al. [91] validated a geometric model of gas exchange xenon as a tracer in a physical model. They reiterated their previous conclusion that "...natural single-ostium sinus ventilation is remarkably slow", and speculated that one functional consequence might be to preserve bacteriostatically high NO levels in the maxillary sinuses. Of note, both studies assumed constant flow (and shear forces) at the ostial opening, thereby minimizing the potential impact of cyclical pressure changes during tidal breathing.

Chung and colleagues [13] utilized a surface-rendered computational model from a normal CT scan of the upper airway and sinuses as the basis of simulating the anatomical alterations resulting from a middle meatal antrostomy (MMA). They then compared modeled NO flux between the two geometric models, taking into account cyclical pressure changes during breathing. As in previous model studies, enlarging of the ostium increased the role of convection, with a consequent increase in NO flux. Cyclic effects were significant, with a larger NO flux predicted during inspiration than expiration.

Kang et al. [47] modeled bidirectional flux of NO in the nose, taking into account both local NO production and reabsorption into the bloodstream. In modeling the phases of respiration separately, their model was able to predict cyclical pressure differences between the paranasal sinuses and nasal airway. Interestingly, however, flow velocities were computed through the maxillary sinuses based on the assumption that the presence of an auxiliary ostium is the normal case. (Although auxiliary ostia are not uncommon, the prevailing literature does not assert that they are the norm.)

To summarize, published CFD studies pertaining to NO flux in the upper airway have utilized either geometrically stylized models or surface-rendered computational models from actual CT scans. In at least one case, a model has been validated utilizing a tracer gas other than NO (i.e., xenon). In no case have individual anatomical models been compared with individual measured nNO levels. Combined in vivo—CFD studies would be a logical extension of existing work in this area.

9.4.5 Conclusions

Nitric oxide (FeNO) is a reliable barometer of inflammation in bronchial asthma, and has achieved accepted clinical status as a tool for managing asthmatic patients. In the upper airway, by contrast, increases in nitric oxide (nNO) due to local mucosal inflammation (e.g., in allergic rhinitis) can be masked by the contribution of NO from the paranasal sinuses. NO flux from the sinuses to nasal airway is paradoxically reduced in chronic rhinosinusitis (particularly when accompanied by polyposis), and in fact obstruction of the OMC by mucosal swelling (and/or nasal polyposis) can lower nNO below the levels found in normal subjects. Quantitative modeling of NO flux in the upper airway is therefore challenging, and simple diffusion models based on OMC diameter may be incomplete. In this situation, CFD may prove useful, particularly in considering such newer metrics as the nNO response to acoustic stimulation.

References

1. J.F. Arnal, P. Flores, J. Rami, M. Murris-Espin, F. Bremont, I.A.M. Pasto, E. Serrano, A. Didier, Nasal nitric oxide concentration in paranasal sinus inflammatory diseases. Eur. Respir. J. **13**(2), 307–312 (1999)
2. T. Asano, M. Takemura, Y. Kanemitsu, M. Yokota, K. Fukumitsu, N. Takeda, H. Ichikawa, H. Hijikata, T. Uemura, O. Takakuwa, H. Ohkubo, K. Maeno, Y. Ito, T. Oguri, A. Nakamura, Y. Maki, Y. Nakamura, M. Suzuki, A. Niimi, Combined measurements of fractional exhaled nitric oxide and nasal nitric oxide levels for assessing upper airway diseases in asthmatic patients. J. Asthma **55**(3), 300–309 (2018)
3. ASTM, *Standard Practice for Determination of Odor and Taste Thresholds By a Forced-Choice Ascending Concentration Series Method of Limits* (American Society for Testing and Materials, Philadelphia, 2011)
4. American Thoracic Society, ATS/ERS recommendations for standardized procedures for the online and offline measurement of exhaled lower respiratory nitric oxide and nasal nitric oxide, 2005. Am. J. Respir. Crit. Care Med. **171**(8), 912–930 (2005)
5. E. Baraldi, N.M. Azzolin, S. Carra, C. Dario, L. Marchesini, F. Zacchello, Effect of topical steroids on nasal nitric oxide production in children with perennial allergic rhinitis: a pilot study. Respir. Med. **92**(3), 558–561 (1998)
6. D.M. Bautista, J. Siemens, J.M. Glazer, P.R. Tsuruda, A.I. Basbaum, C.L. Stucky, S.-E. Jordt, D. Julius, The menthol receptor TRPM8 is the principal detector of environmental cold. Nature **448**(7150), 204 (2007)

7. R. Benoliel, A. Biron, S.Y. Quek, O. Nahlieli, E. Eliav, Trigeminal neurosensory changes following acute and chronic paranasal sinusitis. Quintessence Int. **37**, 6 (2006)
8. L. Bommarito, G. Guida, E. Heffler, I. Badiu, F. Nebiolo, A. Usai, A. De Stefani, G. Rolla, Nasal nitric oxide concentration in suspected chronic rhinosinusitis. Ann. Allergy Asthma Immunol. **101**(4), 358–362 (2008)
9. M.A. Burgos, E. Sanmiguel-Rojas, C. Del Pino, M.A. Sevilla-Garcia, F. Esteban-Ortega, New CFD tools to evaluate nasal airflow. Eur. Arch. Oto-Rhino-Laryngol. **274**(8), 3121–3128 (2017)
10. A. Burrow, R. Eccles, A.S. Jones, The effects of camphor, eucalyptus and menthol vapour on nasal resistance to airflow and nasal sensation. Acta Oto-Laryngol. **96**(1–2), 157–161 (1983)
11. X.B. Chen, H.P. Lee, V.F. Chong, D.Y. Wang, Aerodynamic characteristics inside the rhino-sinonasal cavity after functional endoscopic sinus surgery. Am. J. Rhinol. Allergy **25**(6), 388–392 (2011)
12. N. Chhabra, S.M. Houser, Empty nose syndrome. Recent. Adv. Otolaryngol. Head Neck Surg. **5**, 75 (2016)
13. S.K. Chung, G. Jo, S.K. Kim, Y. Na, The effect of a middle meatal antrostomy on nitric oxide ventilation in the maxillary sinus. Respir. Physiol. Neurobiol. **192**, 7–16 (2014)
14. R. Clarke, A. Jones, The distribution of nasal airflow sensitivity in normal subjects. J. Laryngol. Otol. **108**(12), 1045–1047 (1994)
15. R.W. Clarke, A.S. Jones, P. Charters, I. Sherman, The role of mucosal receptors in the nasal sensation of airflow. Clin. Otolaryngol. Allied Sci. **17**(5), 383–387 (1992)
16. D. Colantonio, L. Brouillette, A. Parikh, G.K. Scadding, Paradoxical low nasal nitric oxide in nasal polyposis. Clin. Exp. Allergy **32**(5), 698–701 (2002)
17. P. Cole, Nose and sinus airflow. Curr. Opin. Otolaryngol. Head Neck Surg. **2**, 16–21 (1994)
18. C. Croce, R. Fodil, M. Durand, G. Sbirlea-Apiou, G. Caillibotte, J.-F. Papon, J.-R. Blondeau, A. Coste, D. Isabey, B. Louis, In vitro experiments and numerical simulations of airflow in realistic nasal airway geometry. Ann. Biomed. Eng. **34**(6), 997–1007 (2006)
19. K.D. Croen, Evidence for an antiviral effect of nitric oxide. J. Clin. Investig. **91**, 2446–52a (1993)
20. Y.G. Dabholkar, A.A. Saberwal, H.K. Velankar, A.K. Shetty, N.P. Chordia, S.R. Budhwani, Correlation of nasal nitric oxide measurement with computed tomography findings in chronic rhinosinusitis. Indian J. Otolaryngol. Head Neck Surg. **66**(1), 92–96 (2014)
21. P. Dalton, D. Dilks, T. Hummel, Effects of long-term exposure to volatile irritants on sensory thresholds, negative mucosal potentials, and event-related potentials. Behav. Neurosci. **120**(1), 180 (2006)
22. M. D'Amato, A. Molino, G. Calabrese, L. Cecchi, I. Annesi-Maesano, G. D'Amato, The impact of cold on the respiratory tract and its consequences to respiratory health. Clin. Transl. Allergy **8**(1), 20 (2018)
23. A. Dayal, J.S. Rhee, G.J. Garcia, Impact of middle versus inferior total turbinectomy on nasal aerodynamics. Otolaryngol.–Head Neck Surg. **155**(3), 518–525 (2016)
24. C. Delclaux, D. Malinvaud, B. Chevalier-Bidaud, E. Callens, B. Mahut, P. Bonfils, Nitric oxide evaluation in upper and lower respiratory tracts in nasal polyposis. Clin. Exp. Allergy **38**(7), 1140–1147 (2008)
25. A.F. Deroee, M. Naraghi, A.F. Sontou, M.R. Ebrahimkhani, A.R. Dehpour, Nitric oxide metabolites as biomarkers for follow-up after chronic rhinosinusitis surgery. Am. J. Rhinol. Allergy **23**(2), 159–161 (2009)
26. P.G. Djupesland, J.M. Chatkin, W. Qian, P. Cole, N. Zamel, P. McClean, H. Furlott, J.S. Haight, Aerodynamic influences on nasal nitric oxide output measurements. Acta Oto-Laryngol. **119**(4), 479–485 (1999)
27. D. Doorly, D. Taylor, R. Schroter, Mechanics of airflow in the human nasal airways. Respir. Physiol. Neurobiol. **163**(1), 100–110 (2008)
28. P.M. Dupuy, S.A. Shore, J.M. Drazen, C. Frostell, W.A. Hill, W.M. Zapol, Bronchodilator action of inhaled nitric oxide in guinea pigs. J. Clin. Invest. **90**(2), 421–428 (1992). https://doi.org/10.1172/JCI115877

29. R. Eccles, Nasal airflow in health and disease. Acta Oto-Laryngol. **120**(5), 580–595 (2000)
30. R. Eccles, S. Morris, N.S. Tolley, The effects of nasal anaesthesia upon nasal sensation of airflow. Acta Oto-Laryngol. **106**(1–2), 152–155 (1988)
31. D.O. Frank, A.M. Zanation, V.H. Dhandha,, K.A. McKinney, G.M. Fleischman, C.S.J. Ebert et al., Quantification of airflow into the maxillary sinuses before and after functional endoscopic sinus surgery. Int. Forum Allergy Rhinol. **3**(10), 834–840 (2013)
32. D.O. Frank-Ito, J.S. Kimbell, A.A.T. Borojeni, G.J.M. Garcia, J.S. Rhee, A hierarchical stepwise approach to evaluate nasal patency after virtual surgery for nasal airway obstruction. Clin. Biomech. (Bristol, Avon) **61**, 172–180 (2019)
33. M. Frendo, K. Hakansson, S. Schwer, A.T. Ravn, H. Meteran, C. Porsbjerg, V. Backer, C. von Buchwald, Exhaled and nasal nitric oxide in chronic rhinosinusitis patients with nasal polyps in primary care. Rhinology **56**(1), 59–64 (2018)
34. C.G. Frostell, Differential ventilation. Acta Anaesthesiol. Scand. **35**, 119–124 (1991). https://doi.org/10.1111/j.1399-6576.1991.tb03408.x
35. M. Gamerra, R.D. Luca, Airflow in paranasal sinuses. Biomed. Sci. Res. **1**(5), 197–201 (2019)
36. G.J.M. Garcia, J.S. Rhee, B.A. Senior, J.S. Kimbell, Septal deviation and nasal resistance: an investigation using virtual surgery and computational fluid dynamics. Am. J. Rhinol. Allergy **24**(1), 46–53 (2010)
37. P. Germann, A. Braschi, G. Della Rocca, A.T. Dinh-Xuan, K. Falke, C. Frostell, L.E. Gustafsson, P. Herve, P. Jolliet, U. Kaisers, H. Litvan, D.J. Macrae, M. Maggiorini, N. Marczin, B. Mueller, D. Payen, M. Ranucci, D. Schranz, R. Zimmermann, R. Ullrich, Inhaled nitric oxide therapy in adults: European expert recommendations. Intensiv. Care Med. **31**(8), 1029–1041 (2005)
38. A.A. Gungor, B.J. Martino, S.C. Dupont, L. Kuo, A human study model for nitric oxide research in sinonasal disease. Am. J. Otolaryngol. **34**(4), 337–344 (2013)
39. A.H. Henriksen, M. Sue-Chu, T.L. Holmen, A. Langhammer, L. Bjermer, Exhaled and nasal no levels in allergic rhinitis: relation to sensitization, pollen season and bronchial hyperresponsiveness. Eur. Respir. J. **13**(2), 301–306 (1999)
40. H.R. Hong, Y.J. Jang, Correlation between remnant inferior turbinate volume and symptom severity of empty nose syndrome. The Laryngoscope **126**(6), 1290–1295 (2016)
41. C.M. Hood, R.C. Schroter, D.J. Doorly, E.J. Blenke, N.S. Tolley, Computational modeling of flow and gas exchange in models of the human maxillary sinus. J. Appl. Physiol. **107**, 1195–1203 (2009)
42. S.M. Houser, Surgical treatment for empty nose syndrome. Arch. Otolaryngol.–Head Neck Surg. **133**(9), 858–863 (2007)
43. E.H. Huizing, J.A.M. de Groot, *Functional Reconstructive Nasal Surgery* (Thieme, Stuttgart, 2003)
44. I. Hörschler, W. Schröder, M. Meinke, On the assumption of steadiness of nasal cavity flow. J. Biomech. **43**, 1081–1085 (2010)
45. K. Inthavong, Z.F. Tian, J.Y. Tu, W. Yang, C. Xue, Optimising nasal spray parameters for efficient drug delivery using computational fluid dynamics. Comput. Biol. Med. **38**(6), 713–726 (2008)
46. J.H. Jeong, H.S. Yoo, S.H. Lee, K.R. Kim, H.J. Yoon, S.H. Kim, Nasal and exhaled nitric oxide in chronic rhinosinusitis with polyps. Am. J. Rhinol. Allergy **28**(1), e11–6 (2014)
47. I. Kang, P. Park, Numerical study on nitric oxide transport inhuman nasal airways. J. Mech. Sci. Technol. **32**(3), 1423–1430 (2018)
48. T. Keck, R. Leiacker, A. Heinrich, S. Khnemann, G. Rettinger, Humidity and temperature profile in the nasal cavity. Rhinology **38**, 167–171 (2001)
49. T. Keck, J. Lindemann, Numerical simulation and nasal air-conditioning. GMS Curr. Top. Otorhinolaryngol. Head Neck Surg. **9**, Doc08–Doc08 (2010)
50. K. Keyhani, P. Scherer, M. Mozell, Numerical simulation of airflow in the human nasal cavity. J. Biomech. Eng. **117**, 429–441 (1995)
51. S.A. Kharitonov, K. Rajakulasingam, B. O'Connor, S.R. Durham, P.J. Barnes, Nasal nitric oxide is increased in patients with asthma and allergic rhinitis and may be modulated by nasal glucocorticoids. J. Allergy Clin. Immunol. **99**(1 Pt 1), 58–64 (1997)

52. R.G. Knowles, S. Moncada, Nitric oxide synthases in mammals. Biochem. J. **298**(Pt 2), 249–258 (1994)
53. H. Kumar, R. Jain, R.G. Douglas, M.H. Tawhai, Airflow in the human nasal passage and sinuses of chronic rhinosinusitis subjects. PLOS ONE **11**(6), e0156379 (2016)
54. D. Laing, Natural sniffing gives optimum odour perception for humans. Perception **12**, 99–117 (1983)
55. J.M. Lee, C.L. McKnight, T. Aves, J. Yip, A.S. Grewal, S. Gupta, Nasal nitric oxide as a marker of sinus mucosal health in patients with nasal polyposis. Int. Forum Allergy Rhinol. **5**(10), 894–899 (2015)
56. K.J. Lee, S.H. Cho, S.H. Lee, K. Tae, H.J. Yoon, S.H. Kim, J.H. Jeong, Nasal and exhaled nitric oxide in allergic rhinitis. Clin. Exp. Otorhinolaryngol. **5**(4), 228–233 (2012)
57. T. Lee, C. Fu, C. Wu, Y. Tam, C. Huang, P. Chang, Y. Chen, M. Wu, Evaluation of depression and anxiety in empty nose syndrome after surgical treatment. The Laryngoscope **126**(6), 1284–1289 (2016)
58. T.S. Lee, P. Goyal, C. Li, K. Zhao, Computational fluid dynamics to evaluate the effectiveness of inferior turbinate reduction techniques to improve nasal airflow. JAMA Facial Plast. Surg. **20**(4), 263–270 (2018)
59. C. Li, A. Farag, G. Maza, S. McGhee, M. Ciccone, B. Deshpande, E. Pribikin, B. Otto, K. Zhao, Investigation of the abnormal nasal aerodynamics and trigeminal functions among empty nose syndrome patients. Int. Forum Allergy Rhinol. **8**(3), 444–452 (2018)
60. C. Li, A.A. Farag, J. Leach, B. Deshpande, A. Jacobowitz, K. Kim, B.A. Otto, K. Zhao, Computational fluid dynamics and trigeminal sensory examinations of empty nose syndrome patients. The Laryngoscope **127**(6), E176–E184 (2017)
61. C. Li, A.A. Farag, G. Maza, S. McGhee, M.A. Ciccone, B. Deshpande, E.A. Pribitkin, B.A. Otto, K. Zhao, Investigation of the abnormal nasal aerodynamics and trigeminal functions among empty nose syndrome patients. Int. Forum Allergy Rhinol. **8**(3), 444–452 (2018)
62. C. Li, J. Jiang, H. Dong, K. Zhao, Computational modeling and validation of human nasal airflow under various breathing conditions. J. Biomech. **64**(7), 59–68 (2017)
63. K.G. Lim, C. Mottram, The use of fraction of exhaled nitric oxide in pulmonary practice. Chest **133**(5), 1232–1242 (2008)
64. C. Liu, M. Zheng, F. He, X. Wang, L. Zhang, Role of exhaled nasal nitric oxide in distinguishing between chronic rhinosinusitis with and without nasal polyps. Am. J. Rhinol. Allergy **31**(6), 389–394 (2017)
65. M. Lu, Y. Liu, J. Ye, H. Luo, Large eddy simulation of flow in realistic human upper airways with obstructive sleep. Procedia Comput. Sci. **29**, 557–564 (2014)
66. J.O. Lundberg, Airborne nitric oxide: inflammatory marker and aerocrine messenger in man. Acta Physiol. Scand. Suppl. **633**, 1–27 (1996)
67. J.O. Lundberg, Nitric oxide and the paranasal sinuses. Anat. Rec. (Hoboken) **291**(11), 1479–1484 (2008)
68. J.O. Lundberg, T. Farkas-Szallasi, E. Weitzberg, J. Rinder, J. Lidholm, A. Anggaard, T. Hokfelt, J.M. Lundberg, K. Alving, High nitric oxide production in human paranasal sinuses. Nat. Med. **1**(4), 370–373 (1995)
69. J.O. Lundberg, M. Maniscalco, M. Sofia, L. Lundblad, E. Weitzberg, Humming, nitric oxide, and paranasal sinus obstruction. JAMA **289**(3), 302–303 (2003)
70. J.O. Lundberg, J. Rinder, E. Weitzberg, J.M. Lundberg, K. Alving, Nasally exhaled nitric oxide in humans originates mainly in the paranasal sinuses. Acta Physiol. Scand. **152**(4), 431–432 (1994)
71. J.O. Lundberg, E. Weitzberg, Nasal nitric oxide in man. Thorax **54**(10), 947–952 (1999)
72. J.O. Lundberg, E. Weitzberg, J. Rinder, A. Rudehill, O. Jansson, N.P. Wiklund, J.M. Lundberg, K. Alving, Calcium-independent and steroid-resistant nitric oxide synthase activity in human paranasal sinus mucosa. Eur. Respir. J. **9**(7), 1344–1347 (1996)
73. J. Malik, C. Li, G. Maza, A.A. Farag, J.P. Krebs, S. McGhee, G. Zappitelli, B. Deshpande, B.A. Otto, K. Zhao, Computational fluid dynamic analysis of aggressive turbinate reductions: is it a culprit of empty nose syndrome? Int. Forum Allergy Rhinol. **9**, 891–899 (2019)

74. J. Malik, A. Thamboo, S. Dholakia, N. Borchard, S. McGhee, C. Li, K. Zhao, J. Nayak, The cotton test redistributes nasal airflow in patients with empty nose syndrome. Int. Forum Allergy Rhinol. in press (2020)
75. R.L. Mancinell, C.P. McKay, Effects of nitric oxide and nitric dioxide on bacterial growth. Appl. Environ. Microbiol. **46**, 198–202 (1983)
76. M. Maniscalco, M. Sofia, E. Weitzberg, L. Carratu, J.O. Lundberg, Nasal nitric oxide measurements before and after repeated humming maneuvers. Eur. J. Clin. Investig. **33**(12), 1090–1094 (2003)
77. M. Maniscalco, M. Sofia, E. Weitzberg, G. De Laurentiis, A. Stanziola, V. Rossillo, J.O. Lundberg, Humming-induced release of nasal nitric oxide for assessment of sinus obstruction in allergic rhinitis: pilot study. Eur. J. Clin. Investig. **34**(8), 555–560 (2004)
78. M. Maniscalco, E. Weitzberg, J. Sundberg, M. Sofia, J.O. Lundberg, Assessment of nasal and sinus nitric oxide output using single-breath humming exhalations. Eur. Respir. J. **22**(2), 323–329 (2003)
79. G. Maza, C. Li, J.P. Krebs, B.A. Otto, A.A. Farag, R.L. Carrau, K. Zhao, Computational fluid dynamics after endoscopic endonasal skull base surgery possible empty nose syndrome in the context of middle turbinate resection. Int. Forum Allergy Rhinol. **9**(2), 204–211 (2019)
80. L. McKinlay, S. Vaidyanathan, P.A. Williamson, B.J. Lipworth, Nasal nitric oxide as a measure of osteomeatal complex patency in nasal polyps. Ann. Allergy Asthma Immunol. **107**(2), 179–180 (2011)
81. U. Mercke, C.H. Hakansson, N.G. Toremalm, The influence of temperature on mucociliary activity. temperature range 20 degrees c-40 degrees c. Acta Oto-Laryngol. **78**(5–6), 444–450 (1974)
82. T. Meusel, S. Negoias, M. Scheibe, T. Hummel, Topographical differences in distribution and responsiveness of trigeminal sensitivity within the human nasal mucosa. Pain **151**(2), 516–521 (2010)
83. G. Mylavarapu, S. Murugappan, M. Mihaescu, M. Kalra, S. Khosla, E. Gutmark, Validation of computational fluid dynamics methodology used for human upper airway flow simulations. J. Biomech. **42**(10), 1553–1559 (2009)
84. S. Naftali, M. Rosenfeld, M. Wolf, D. Elad, The air-conditioning capacity of the human nose. Ann. Biomed. Eng. **33**, 545–553 (2005)
85. M. Naraghi, A.F. Deroee, M. Ebrahimkhani, S. Kiani, A. Dehpour, Nitric oxide: a new concept in chronic sinusitis pathogenesis. Am. J. Otolaryngol. **28**(5), 334–337 (2007)
86. V.S. Nesic, V.Z. Djordjevic, V. Tomic-Spiric, Z.R. Dudvarski, I.A. Soldatovic, N.A. Arsovic, Measuring nasal nitric oxide in allergic rhinitis patients. J. Laryngol. Otol. **130**(11), 1064–1071 (2016)
87. J.D. Oliver, K.G. Lim, E.K. O'Brien, Correlation of exhaled nasal nitric oxide with sinus computed tomography and sinonasal outcome test scores: A cross-sectional pilot study. Am. J. Rhinol. Allergy **32**(6), 533–538 (2018)
88. J.P. Palm, K. Alving, J.O. Lundberg, Characterization of airway nitric oxide in allergic rhinitis: the effect of intranasal administration of l-name. Allergy **58**(9), 885–892 (2003)
89. S.M. Ragab, V.J. Lund, H.A. Saleh, G. Scadding, Nasal nitric oxide in objective evaluation of chronic rhinosinusitis therapy. Allergy **61**(6), 717–724 (2006)
90. C. Rennie, K. Gouder, D. Taylor, N. Tolley, R. Schroter, D. Doorly, Nasal inspiratory flow: at rest and sniffing. Int. Forum Allergy Rhinol. **1**, 128–135 (2011)
91. C.E. Rennie, C.M. Hood, E.J. Blenke, R.S. Schroter, D.J. Doorly, H. Jones, D. Towey, N.S. Tolley, Physical and computational modeling of ventilation of the maxillary sinus. Otolaryngol. Head Neck Surg. **145**(1), 165–170 (2011)
92. F.L. Ricciardolo, P.J. Sterk, B. Gaston, G. Folkerts, Nitric oxide in health and disease of the respiratory system. Physiol. Rev. **84**(3), 731–765 (2004)
93. T. Runer, A. Cervin, S. Lindberg, R. Uddman, Nitric oxide is a regulator of mucociliary activity in the upper respiratory tract. Otolaryngol. Head Neck Surg. **119**, 278–287 (1998)
94. T. Runer, S. Lindberg, Ciliostimulatory effects mediated by nitric oxide. Acta Oto-Laryngol. **119**(7), 821–825 (1999)

95. M.O. Scheithauer, Surgery of the turbinates and "empty nose" syndrome. GMS Curr. Top. Otorhinolaryngol. Head Neck Surg. **9** (2010)
96. D. Shusterman, The nasal nitric oxide response to external acoustic energy: a pilot study of sampling dynamics. Sinusitis **1**, 1 (2016)
97. D. Shusterman, M. Murphy, Nasal hyperreactivity in allergic and non-allergic rhinitis: a potential risk factor for non-specific building-related illness. Indoor Air **17**(4), 328–333 (2007)
98. D.J. Shusterman, K. Jansen, E.M. Weaver, J.Q. Koenig, Documentation of the nasal nitric oxide response to humming: Methods evaluation. Eur. J. Clin. Investig. (2007)
99. D.J. Shusterman, E.M. Weaver, A.N. Goldberg, S.F. Schick, H.H. Wong, J.R. Balmes, Pilot evaluation of the nasal nitric oxide response to humming as an index of osteomeatal patency. Am. J. Rhinol. Allergy **26**(2), 123–126 (2012)
100. J. Siu, J. Dong, K. Inthavong, Y. Shang, R.G. Douglas, Quantification of airflow in the sinuses following functional endoscopic sinus surgery. Rhinology Accepted for publication Jan 10 2020 (2020)
101. J. Siu, K. Shrestha, K. Inthavong, Y. Shang, R. Douglas, Particle deposition in the paranasal sinuses following endoscopic sinus surgery. Comput. Biol. Med. **116**, 103573 (2020)
102. W.M. Smith, T.M. Davidson, C. Murphy, Toxin-induced chemosensory dysfunction: a case series and review. Am. J. Rhinol. Allergy **23**(6), 578–581 (2009)
103. G.M. Sokol, G.G. Konduri, K.P. Van Meurs, Inhaled nitric oxide therapy for pulmonary disorders of the term and preterm infant. Semin. Perinatol. **40**(6), 356–369 (2016)
104. J. Sozansky, S.M. Houser, The physiological mechanism for sensing nasal airflow: a literature review. Int. Forum Allergy Rhinol. **4**(10), 834–838 (2014)
105. C.D. Sullivan, G.J.M. Garcia, D.O. Frank-Ito, J.S. Kimbell, J.S. Rhee, Perception of better nasal patency correlates with increased mucosal cooling after surgery for nasal obstruction. Otolaryngol.–Head Neck Surg. **150**(1), 139–147 (2013)
106. H. Suojalehto, T. Vehmas, I. Lindstrom, D.W. Kennedy, M. Kilpelainen, T. Plosila, S. Savukoski, J. Sipila, M. Varpula, H. Wolff, H. Alenius, E. Toskala, Nasal nitric oxide is dependent on sinus obstruction in allergic rhinitis. Laryngoscope **124**(6), E213–8 (2014)
107. D. Takahara, T. Kono, S. Takeno, T. Ishino, T. Hamamoto, K. Kubota, T. Ueda, Nasal nitric oxide in the inferior turbinate surface decreases with intranasal steroids in allergic rhinitis: A prospective study. Auris Nasus Larynx **46**(4), 507–512 (2019)
108. S. Takeno, H. Yoshimura, K. Kubota, T. Taruya, T. Ishino, K. Hirakawa, Comparison of nasal nitric oxide levels between the inferior turbinate surface and the middle meatus in patients with symptomatic allergic rhinitis. Allergol. Int. **63**(3), 475–483 (2014)
109. J. Tan, D. Han, J. Wang, T. Liu, T. Wang, H. Zang, Y. Li, X. Wang, Numerical simulation of normal nasal cavity airflow in chinese adult: a computational flow dynamics model. Eur. Arch. Oto-Rhino-Laryngol. **269**(3), 881–889 (2012)
110. A. Thamboo, N. Velasquez, A.R. Habib, D. Zarabanda, H. Paknezhad, J.V. Nayak, Defining surgical criteria for empty nose syndrome: validation of the office-based cotton test and clinical interpretability of the validated empty nose syndrome 6-item questionnaire. The Laryngoscope **127**, 1746–1752 (2017)
111. S. Vaidyanathan, P. Williamson, K. Anderson, B. Lipworth, Effect of systemic steroids on humming nasal nitric oxide in chronic rhinosinusitis with nasal polyposis. Ann. Allergy Asthma Immunol. **105**(6), 412–417 (2010)
112. N. Velasquez, A. Thamboo, A.R. Habib, Z. Huang, J.V. Nayak, The empty nose syndrome 6-item questionnaire (ens6q): a validated 6-item questionnaire as a diagnostic aid for empty nose syndrome patients. Int. Forum Allergy Rhinol. **7**(1), 64–71 (2017)
113. T. Wang, D. Chen, P. Wang, J. Chen, J. Deng, Investigation on the nasal airflow characteristics of anterior nasal cavity stenosis. Braz. J. Med. Biol. Res. **49**, 9 (2016)
114. Y. Wang, S. Elghobashi, On locating the obstruction in the upper airway via numerical simulation. Respir. Physiol. Neurobiol. **193**, 1–10 (2014)
115. P. Webb, Air temperatures in respiratory tracts of resting subjects in cold. J. Appl. Physiol. **4**(5), 378–382 (1951)

116. E. Weitzberg, J.O. Lundberg, Humming greatly increases nasal nitric oxide. Am. J. Respir. Crit. Care Med. **166**(2), 144–145 (2002)
117. J. Wen, K. Inthavong, J. Tu, S. Wang, Numerical simulations for detailed airflow dynamics in a human nasal cavity. Respir. Physiol. Neurobiol. **161**(2), 125–135 (2008)
118. P.A. Williamson, S. Vaidyanathan, K. Clearie, M. Stewart, B.J. Lipworth, Relationship between fractional exhaled nitric oxide and nasal nitric oxide in airways disease. Ann. Allergy Asthma Immunol. **105**(2), 162–167 (2010)
119. M.R. Wofford, J.S. Kimbell, D.O. Frank-Ito, V. Dhandha, K.A. McKinney, G.M. Fleischman et al., A computational study of functional endoscopic sinus surgery and maxillary sinus drug delivery. Rhinology **53**(1), 41–48 (2015)
120. B.B. Wrobel, A.G. Bien, E.H. Holbrook, G.E. Meyer, N.A. Bratney, J. Meza, D.A. Leopold, Decreased nasal mucosal sensitivity in older subjects. Am. J. Rhinol. **20**(3), 364–368 (2006)
121. J. Xi, J. Kim, X.A. Si, Effects of nostril orientation on airflow dynamics, heat exchange, and particle depositions in human noses. Eur. J. Mech.-B/Fluids **55**, 215–228 (2016)
122. G. Xiong, J. Zhan, K. Zuo, J. Li, L. Rong, G. Xu, Numerical flow simulation in the post-endoscopic sinus surgery nasal cavity. Med. Biol. Eng. Comput. **46**(11), 1161–1167 (2008)
123. Z. Zhang, C. Kleinstreuer, Laminar-to-turbulent fluid-nanoparticle dynamics simulations: model comparisons and nanoparticle-deposition applications. Int. J. Numer. Methods Biomed. Eng. **27**(12), 1930–1950 (2011)
124. K. Zhao, K. Blacker, Y. Luo, B. Bryant, J. Jiang, Perceiving nasal patency through mucosal cooling rather than air temperature or nasal resistance. PLoS One **6**(10), e24618 (2011)
125. K. Zhao, P. Dalton, G.C. Yang, P.W. Scherer, Numerical modeling of turbulent and laminar airflow and odorant transport during sniffing in the human and rat nose. Chem. Senses **31**(2), 107–118 (2006)
126. K. Zhao, J. Jiang, What is normal nasal airflow? a computational study of 22 healthy adults. Int. Forum Allergy Rhinol. **4**(6), 435–446 (2014)
127. K. Zhao, E.A. Pribitkin, B.J. Cowart, D. Rosen, P.W. Scherer, P. Dalton, Numerical modeling of nasal obstruction and endoscopic surgical intervention: outcome to airflow and olfaction. Am. J. Rhinol. **20**(3), 308–316 (2006)
128. K. Zhao, P.W. Scherer, S.A. Hajiloo, P. Dalton, Effect of anatomy on human nasal air flow and odorant transport patterns: implications for olfaction. Chem. Senses **29**(5), 365–379 (2004)

Chapter 10
Clinical CFD Applications 2

Yu Feng, Hamideh Hayati, Alister J. Bates, Koch Walter, Lehner Matthias, Benda Odo, Ortiz Ramiro, and Koch Gerda

Abstract This chapter is the second of the two chapters demonstrating the wide variety of CFD studies in clinical applications presented from leading researchers in their respective fields. This chapter covers the latest research techniques and outcomes in whole lung modelling; Modeling the Effect of Airway Motion Using Dynamic Imaging; and Automatic reconstruction of the nasal geometry from CT scans.

10.1 Whole-Lung Modelling

Yu Feng, Hamideh Hayati

10.1.1 Introduction

To accurately evaluate the lung dose of inhaled particulate matters from the nose, the influence of lower respiratory tract on the upstream airflow patterns in human nose needs to be modeled. Specifically, the presence of lower respiratory tract will alter the impedance of the expiratory airflows and the resultant particle transport dynamics backflowing into the nasal cavity. Therefore, it is necessary to improve predictive capabilities of computational fluid-particle dynamics (CFPD), i.e., the whole-lung modeling strategy, to precisely estimate the aerosol deposition from nasal cavity to alveoli with full inhalation-exhalation breathing cycles, which is difficult to acquire

Y. Feng (✉) · H. Hayati
Oklahoma State University, Stillwater, USA
e-mail: yu.feng@okstate.edu

A. J. Bates
Cincinnati Children's Hospital, Cincinnati, USA

K. Walter · L. Matthias · B. Odo · O. Ramiro · K. Gerda
AIT Angewandte Informationstechnik Forschungsgesellschaft mbH, Graz, Austria

K. Inthavong et al. (eds.), *Clinical and Biomedical Engineering in the Human Nose*, Biological and Medical Physics, Biomedical Engineering,
https://doi.org/10.1007/978-981-15-6716-2_10

via in vitro and in vivo methodologies [13, 21, 37]. The overall complexity of the lung with bifurcating architecture, hierarchical cellular composition, and continual dynamic motion results in complex fluid phenomena which perplex deposition of inhaled aerosols.

Whole-lung modeling capabilities are not only able to provide a full map of regional depositions of inhaled aerosols, but also enable the accurate simulation of expiratory particle dynamics after the inhalation. Specifically, simulations using the truncated subject-specific airway geometries reconstructed from CT/MRI data only contain airways up to generation 9 (G9). The missing lower airways force the researchers to make assumptions on how many particles will be able to "re-enter" the flow domain from the truncated airway outlets during the exhalation after the inhalation. With the whole-lung modeling strategies, the lung aerosol dynamics during the full inhalation-exhalation cycles can be predicted without such ad-hoc simplifications. Whole-lung modeling efforts have been made to demonstrate that because of their small size, micro-to-nano scale toxic and/or therapeutic aerosols that vastly exist in the ambient environment are respirable and can penetrate the peripheral lung and enter the systemic region via the air-blood barriers [13, 15, 16, 22, 27, 35–40, 47, 48, 74, 75, 78, 81].

Due to the morphological complexity of the human respiratory system, and the limited resolution of medical imaging techniques [15, 20, 37, 59], identified challenges to develop whole-lung models are: (1) the lack of small-airway CT/MRI data for airway reconstructions [20], and (2) the extremely high computational cost of full-scale CFPD simulations of all airways in entire 23 generations [36, 37]. Accordingly, research efforts have been made to develop alternative whole-lung modeling strategies to optimize the balance between computational accuracy and efficiency [13, 15, 35–40, 47, 48, 74, 75, 78, 81], which are discussed in the next section.

10.1.2 Existing Whole-Lung Modeling Strategies

Realistic airway configurations are currently restricted to the few upper air-way generations since lower airway configurations are not easy to reconstruct due to the limited resolution of the scanned images. Moreover, even if the entire lung was fully segmentable, it is not computationally feasible to simulate the full lung tree accurately, because it requires billions of 3D mesh elements [15, 37, 59]. Therefore, several common strategies have been employed in existing whole-lung models: (1) using equivalent geometry (e.g., trumpet) to represent the entire tracheobronchial tree, (2) reducing the degree of freedom (DOF) of the whole lung airway trees by truncating airways and applying advanced coupled boundary conditions [36, 37, 59, 74, 75, 78], and (3) by simplifying small airways using 1D pipelines or 2D in-plane airway models instead of 3D geometries [74, 81].

Specifically, popular approaches for simulating inhaled particle deposition are 'trumpet' model [73], deterministic single and multi-path models [1, 13, 48, 54], and stochastic multi-path models [33, 34, 47]. The stochastic model of human lung consider asymmetry and randomness of the airway system, as well as, variation in parameters (inter-subject variation), i.e., diameter and length of the tubes, branching and gravity angles and the number of bifurcation in each bronchial pathway [34]. In contrast, deterministic approaches do not take the inter-subject variability into account [37].

10.1.2.1 Trumpet Models

The trumpet model [73] is one of the simplified deterministic models that consider the tracheobronchial tree as an axisymmetric trumpet matching the morphometric data of the deterministic Weibel [80] model. Choi and Kim [13] developed a single-path mathematical model, i.e., the two-dimension (2D) trumpet model with connected alveoli [73], to estimate the whole-lung particle deposition. They considered both axial and lateral transport of inhaled particles with the contraction and expansion of alveoli walls. Figure 10.1a demonstrates the 2D trumpet geometry to represent the human lung. Specifically, the alveolar region present G18 to G23.The trumpet is designed based on the fact that the volume of each generation increases from G1 to G23. Although the computational efficiency using the 2D trumpet model is high, the regional deposition results are not predictive, because of the simple inlet conditions and semi-analytical correlations for the deposition mechanisms. Such a 2D trumpet model may not be able to predict lung aerosol dynamics, especially related to the dense particle suspensions from dry powder inhalers (DPIs), in which the interactions between particles may significantly affect particle trajectories in the lung, but cannot be modeled using the 2D trumpet model. Kolanjiyil and Kleinstreuer [37] further developed a 3D trumpet modeling approach, i.e., the whole-lung airway model (WLAM). Considering the fact that the extra-thoracic airway has significant effects on the pulmonary air-flow patterns and the resultant particle transport and deposition, they connected an idealized upper airway model with the 3D trumpet, which can be replaced by subject-specific nasal cavity model (see Fig. 10.1b). By adding an artificial force in the radial directions in the trumpet, WLAM is able to provide reasonably accurate depositions compared to experimental data within the full inhalation-exhalation cycle. The trumpet was divided into multiple regions to represent different airway generations. Specifically, the division is done to equalize the regional volumes to the volumes of certain generations. Furthermore, the lung deformation effect was also approximately achieved by the deformation of the wall at the bottom of the trumpet. However, since the trumpet in WLAM is still axisymmetric, the effect of realistic asymmetric airway features were not able to be captured. Another limitation of the model is that it is only accurate to predict particles larger than 400 nm.

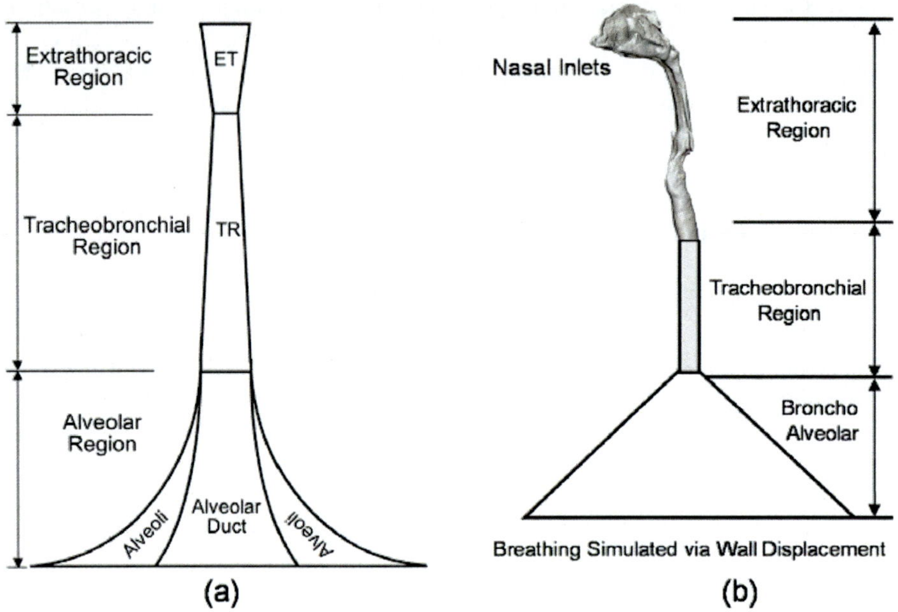

Fig. 10.1 Flow-domain schematics of trumpet models: **a** the 2D trumpet model [13], and **b** the whole-lung airway model [37] (Reprinted from [13, 37], with permission from Taylor and Francis as well as Elsevier)

10.1.2.2 Single-Path and Multi-path Models with Truncations

To simultaneously capture the asymmetrical effect of the airway trees and keep the computational efficiency high, another set of whole-lung modeling strategies is to truncate the branches of the lung geometry and solve the governing equations for one single branch coupling with appropriate boundary conditions at those truncated airways. Modeling the pulmonary routes from mouth/nose to G9 can be achieved by geometry reconstructions using clinical data. However, due to the limitation of medical imaging resolution, small airway reconstructions have been done by connecting idealized bifurcation unit with different geometric dimensions [35, 36, 47, 48, 75, 78]. Specifically, Zhang et al. [81] made the first effort to decompose the complex airway network from the trachea to G17 into a series of detached multiple triple bifurcation units (TBU) [80]. Using an improved modeling strategy based on Zhang et al. [81], Longest et al. [47, 48] developed the stochastic individual paths (SIP) and multiple stochastic individual paths (MSIP) approaches to represent the whole-lung geometry. Compared to the coupling method applied on detached airway geometries [80], this model is able to simulate lung aerosol dynamics with transient inhalation-exhalation breathing cycles.

Also, to model the entire conductive zone, researchers developed a deterministic reduced lung geometry model up to G16 [75, 78]. Figure 10.2 shows the details about

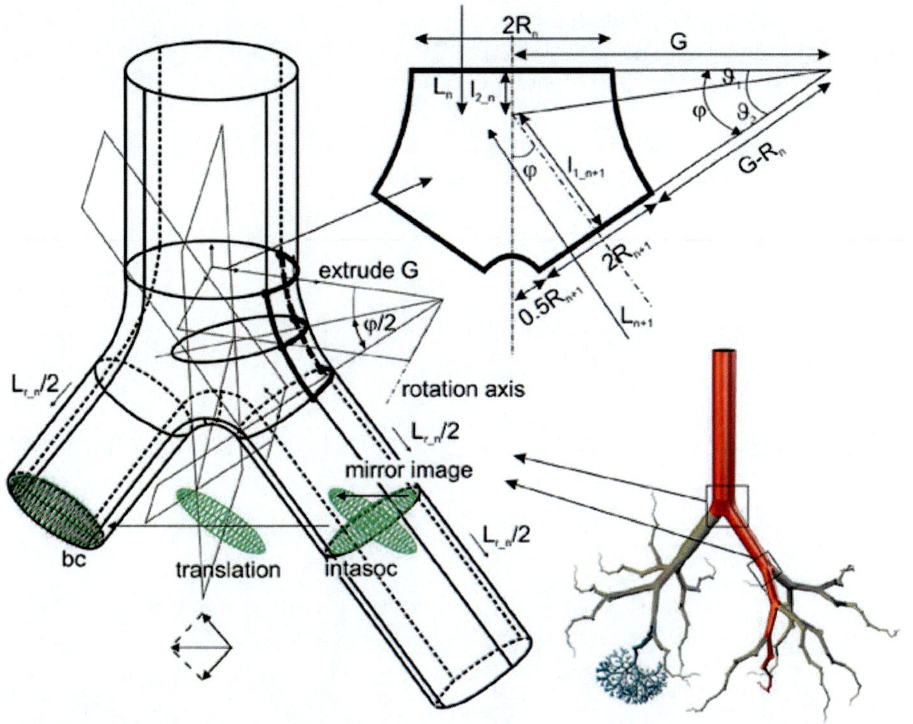

Fig. 10.2 Construction of the reduced lung geometry model (Reprinted from [75], with permission from Wiley)

how the reduced lung geometry model was generated and how the boundary condition was defined at the truncated outlets. Specifically, coupled boundary conditions are applied at the truncated branches, obtained from the complete airway routes. The flow variables were mapped from the resolved pathway to the unresolved truncated branch.

In addition, for the separate effort done for modeling the deep lung and the respiratory zone has been done by Koullapis et al. [40] using an idealized 10-generation bifurcation tree and acini models. The model is shown in Fig. 10.3. This model is capable of tracking the particles that remained suspending during the inhalation phase in the deep lung, and their backflow motion during the exhalation with no assumptions made. However, the disadvantage of this model is lack of realistic randomness of the airway morphology. Indeed, the whole airway tree was constructed by connecting scaled single bifurcations from the same single bifurcation geometry. Therefore, the branching angle and ratios of parent-to-daughter tubes diameter and length-to-diameter are constant for all the generations, which is not physiologically true. The rigid-wall assumption is another disadvantage of the aforementioned model. Furthermore, the acinar model employed in this model represented about 12% of the total acinus, which led to under-estimated acinar deposition of particles, since there are two additional acinar generations in the human lung.

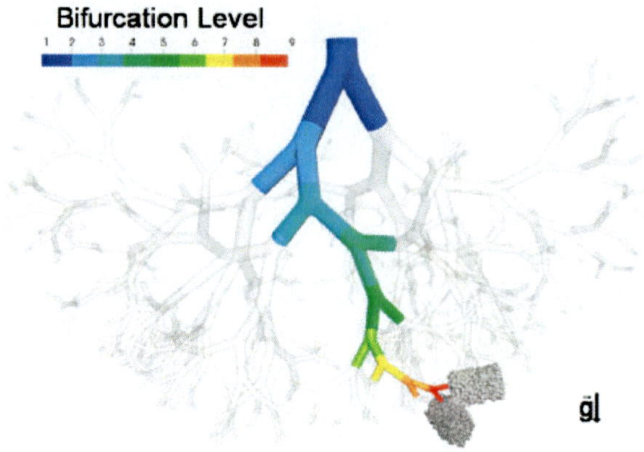

Fig. 10.3 The developed simplified deep lung model (Reprinted from [40], with permission from Elsevier)

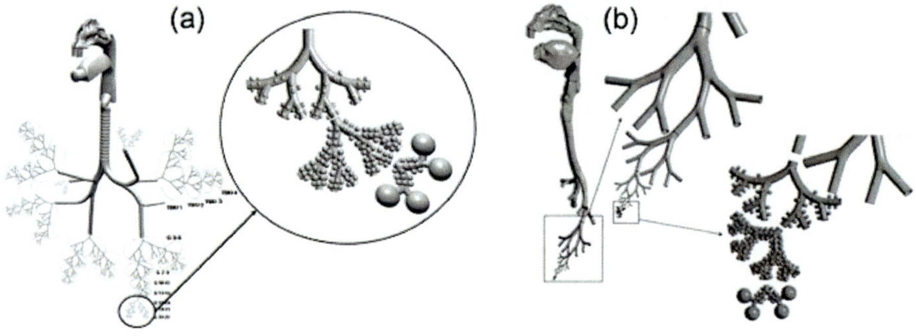

Fig. 10.4 Schematics of (**a**) the whole-lung model using an idealized human upper air-way model and TBU cascades [36], and (**b**) the whole-lung dual-path model using an subject-specific human upper airway model and TBU cascades

Moreover, Kolanjiyil and Kleinstreuer [36] combined an idealized upper airway model and a series of TBUs to generate the representative whole-lung model (see Fig. 10.4a). Using the fluid-structure interaction (FSI) approach, this whole-lung model [15] is able to study the effect of alveolar deformations on the pulmonary airflow pattern and the resultant particle dynamics. However, the uniform zero-gauge pressure boundary conditions applied at truncated outlets may introduce errors on pulmonary airflow predictions. In addition, the asymmetrical and non-planar nature of the airway, which influence both fluid flow and particle deposition were not considered. Later, Kolanjiyil and Kleinstreuer [38] built another dual-path whole lung model (see Fig. 10.4b). Instead of using the idealized upper airway geometry, they employed a subject-specific human upper airway model widely used in computational and numerical lung aerosol dynamics research [19, 28, 70, 71].

10.1.2.3 Full 3D-1D Airway-Tree Models Without Truncations

Efforts have been also made to achieve the whole-lung modeling capability with reasonable computational efficiency via reducing the DOF of airways without truncating a portion of the tracheobronchial tree. As the prototype of this type of whole-lung simulation strategy, the impedance model was developed which simplifies the whole lung airway structure using one-dimensional (1D) pipelines [72]. However, due to the existence of strong secondary flows in lateral directions from G1 to G6, the airflow field cannot be captured accurately by solving the 1D governing equation systems. To address such disadvantages, a fully coupled 3D-1D approach was developed based on asymmetric physiological characteristics of the realistic lung airway trees [15, 51, 57, 74, 76]. Specifically, Comerford et al. [15] utilized coupled 3D-1D modeling approach to study pulmonary mechanics of the entire conducting zone under mechanical ventilation. By using this method, the conducting zone of the airway can be modeled efficiently and results in finding important pressure and flow data. In this approach, pressure domain in down-stream (1D) domain obtained from the impedance model used to find the time dependence pressure at the outflow of the 3D domain. To couple 1D model to 3D model, Dirichlet or Neumann boundary conditions were employed [76]. In order to estimate the lung input impedance, Ma and Lutchen [51] applied a model combining 3D CFD simulation in upper and central airway and 1D transmission line-like model for the small airways, which they called hybrid computational model for the human lung. They simulated unsteady flow in upper airway and lower airway up to G6 with impedance (i.e., time-dependent pressure) boundary conditions. The small airways from G6 to alveoli were represented by corresponding small airway trees in the hypothetical airway tree model [74]. Viscoelastic alveoli tissue unit was attached to the end of the conducting zone. The 3D-1D whole-lung models are able to efficiently simulate the pulmonary airway fields in the entire pulmonary system. However, this modeling approach has not been further developed to predict the transport and deposition of inhaled particles.

10.1.3 The Future of Whole-Lung Modeling

For patients with pre-existing lung diseases, including emphysema, asthma, cystic fibrosis (CF), and idiopathic pulmonary fibrosis (IPF), etc., lung dynamic motion, breathing pattern, patient-specific geometries, age, and disease state are all known to contribute to the alteration of airflow and inhaled particle dynamics in the pulmonary system, yet no computational whole-lung models currently incorporates any one of these variables to successfully predict aerosol transport and deposition. Existing whole-lung models are not able to cope with the challenges to:

- Recover the realistic real-time lung deformations considering fluid-structure interaction (FSI);
- Consider soft palate and glottis motions;

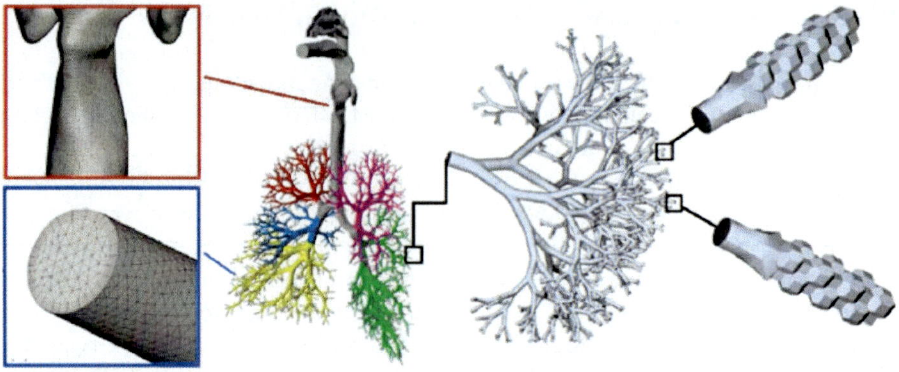

Fig. 10.5 The next-generation 3D elastic whole-lung model covering the entire conducting and respiratory zones

- Capture the disease-specific mucus movement and clearance of inhaled aerosols driven by cilia; and
- Predict the translocation of inhaled drugs or toxicants to clinical endpoints.

Therefore, the multiphase flow phenomena (air-mucus-aerosol transport) in the healthy and diseased human respiratory systems are still not comprehensively modeled and studied. Hence, there is still a gap among the whole-lung modeling capabilities, to sufficiently investigate how pre-existing lung disease including lung morphology and stiffness changes affect the advection and diffusion in pulmonary airways, the mucus movement and clearance, and the resultant variations of the inhaled therapeutic drug delivery efficiency. Thus, the future work to build the next-generation whole-lung model is to enable the development of per-sonalized aerosol therapeutics optimized for patient-specific variables including patient age, anatomy, breathing maneuvers, and pathophysiology. The next-generation whole-lung model should be able to answer the fundamental ques-tion of biological fluid dynamics on how the physiological and biomechanical feature variations between the healthy and diseased airways will affect pulmonary airflow dynamics, mucus movement, and the advection and diffusion of tracing particles/gases into small airways. The next-generation of whole-lung model, i.e., "living lung model", should provide a scientific basis for further understanding altered characteristics of diseased airways and their implications to clinical diagnosis and treatment as well as evaluation of drug delivery. The most straightforward methodology is to integrate the elastic lung model [76] into the modeling ideas for conducting and respiratory zones [40, 75], to form the proto-type of the next-generation of virtual whole-lung model (see Fig. 10.5).

10.2 Modeling the Effect of Airway Motion Using Dynamic Imaging

Bates, Alister

10.2.1 Introduction and Clinical Significance

Many airway diseases involve abnormal motion of the airway wall. This motion may result in the airway lumen narrowing or entirely collapsing, either of which will affect the patient's ability to breathe. Abnormal airway motion may occur in various portions of the airway. Examples include obstructive sleep apnea (OSA), a condition in which the upper airway collapses during sleep, often because the soft palate and/or tongue collapse against the posterior wall of the pharynx. Vocal cord dysfunction occurs when the vocal cords close abnormally during breathing. The larynx, trachea, and bronchi collapse in laryngomalacia, tracheomalacia, and bronchomalacia, respectively. Even in otherwise healthy subjects, the nasal valve can collapse during rapid inhalation.

Airway motion occurs for a variety of reasons. Firstly, the air pressure within the airway varies throughout the breathing cycle from below atmospheric pressure during inhalation to above atmospheric pressure during exhalation. These pressure forces act on the airway walls and can cause them to collapse inwards under low pressures or expand outwards under high pressure.

Secondly, the airway below the thoracic inlet (approximately one third of the way down the trachea) is within the thoracic cavity. Therefore, the intrathoracic pressures generated by the diaphragm and intercostal muscles to expand and contract the lungs also act on the exterior of the intrathoracic airway walls. This effect leads to conditions such as tracheomalacia, in which the intrathoracic pressure generated to exhale air from the lungs is enough to collapse the trachea. The trachea may collapse if the cartilaginous tissues that provide its structure are abnormally soft and are unable to support the membrane in the posterior tracheal wall, the trachealis. However, a healthy trachea may also collapse if the intrathoracic pressure becomes unusually high, due to respiratory demands or lung parenchyma abnormalities. In this latter case, the condition is known as excessive dynamic airway collapse (EDAC).

The third mechanism for airway motion is neuromuscular control of the structures surrounding the airway, many of which are muscular, such as the tongue, soft palate, and pharyngeal walls. Opening the jaw or changing head position also alter the airway shape and size. Patients control these structures in order to maintain airway patency, but their ability to do so depends on their anatomy, ability to sense airflow, muscle tone, and the pressure forces they are working against. This factor also varies according to the subject's level of arousal, leading to sleep related conditions such as OSA.

Regardless of the location or cause of airway collapse, dynamic narrowing will increase the amount of energy required to respire, which is known as the work of breathing. The airway resistance will also increase during the portion of the breath in which the airway is narrowed, thereby increasing the pressure difference between the lungs and atmospheric pressure or limiting peak flow rates. This effect may, in turn, alter the patient's respiratory flow waveform. Clinically, the important question is whether airway motion affects the patient's symptoms or wellbeing due to increased airway resistance and/or work of breathing. CFD simulations have the potential to quantify these factors but require realistic airway motion to be incorporated as a boundary condition.

Two approaches have been used for modelling airway motion. The first is fluid structure interaction (FSI). These models simulate both the airflow and the structures surrounding the airway by modeling structural deformations using material properties such as Young's Modulus. They have been employed to model passive motion of the airway (i.e. when the airway moves solely due to air pressure in the lumen) [2, 62, 79, 82–84]. However, tissue properties may be difficult to calculate in vivo and vary significantly across the tissues surrounding the upper airway and between subjects with various pathological conditions [12]. Furthermore, neuromuscular control of structures surrounding the airway is difficult to predict and incorporate into FSI models. The tissues may move out-of-phase with the breath profile, or may move differently from breath to breath [58, 68]. Therefore, a second technique has been implemented to model conditions where neuromuscular control of the airway plays a significant role in airway motion, such as OSA. In this technique the in vivo motion of the airway is extracted from cine imaging and applied to the surface of the CFD model, causing the modeled airway surface to move as the real airway moves in vivo. This technique will be described in more detail below.

10.2.2 3D Cine Imaging

Three-dimensional, time varying CFD simulations require information on the airway motion in all directions and across the airway surface throughout the period of interest. This information can be obtained from four-dimensional (4D—time and three spatial dimensions) cine image volumes acquired from either computed tomography (CT) or magnetic resonance imaging (MRI). CT has the advantage over MRI of a higher contrast to noise ratio and is therefore easier to determine the position of the airway at each time-point in a cine image sequence. CT also allows for high spatial resolution (approximately $0.4 \times 0.4 \times 0.4\,\text{mm}$) which yields accurate reconstruction of the airway shape and motion, and lower dynamic times (i.e. the amount of time between one image and the next in the sequence—as low as 135 ms, which is half the gantry rotation time), making it more suitable for fast moving structures. An example of airway motion captured with CT is shown in Fig. 10.6a. However, the disadvantage of this modality is that CT imaging exposes subjects to ionizing radiation [61].

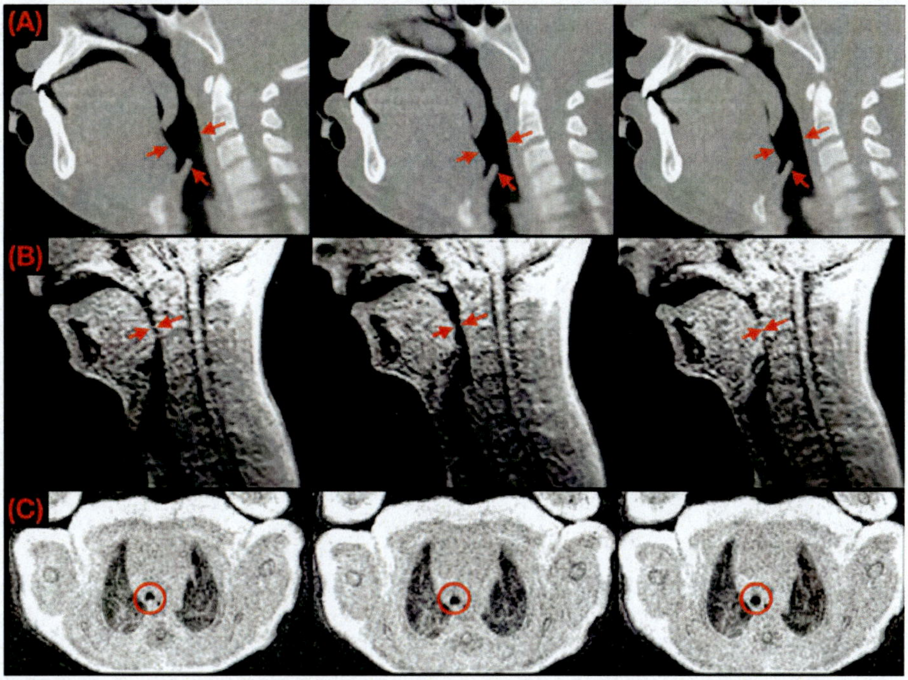

Fig. 10.6 **a** A sagittal slice through the upper airway of an 18-year-old patient with OSA taken from a 4D dynamic CT image. The slice is shown at three instants through a breath. There is a subtle change in caliber of the retroglossal airway and the epiglottis moves anteriorly (red arrows). **b** A sagittal slice through the upper airway of an 11-year-old patient with OSA at three instants taken from real-time cine MRI. The retroglossal airway caliber starts narrow (left image, red arrows), widens (central image), and then collapses completely (right image). **c** An axial slice showing the intrathoracic trachea in a neonatal patient with tracheomalacia at three instants. In the first image (left), the rear wall of the trachea has collapsed inwards creating a "D" shaped lumen (red circle). In the central image, the rear wall has started to move posteriorly, and it keeps moving until the trachea is almost circular (right image)

An alternative modality is MRI, which does not employ ionizing radiation and therefore can be used in vulnerable populations, such as children, and in serial studies in which patients are imaged before and after treatment or to assess disease progression. Patients may also be imaged for long periods in order to capture particular events, such as airway collapse in OSA, which may only occur a few times per hour. MRI provides two techniques for achieving three-dimensional imaging of dynamic anatomy. The first is real-time cine imaging. Fast cine MRI techniques, such as enhanced T1 high-resolution isotropic volume excitation (THRIVE), can produce 3D images of the airway every 300 ms, with spatial resolution of $2.0 \times 2.0 \times 3.0$ mm [8]. This temporal resolution is not high enough to capture fine details of the airway motion, such as oscillations of the uvula due to snoring, but does capture coarse airway motion. The value of this technique is the high temporal resolution and real-time capture of events that do not occur every breath. Figure 10.6b shows real-time MRI of complete collapse of the retroglossal airway in a patient with OSA.

The second MRI approach for three-dimensional imaging of dynamic anatomy utilizes retrospective gating of the raw MRI data, which is modulated by respiratory motion due to imaging physics. High-resolution imaging data is continually obtained for several minutes while the subject performs tidal breathing. The data is then sorted into bins depending on when in the respiratory cycle it was captured. Images are reconstructed from each bin of data, providing average images of the typical airway shape during that portion of the breath over the imaging period [29]-[17]. For example, data could be sorted into inspiratory and expiratory bins or into many shorter periods throughout the respiratory cycle. Using more bins increases the temporal resolution of the resultant images at the expense of the images' signal to noise ratio [25]. The advantage of this method over real time cine MRI is that higher spatial resolution images are obtained (a CT-like submillimeter resolution, such as $0.7 \times 0.7 \times 0.7$ mm) and higher temporal resolution (e.g. 90 ms) is possible depending on the required signal to noise ration and imaging duration. The disadvantage is that the images represent averages of the airway shape during particular periods of the breathing cycle rather than the instantaneous "snap shots" provided by real-time cine MRI. This renders the technique suitable for rhythmic airway motion such as tracheal collapse in patients with tracheomalacia, but unsuitable for the occasional upper airway collapses in patients with OSA. An example of retrospectively gated MRI demonstrating tracheomalacia in a premature neonate is shown in Fig. 10.6c. This technique has been shown to quantify similar levels of airway dynamics as bronchoscopy [7].

There is significant clinical value in geometric analysis of these moving airway surfaces in addition to analysis of the respiratory airflow within them. Analysis of these airway surfaces has been used in surgical planning in neonates born with tracheoesophageal defects [24] and used to predict which infants with bronchopulmonary dysplasia (respiratory disease of prematurity) will require a tracheotomy [23].

10.2.3 Extracting Airway Motion from Cine Images

Moving airway surfaces are extracted from images through segmentation and registration. The first image in the cine image sequence of interest must be segmented as described in Chap. 5 segmentation) and an initial airway surface generated from the segmentation. If the images are real-time cine MRI, these images may not be high enough spatial resolution to yield a good quality surface. Therefore, a high-resolution static MRI can be acquired and segmented instead and registered to the initial cine image, [9] as described below.

The goal of the registration is to produce a high-resolution, three-dimensional map of how the airway moves from each image to the next in the cine-image sequence. This can be achieved through image registration, surface registration, or a combination of both. While any non-rigid registration technique that yields suitable motion maps can be employed, an extension of the four dimensional joined, deformable motion-

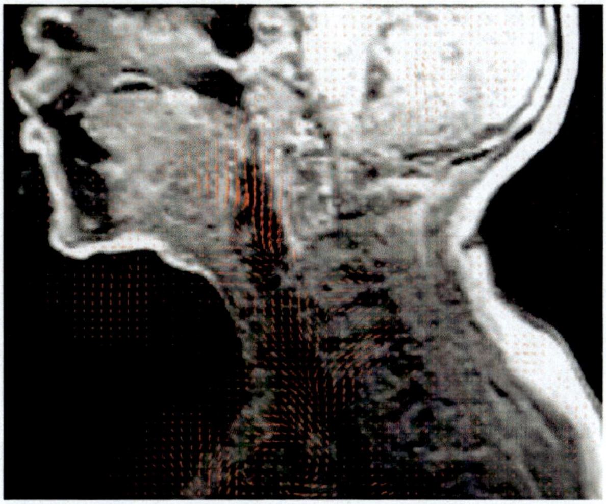

Fig. 10.7 Motion vectors between 2 frames of real-time cine MRI (orange) in a pediatric patient with OSA. Between these image frames, the patient's soft palate moves down, as does the jaw. The chest expands outwards and there is some airway collapse around the larynx. Each of these motions is illustrated by large vectors in these regions

tracking algorithm originally described by Rueckert et al. [66] has been used in the examples shown in this section and has been applied to the airways previously [8, 9]. Image registration entails seeding the initial image with control points. These control points are then moved so that they are surrounded by a similar pattern of voxel intensities in the second image as they were in the first image. Adjacent control points are prevented from moving too far apart by a bending stiffness term. Surface registration requires that airway surfaces are produced based on segmentations of each image in the image sequence. Using these surfaces, registration moves control points to minimize the difference in node positions and/or surface normal vectors between the surface at one time point and the next.

The registration algorithm then minimizes the sum of the dissimilarity between consecutive images, consecutive meshes, and the bending stiffness caused by moving control points. Figure 10.7 shows airway motion vectors between one image and the next in a pediatric subject with OSA. Typical CFD time-steps for transient (i.e. temporally varying) simulations are much shorter than the period between images (<1 ms vs (O)100 ms). Therefore, the motion between each image must be interpolated to much finer temporal resolution to provide smooth motion between CFD time-steps. Figure 10.8 shows airway surfaces from a pediatric patient with sleep apnea at peak inhalation and peak exhalation. The differences between the surfaces show how the airway has moved between these timepoints.

10.2.4 Dynamic CFD Simulations

Dynamic CFD simulations require the following boundary conditions: (1) the airway surface at the initial time point; (2) the motion of this surface through the period of interest; and (3) the pressure or flow inlet and outlet boundary conditions that

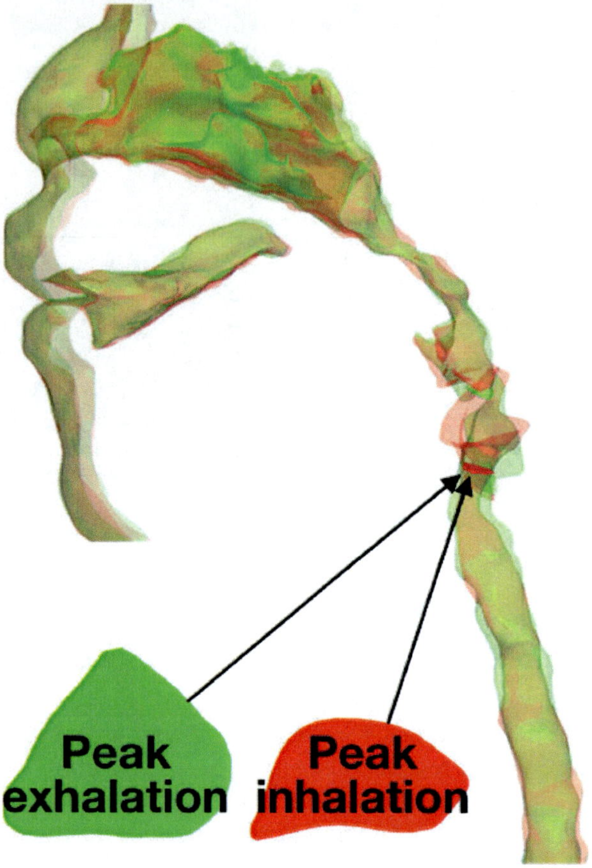

Fig. 10.8 Airway surfaces at peak exhalation (green) and peak inhalation (red). The motion of the airway between these timepoints is shown where these surfaces do not overlap. A cross-sectional plane in the larynx from each timepoint is shown in the lower left

vary through the period of interest synchronously with airway motion. Flow rates throughout the breath cycle can be captured synchronously with imaging by having the subject breathe through a pneumotach. Recording these flow measurements synchronously with a trigger signal denoting each image acquisition allows alignment of airflow and motion boundary conditions, as shown in Fig. 10.9. The high frequency oscillations in the flow waveform represent snoring, which occurs faster than the temporal resolution of the MRI acquisition. However, the temporal resolution does allow changes to the airway between inhalation and exhalation to be captured.

Once these boundary conditions have been obtained, a volume mesh must be generated based on the initial surface. The transient (i.e. boundary conditions vary with time [4, 6]) CFD simulations can then be performed, as described in Chap. 7 (transient solutions). However, the Navier-Stokes equations must be adapted to incorporate the mesh movement. The continuity and momentum equations become

$$\frac{\partial}{\partial t}\int_{\widetilde{V}} \rho \widetilde{V} + \oint \rho\left(\mathbf{v}-\mathbf{v}_{\mathbf{g}}\right) \cdot d\mathbf{a} = 0 \tag{10.1}$$

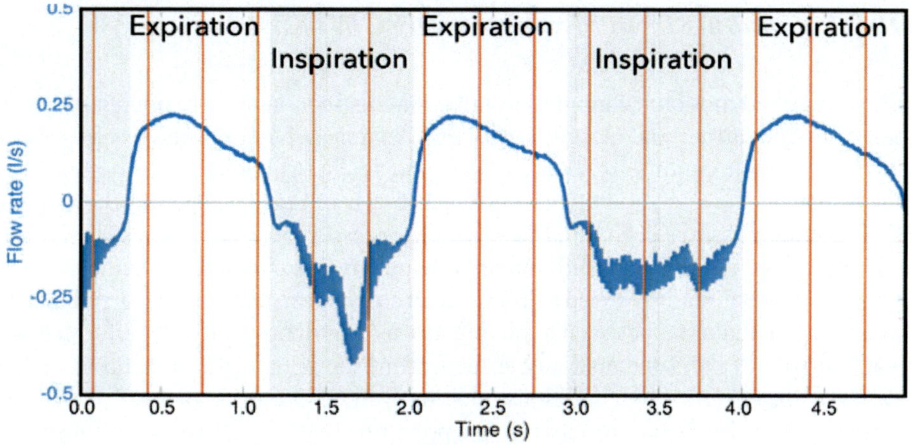

Fig. 10.9 Respiratory flow waveform (blue) and vertical red lines denoting the start time for each cine MRI acquisition (every 320 ms in this case)

and

$$\frac{\partial}{\partial t}\int_{\widetilde{V}} \rho \mathbf{v} d\widetilde{V} + \oint \left(\rho\left(\mathbf{v}-\mathbf{v_g}\right)\otimes \mathbf{v}\right)\cdot d\mathbf{a} = -\oint p\mathbf{I}\cdot d\mathbf{a} + \oint \mathbf{T}\cdot d\mathbf{a} \tag{10.2}$$

respectively, where t is time, $\tilde{V}$ is the cell volume, ρ is the air density, $\mathbf{v}$ is the airflow velocity, $\mathbf{v_g}$ is the mesh velocity as calculated from the registration, $\mathbf{a}$ is a vector representing the surface of each mesh cell, $\mathbf{I}$ is the identity matrix, and $\mathbf{T}$ is the viscous stress tensor.

Once the Navier-Stokes equations are solved for the current time-step, the mesh must be morphed to its new position. This motion may affect the quality of the mesh, which is therefore checked at each time-step. Should the mesh quality fall below the requirements for the simulation (see Chap. 6—Meshing), the current geometry should be remeshed and the pressure and flow fields interpolated onto the new mesh. Frequent remeshing can greatly extend the time taken to perform simulations, so care should be taken to produce a high-quality mesh at the initial time point. Once a mesh of suitable quality is obtained, the inlet and outlet boundary conditions are updated to the flow rates and pressures presenting the new time-step, and the Navier-Stokes equations are solved again, starting from the solution at the previous time-step. This process is repeated until the entire period of interest has been simulated.

10.2.5 Outputs from Moving Airway CFD Simulations

The prescribed motion techniques described in this section provide boundary conditions for simulations that closely follow the dynamic behavior of the in vivo airway in patients with conditions such as OSA and tracheomalacia. Velocity and local resistance fields for a patient with OSA are shown in Fig. 10.10A and B. Standard CFD output measures such as airflow velocity, pressure loss, airway resistance, and energy flux are significantly different in moving airways compared to static airways. For example, one study showed a difference in airway resistance of 76% in inhalation and 84% in exhalation between a moving airway simulation and one performed in a static airway of the same anatomy at the start of the breath [8]. In addition to providing realistic boundary conditions, moving wall CFD allows analysis of whether airway movement is due to passive (air pressure) motion, active (neuromuscular) motion, or a combination of both. For example, in OSA, muscle tone in the upper airway is an important factor in a patient's ability to prevent airway collapse.

CFD simulations calculate the pressure forces and wall shear stress acting on the airway wall. These can be combined to show the total force acting on each face on the airway wall mesh due to the airflow, $\mathbf{F_w}$, as shown in equation E3, where A is the area of the face, p is the pressure force, **n**, the normal vector to the face, and $\boldsymbol{\tau}$ the wall shear stress vector.

$$\mathbf{F_w} = A\,(p\mathbf{n} + \boldsymbol{\tau}) \qquad (10.3)$$

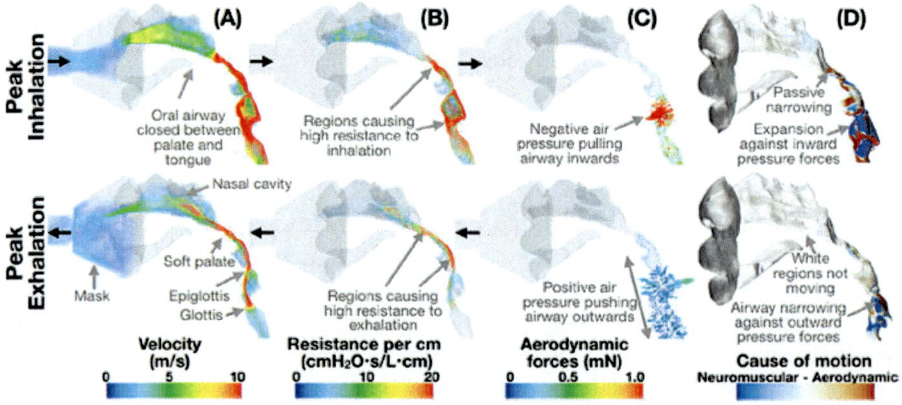

Fig. 10.10 Results from a CFD simulation in a patient with OSA with airway wall motion incorporated. **A** CFD calculations of airflow velocity showing high velocities in narrow portions of the airway. **B** Resistance per cm of airway traversed by the air-flow highlighting portions of the airway responsible for high resistance. **C** Aerodynamic force vectors (pressure plus wall shear stress). Internal pressure forces pull the airway inwards during inhalation and push outwards during exhalation. **D** The cause of airway motion determined by comparison of the direction of aerodynamic forces and the direction of motion of the airway wall. Causes of motion are due to neuromuscular control (blue), aerodynamic forces (red), or no motion (white)

Figure 10.10c shows the forces acting on the airway wall in a patient with OSA. Taking the vector dot product of this force with the prescribed airway wall velocity $\mathbf{v_w}$ at each airway wall face allows the power transfer, P, between the airway wall and the airflow to be calculated (Eq.10.3).

$$P = \mathbf{v_w} \cdot \mathbf{F_w} \tag{10.4}$$

This calculation determines the cause of motion between motion due to air pressure forces (passive motion) and motion caused by the airway through neuromuscular control. When the value is positive, the air pressure force is acting in the same direction as the motion of the airflow (for example, airway collapse due to sub-atmospheric air pressure in the airway), and therefore the cause of airway motion is likely to be due to airflow. However, when the airway motion and air pressure forces are in opposite directions or orthogonal to one another, the air pressure force cannot be the cause of motion, and P will be negative. Figure 10.10D shows the instantaneous power transfer between the air and the airway wall at peak inhalation and peak exhalation in a patient with OSA. This map shows regions where the airway is able to maintain patency due to neuromuscular control (blue) and where it is collapsing due to air pressure force acting on it (red).

10.2.6 Clinical Implications of Moving CFD Simulations

CFD simulations have previously revealed how airflow characteristics influence patient symptoms in patients with conditions such as stenotic tracheas, [52, 55] compressive goiters, [3, 5]. CFD simulations with realistic motion determined from cine imaging allows respiratory CFD to be applied to a wider set of airway conditions such as OSA and provides extra information that may aid in clinical decision-making. For example, surgical success rates in OSA are highly variable, as the location, cause, and degree of airway collapse varies between patients. CFD modeling may allow surgeries to be better matched to patients by calculating the regions of highest resistance (e.g. Fig. 10.10b), which indicate candidate sites for surgeries. The causes of airway collapse can also be separated into passive and neuromuscular, thereby revealing the degree of muscle tone and neuromuscular control in the patient. Finally, the airway component of the work of breathing can be calculated, allowing the energy expense of airway abnormalities to be quantified. Clinically, this can be used to develop thresholds to determine when treatments should be applied. Basing these decisions on energy expenditure rather than qualitative anatomic observations may allow treatment strategies to better match patient symptoms. Furthermore, in patients with various respiratory comorbidities, symptoms due to airway abnormalities can be separated from parenchymal respiratory issues, allowing treatments to be targeted to the regions of the respiratory symptom most responsible for symptoms

10.3 Automatic Reconstruction of the Nasal Geometry from CT Scans

Koch Walter, Lehner Matthias, Benda Odo, Ortiz Ramiro, Koch Gerda

10.3.1 Introduction

Accurate reconstruction of human nasal geometry is a critical prerequisite for realistic CFD simulations of nasal airflow. Fur-thermore, it is desirable that reconstruction can be performed rapidly and with minimal human intervention. Ideally, the reconstruction process can be automated completely while yielding anatomically plausible results immediately. From a clinical perspective, fast and accurate reconstruction would enable physicians as well as patients themselves to examine the nasal cavity and paranasal sinuses in an intuitive three-dimensional manner. Moreover, quickly obtained 3D models of air-filled cavities and bone provide a means for automated pathology detection, an extensive topic in its own right which will not be addressed in this chapter. From a scientific and engineering perspective, minimally time-intensive yet anatomically truthful 3D reconstructions enable the CFD specialist to focus on the investigation of nasal aerodynamics. Instead of having to expend excessive time and effort on image segmentation and 3D modelling, once a viable automated solution is available, researchers can concentrate on the in silico analysis of airflow and the verification of simulation results using experimental data. The latter can be obtained from sensors placed on 3D-printed models and measurements performed on patients.

In recent years, deep learning [41] algorithms have led to significant improvements in computer vision by outperforming techniques which were previously considered state-of-the-art [77]. A particular type of deep learning algorithm called a Convolutional Neural Network (CNN), which is most commonly used for image analysis (e.g. handwriting recognition [42, 43], outperforms classical approaches in many computer vision tasks. Medical image segmentation is one such task where promising results have been obtained in recent years. G. Litjens et al. [46] conducted a survey which included over 300 publications on applications of deep learning to medical image analysis published from 2012 to 2017. The survey showed that CNNs have become the prevailing deep learning algorithm in medical image analysis deduced by the number of papers published in the field which used them. In addition, image segmentation turned out to be the task they were most commonly applied to, with MR and CT images being two of the imaging modalities predominantly used. CNN architectures have, for instance, successfully been applied to segment brain tumors [30], brain lesions [31] and different brain tissues [69] in MRI, fluid abnormalities in OCT images [67], heart substructures in cardiac US [60] as well as pancreas [64], bladder [11], liver [50] and liver tumors [14] in CT images (cf. overviews of publications by topic and imaging modality in G. Litjens et al. [46]).

Based on previous work by Maninis et al. (2016) on retinal vasculature and optic disc segmentation in eye fundus images [53], we have developed CNN architectures which automate accurate segmentation of CT scans of the sinonasal cavities. The highly accurate results obtained for the thin blood vessels of the retina served as a good starting point for our work as the nasal cavity, ethmoid air cells and bones surrounding the cavities are similarly delicate (the frontal, maxillary and sphenoid sinuses are easier to segment because they constitute comparatively large, connected areas of pixels with similar intensity values).

Our CNNs work in an end-to-end fashion. This means that the CNN acts as a function which receives minimally preprocessed CT images as input and directly returns the corresponding 2D segmentations as output. The CNN assigns a class (e.g. air or not air, where 1 corresponds to air and 0 corresponds to not air) to each pixel. From the 2D segmentations, 3D models can then be generated. For instance, triangular surface meshes can be created using the Marching Cubes algorithm [49]. Note that it is also possible to implement CNNs which directly classify each voxel in a CT volume, i.e. to perform 3D segmentation [56]. We started with a 2D approach which is easier to implement and retained it because we obtained the desired results. In an initial phase, we developed separate CNNs which perform 2D segmentations of either air-filled regions or bone only, i.e. binary segmentations where white is either air or bone and black is everything else. Figure 10.11 shows two sample CT slices, the air and bone segmentation our CNNs computed for these and the 3D model generated from the segmentations of all slices of this patient's CT volume. Note that the air surrounding the patient's head has been removed in a post processing step and is therefore not visible in the air 3D model.

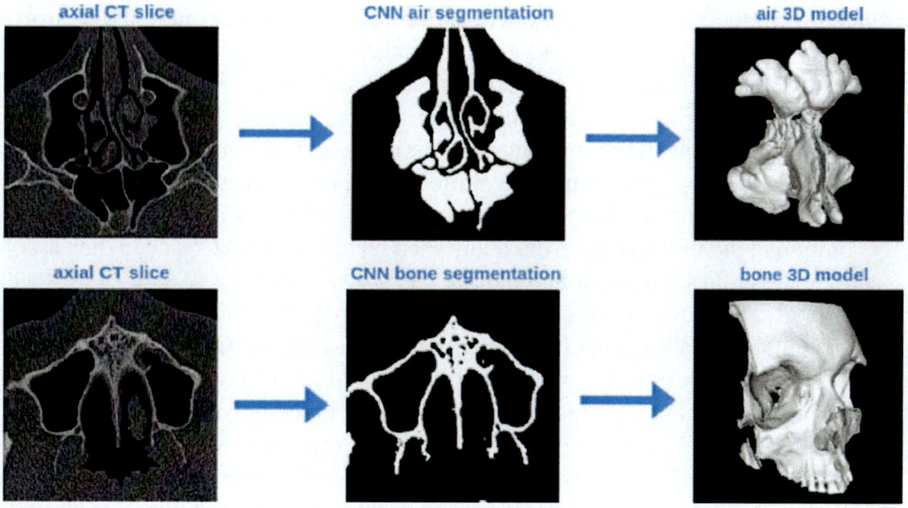

Fig. 10.11 CT slices, corresponding automatic segmentations and 3D models generated from these

10.3.2 How Does a CNN Work and How Can a CNN Be Trained to Segment CT Images?

This section will give an overview of what a CNN does and how it can be used for image segmentation. A CNN is a type of Artificial Neural Network (ANN) which processes images or other data represented by a grid-like topology. Digital images can be represented as matrices of real-valued pixel intensity values. In the case of grayscale CT images, each pixel has a single intensity value which represents the brightness of the pixel (shade of gray). We want the CNN to learn to infer whether a given pixel should be classified as either air or not air (or as either bone or not bone, i.e. as one class or the other). A CNN can achieve this goal by using several layers which perform different mathematical operations. These operations allow a CNN to extract information about the neighbourhood of a given pixel to decide if the pixel should belong to one class or the other.

Our CNN architecture consists of multiple Convolution Layers, each followed by a Nonlinearity (ReLU Layer) and a Max Pooling Layer. The most important layer, the one which gives the CNN its name, performs a (discrete) convolution and is hence called a Convolution Layer. It performs an element-wise multiplication of pixel intensity values in a neighbourhood of each pixel of the input image with the scalar elements of a so called filter matrix, followed by addition of the re-sulting terms (linear combination) [18]. This is how image filters (kernels) work. The result is a new image (matrix) called a feature map. Applying a convolution matrix to an image filters it to extract certain features. Note that in mathematics, the described operation is called a cross-correlation, but in the context of image processing the term convolution is used. A CNN's architecture is loosely inspired by the workings of neurons in the human visual system as the neighbourhood around a given pixel can be viewed as a receptive field and the connectivity pattern of the neurons of a CNN bears resemblance to visual cortex organization [26].

Typically, a nonlinear function (most often a so-called rectified linear unit, ReLU) is applied to the output of the convolution. This makes the output a non-linear function of the inputs. By applying many convolutions to an input image, a CNN can extract different features from an image. Each convolution results in a different feature map. By combining the information from many such feature maps, a CNN can then assign a certain class to each pixel of an input image.

Furthermore, convolutions can also be applied to the feature maps themselves to obtain additional feature maps. By using multiple successive Convolutional Layers, more abstract features can be extracted (e.g. an initial layer may provide information about simple geometric features like edges, a deeper layer may involve more abstract features like the general shape of a given cavity, i.e. one obtains a feature hierarchy). Using several layers in this way makes a neural network deeper, hence the term deep learning. In our CNNs, a volume of feature maps is linearly combined to obtain a single prediction matrix containing real values. The values in this matrix can then be mapped to probabilities in (0, 1) by applying a so called activation function (e.g. a sigmoid function). In a final step, a threshold is applied to this probability matrix.

This sets each value greater than or equal to 0.5 to 1 (air) and each value less than 0.5 to 0 (not air), thereby obtaining a binary segmentation.

The elements of a convolution matrix are called weights. In traditional image processing, the programmer specifies the weights of a convolution matrix to achieve a desired effect (e.g. sharpening of an image). Using a CNN, the task of assigning a class to each pixel in an image can be formulated as a mathematical optimization problem where a cost function is minimized. Weights can be initialized randomly and are improved over time rather than being manually specified by the programmer at the outset. The CNN is fed with pairs of CT images and manual segmentations as inputs. Each input helps the CNN learn better values for the weights, where better means that the cost becomes smaller, which itself means that the CNN outputs a segmentation which is more similar to the manual segmentation it is provided with (which serves as an example for what the desired output looks like). This way the CNN becomes progressively better at segmenting images by learning from examples. Figure 10.12 shows an illustration of the basic architecture of our CNN (for 2 Convolutional Layers, we use up to 4, but some of our best CNNs only use a single layer).

A CNN is trained with so-called supervised learning. This means that a CNN is provided with training data in the form of input-output pairs (CT images paired with segmentations which are manually created by a human expert) where the provided outputs tell the CNN what the desired output is supposed to look like for new inputs. This allows the CNN to learn from the provided examples. Once fully trained, the CNN can segment CT images it has not seen before. Segmentation and 3D model generation for an entire CT volume of e.g. 1000 slices is then reduced to 3 min instead of several days. The so-called labelled data consists of CT images of different patients (DICOM files obtained from partnering clinics) and manual segmentations created for them. We use 256×256 px partial images (our CT images have a resolution of 512×512 px, we use only a region of interest) of axial slices. The Hounsfield unit range is mapped to [0, 1] as a pre-processing step. Rotated and scaled versions of these axial CT slices can also be included in the dataset to create more data artificially (data augmentation). Segmentations for a subset of patients were created manually, this took several days. The manual segmentations were then inspected visually and checked for accuracy and anatomical plausibility by a practicing ENT (Ear-Nose-Throat) surgeon. Three datasets (training, validation and test set) were created from these images, where each dataset consists of both, CT images and their corresponding manual segmentations. First, the training set was used to train many CNNs with different so-called hyperparameters (described in the following paragraphs). Second, the validation set was used to select the top-performing CNN from these. Third, since this selection process can lead to overfitting on the validation set, the performance was confirmed by assessing the CNN's performance on the test set. As a heuristic, training, validation and test set often constitute 60%, 20% and 20% of the entire, randomized dataset.

In each iteration a single CT image is passed through the CNN to generate segmentations (forward pass). The cost, for instance the cross entropy between the probability matrix and the human segmentation, averaged over all images used, is then propagated back to update the weights (backward pass, back propagation of

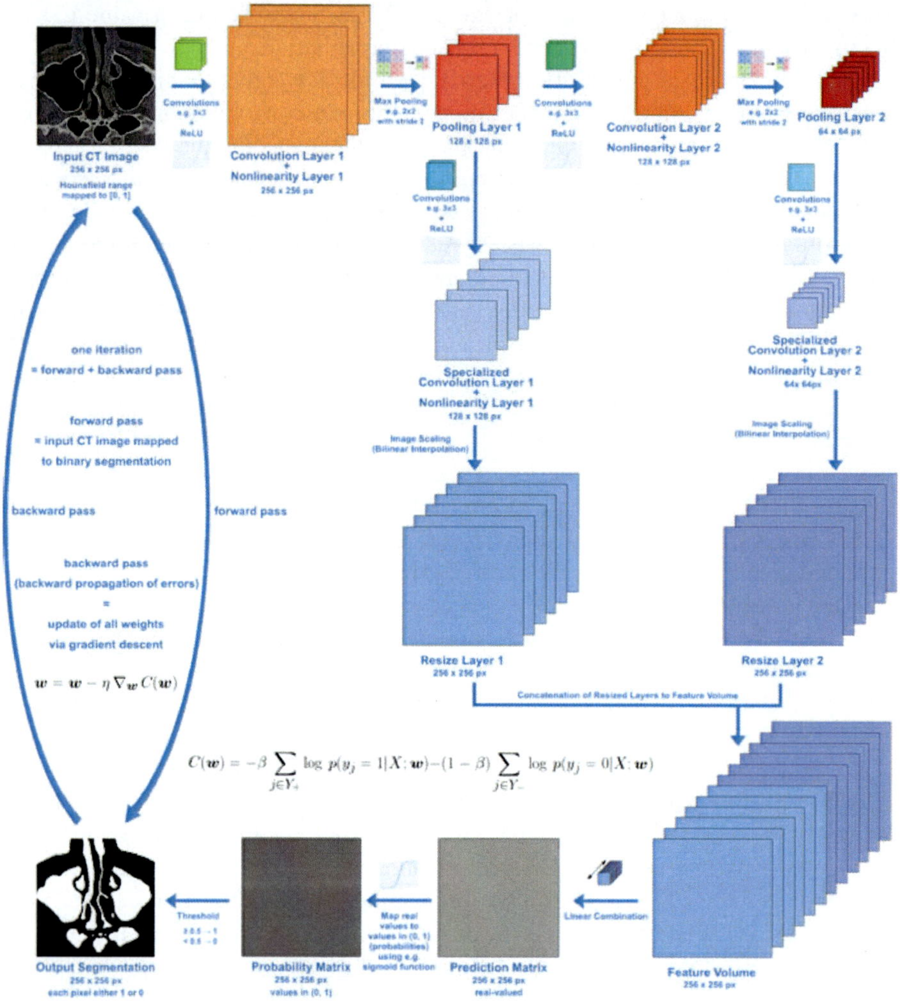

Fig. 10.12 Basic architecture of our binary segmentation CNNs (for either air or bone segmentation)

errors). This process is called stochastic gradient descent (if one training image is used each iteration and gradient descent is taken to perform updates of the weights). A step size (learning rate) is set to specify how much the weights should be adjusted within each iteration. Performing this process iteratively improves the weights and hence the obtained segmentations. The learning process is stopped once certain metrics are met, for instance once the accuracy (percentage of pixels which are the same in both CNN and human segmentation) is high enough. CNNs with different hyper-parameters (number of layers, number and size of filter matrices, step sizes etc.) are trained and the top performing ones selected using the validation and test sets. We obtained good results for a constant step size of 0.01 using up to 10 filter matrices

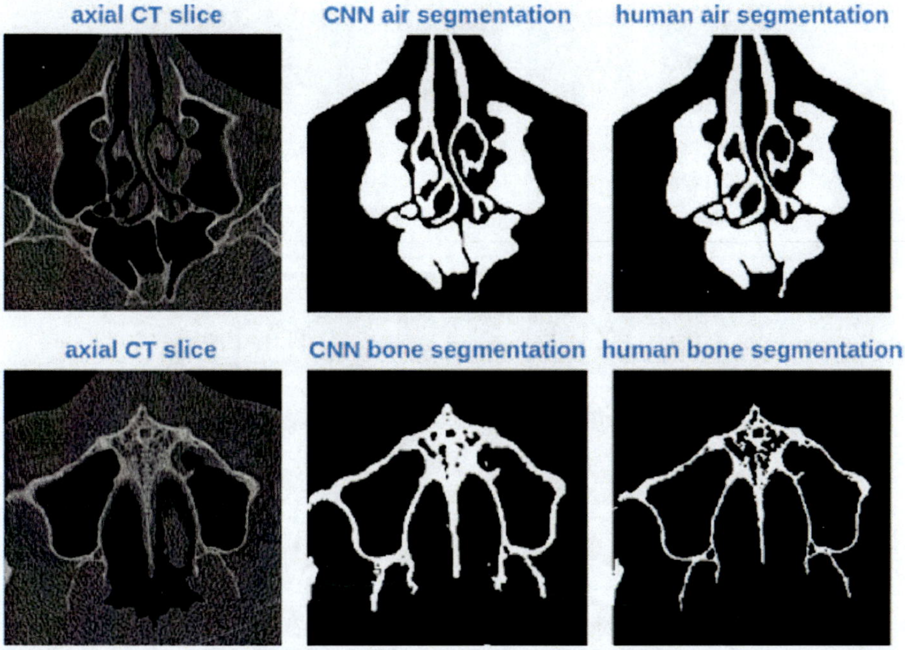

Fig. 10.13 CNN air and bone segmentation results compared to segmentations generated by a human

of size 3×3 or 5×5 for each Convolution Layer, 1 to 4 layers in total, mean cross entropy as the cost function and either Adam or Momentum as optimizer to perform stochastic gradient descent [32, 65]. The CNNs were trained for about 1000 to 5000 iterations on an NVIDIA GTX 1080 GPU and stopped once requirements for metrics such as accuracy, overlap (Jaccard index) and similar number of connected components were met.

Figure 10.13 shows a comparison of automated and human segmentations obtained for 2 sample CT images. Our top performing CNNs achieve an accuracy of 99–96% for air segmentation and an accuracy of 95–96% for bone segmentation. The latter is more difficult because bones and surrounding tissue often have very similar intensity values and are often hard to distinguish even for human experts due to the noise in the CT images. Figure 10.14 illustrates how the segmentation outputs of a CNN improve over time.

In a second phase, motivated by the good results for binary segmentations, we developed a CNN which, instead of predicting for a single pixel whether it belongs to a single class or not, predicts whether the pixel belongs to one of multiple classes. This is achieved by performing a different linear combination of the feature map volume for each class, which results in a vector of class probabilities for each pixel. The predicted class is then given by the largest element of this vector. We trained CNNs for either 11 classes (left/right frontal/sphenoid/maxillary sinus, nasal cavity, oral cavity, air outside, bone and tissues) or 9 classes (left/right frontal/sphenoid

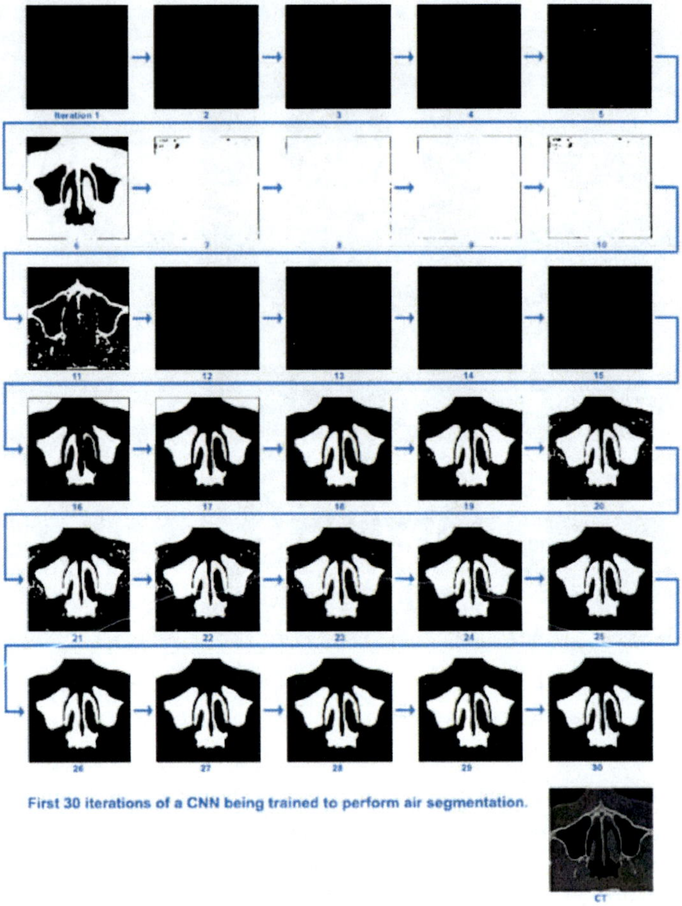

Fig. 10.14 CNN segmentation results for a single CT image during the first 30 iterations of training (CT image is taken from validation set; to update weights during training, a different CT image is used in each iteration)

sinus, left and right maxillary sinuses combined, nasal and oral cavity combined, air outside, bone and tissues). The maxillary sinuses are not adjacent to each other as they are located on opposite sides of the head. They can therefore be combined into a single layer and are easily separated later once the 3D model has been generated. Oral and nasal cavities need not necessarily be separate segments if one does not want to distinguish between the two. In general, a lower number of classes is easier to process accurately for the CNN. Figure 10.15 shows preliminary results obtained for 2 sample CT slices by a CNN trained to distinguish 11 different classes. By learning a distinct class for each type of cavity and tissue, the CNN also learns to distinguish air from bone more accurately and there is no overlap between any of the segments.

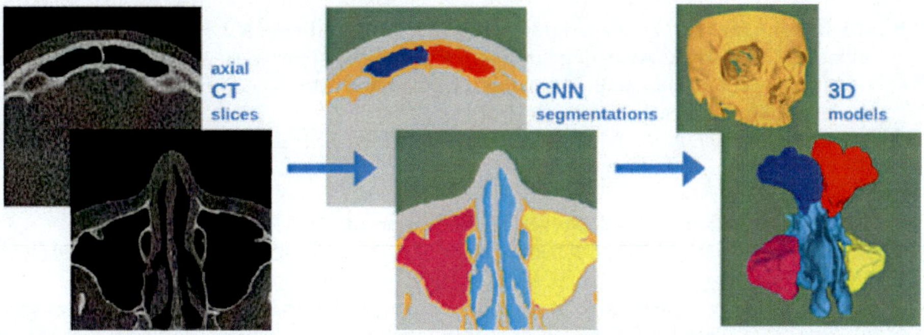

Fig. 10.15 Axial CT slices, the multi-class segmentations computed by a CNN, and the 3D models generated from these

As some classes are found in more CT images than others, form components of different size than others and make up different percentages of the total number of pixels, class balancing techniques may be required (e.g. focal loss [44] or an adequate sampling strategy [10, 63]). At the moment, the CNNs we have trained do not learn each class similarly well. Hence we are currently in the process of optimizing our results for multi-class segmentation by experimenting with different approaches to handle imbalanced classes. Once our approach has been refined, accurate 3D models of bone, tissues and individual cavities shall be obtained for new patients within minutes. Currently, this research work is partly funded by the COIN/IraSME project Rhinodiagnost, Morphological and Functional Precision Diagnostics for the Nose [45].

References

1. B. Asgharian, W. Hofmann, R. Bergmann, Particle deposition in a multiple-path model of the human lung. Aerosol Sci. Technol. **34**(4), 332–339 (2001)
2. S. Ashaat, A.M. Al-Jumaily, Reducing upper airway collapse at lower continuous positive airway titration pressure. J. Biomech. **49**(16), 3915–3922 (2016)
3. A. Bates, R. Cetto, D. Doorly, R. Schroter, N. Tolley, A. Comerford, The effects of curvature and constriction on airflow and energy loss in pathological tracheas. Respir. Physiol. Neurobiol. **234**, 69–78 (2016)
4. A. Bates, A. Comerford, R. Cetto, D. Doorly, R. Schroter, N. Tolley, Computational fluid dynamics benchmark dataset of airflow in tracheas. Data Brief **10**, 101–107 (2017)
5. A. Bates, A. Comerford, R. Cetto, R. Schroter, N. Tolley, D. Doorly, Power loss mechanisms in pathological tracheas. J. Biomech. **49**(11), 2187–2192 (2016)
6. A.J. Bates, D.J. Doorly, R. Cetto, H. Calmet, A. Gambaruto, N. Tolley, G. Houzeaux, R. Schroter, Dynamics of airflow in a short inhalation. J. R. Soc. Interface **12**(102), 20140880 (2015)
7. A.J. Bates, N.S. Higano, E.B. Hysinger, R.J. Fleck, A.D. Hahn, S.B. Fain, P.S. Kingma, J.C. Woods, Quantitative assessment of regional dynamic airway collapse in neonates via retrospectively respiratory-gated 1h ultrashort echo time mri. J. Magn. Reson. Imaging **49**(3), 659–667 (2019)

8. A.J. Bates, A. Schuh, G. Amine-Eddine, K. McConnell, W. Loew, R.J. Fleck, J.C. Woods, C.L. Dumoulin, R.S. Amin, Assessing the relationship between movement and airflow in the upper airway using computational fluid dynamics with motion determined from magnetic resonance imaging. Clin. Biomech. **66**, 88–96 (2017)
9. A.J. Bates, A. Schuh, K. McConnell, B.M. Williams, J.M. Lanier, M.M. Willmering, J.C. Woods, R.J. Fleck, C.L. Dumoulin, R.S. Amin, A novel method to generate dynamic boundary conditions for airway cfd by mapping upper airway movement with non-rigid registration of dynamic and static mri. Int. J. Numer. Methods Biomed. Eng. **34**(12), e3144 (2018)
10. G.E. Batista, R.C. Prati, M.C. Monard, A study of the behavior of several methods for balancing machine learning training data. ACM SIGKDD Explor. Newsl. **6**(1), 20–29 (2004)
11. K.H. Cha, L.M. Hadjiiski, R.K. Samala, H.-P. Chan, R.H. Cohan, E.M. Caoili, C. Paramagul, A. Alva, A.Z. Weizer, Bladder cancer segmentation in ct for treatment response assessment: application of deep-learning convolution neural networkâĂăa pilot study. Tomography **2**(4), 421 (2016)
12. S. Cheng, S. Gandevia, M. Green, R. Sinkus, L. Bilston, Viscoelastic properties of the tongue and soft palate using mr elastography. J. Biomechan. **44**(3), 450–454 (2011)
13. J.I. Choi, C.S. Kim, Mathematical analysis of particle deposition in human lungs: An improved single path transport model. Inhal. Toxicol. **19**, 925–939 (2007)
14. P.F. Christ, M.E.A. Elshaer, F. Ettlinger, S. Tatavarty, M. Bickel, P. Bilic, M. Rempfler, M. Armbruster, F. Hofmann, M. DÂĂŹAnastasi et al., Automatic liver and lesion segmentation in ct using cascaded fully convolutional neural networks and 3d conditional random fields, in *International Conference on Medical Image Computing and Computer-Assisted Intervention* (Springer, Berlin, 2016), pp. 415–423
15. A. Comerford, C. Forster, W.A. Wall, Structured tree impedence outflow boundary conditions for 3d lung simulations. J. Biomechan. Eng. **132**, 10 (2010)
16. J. Dong, Y. Shang, K. Inthavong, H.-K. Chan, J. Tu, Numerical comparison of nasal aerosol administration systems for efficient nose-to-brain drug delivery. Pharm. Res. **35**(1), 5 (2018)
17. G. Dournes, D. Grodzki, J. Macey, P.-O Girodet, M. Fayon, J.-F. Chateil, M. Montaudon, P. Berger, F. Laurent (2015) Quiet submillimeter mr imaging of the lung is feasible with a petra sequence at 1.5 t. Radiology **276**(1), 258–265
18. V. Dumoulin, F. Visin, A guide to convolution arithmetic for deep learning. arXiv:1603.07285 (2016)
19. Y. Feng, C. Kleinstreuer, Analysis of non-spherical particle transport in complex internal shear flows. Phys. Fluids **25**(9), 091904 (2013)
20. Y. Feng, Z. Xu, A. Haghnegahdar, Computational fluid-particle dynamics modeling for unconventional inhaled aerosols in human respiratory systems. *Aerosols-Science and Case Studies* (2016)
21. Y. Feng, J. Zhao, C. Kleinstreuer, Q. Wang, J. Wang, D.H. Wu, J. Lin, An in silico intersubject variability study of extra-thoracic morphology effects on inhaled particle transport and deposition. J. Aerosol Sci. **123**, 185–207 (2018)
22. A. Haghnegahdar, Y. Feng, X. Chen, J. Lin, Computational analysis of deposition and translocation of inhaled nicotine and acrolein in the human body with e-cigarette puffing topographies. Aerosol Sci. Technol. **52**(5), 483–493 (2018)
23. N. Higano, A. Bates, E. Hysinger, I. St. Onge, R. Fleck, P. Kingma, J. Woods, Dynamic tracheal collapse and correlation to later tracheostomy in neonates with bronchopulmonary dysplasia via quantitative ultrashort echo-time mri, in *C16. Clinical Studies in Bronchopulmonary Dysplasia* (American Thoracic Society, New York, 2019), pp. A4264–A4264
24. N.S. Higano, A.J. Bates, J.A. Tkach, R.J. Fleck, F.Y. Lim, J.C. Woods, P.S. Kingma, Pre- and post-operative visualization of neonatal esophageal atresia/tracheoesophageal fistula via magnetic resonance imaging. J. Pediatr. Surg. Case Rep. **29**, 5–8 (2018)
25. N.S. Higano, A.D. Hahn, J.A. Tkach, X. Cao, L.L. Walkup, R.P. Thomen, S.L. Merhar, P.S. Kingma, S.B. Fain, J.C. Woods, Retrospective respiratory self-gating and removal of bulk motion in pulmonary ute mri of neonates and adults. Magn. Reson. Med. **77**(3), 1284–1295 (2017)

26. D.H. Hubel, T.N. Wiesel, Receptive fields and functional architecture of monkey striate cortex. J. Physiol. **195**(1), 215–243 (1968)
27. K. Inthavong, Z. Tian, J. Tu, W. Yang, C. Xue, Optimising nasal spray parameters for efficient drug delivery using computational fluid dynamics. Comput. Biol. Med. **38**(6), 713–726 (2008)
28. T. Janke, P. Koullapis, S. Kassinos, K. Bauer, Piv measurements of the siminhale benchmark case. Eur. J. Pharm. Sci. **133**, 183–189 (2019)
29. K.M. Johnson, S.B. Fain, M.L. Schiebler, S. Nagle, Optimized 3d ultrashort echo time pulmonary mri. Magn. Reson. Med. **70**(5), 1241–1250 (2013)
30. K. Kamnitsas, E. Ferrante, S. Parisot, C. Ledig, A.V. Nori, A. Criminisi, D. Rueckert, B. Glocker, Deepmedic for brain tumor segmentation, in *International workshop on Brainlesion: Glioma, multiple sclerosis, stroke and traumatic brain injuries* (Springer, Berlin, 2016), pp. 138–149
31. K. Kamnitsas, C. Ledig, V.F. Newcombe, J.P. Simpson, A.D. Kane, D.K. Menon, D. Rueckert, B. Glocker, Efficient multi-scale 3d cnn with fully connected crf for accurate brain lesion segmentation. Med. Image Anal. **36**, 61–78 (2017)
32. D.P. Kingma, J. Ba, Adam: A method for stochastic optimization. arXiv:1412.6980 (2014)
33. L. Koblinger, W. Hofmann, Analysis of human lung morphometric data for stochastic aerosol deposition calculations. Phys. Med. Biol. **30**(6), 541–556 (1985)
34. L. Koblinger, W. Hofmann, Monte carlo modeling of aerosol deposition in human lungd. part i: Simulation of particle transport in a stochastic lung structure. J. Aerosol Sci. **21**(5), 661–674 (1990)
35. A.V. Kolanjiyil, C. Kleinstreuer, Nanoparticle mass transfer from lung airways to systemic regions—part i: Whole-lung aerosol dynamics. J. Biomechan. Eng. **135**(12), 11 (2013)
36. A.V. Kolanjiyil, C. Kleinstreuer, Nanoparticle mass transfer from lung airways to systemic regions—part ii: Multi-compartmental modeling. J. Biomech. Eng. **135**, 12 (2013)
37. A.V. Kolanjiyil, C. Kleinstreuer, Computationally effecient analysis of particle transport and deposition in a human whole-airway model. part i: theory and model validation. Comput. Biol. Med. **76**, 193–204 (2016)
38. A.V. Kolanjiyil, C. Kleinstreuer, Computational analysis of aerosol-dynamics in a human whole-lung airway model. J. Aerosol Sci. **114**, 301–316 (2017)
39. A.V. Kolanjiyil, C. Kleinstreuer, R.T. Sadikot, Computationally efficient analysis of particle transport and deposition in a human whole-lung-airway model. part ii: Dry powder inhaler application. Comput. Biol. Med. **2017**, 247–253 (2017)
40. P. Koullapis, P. Hofemeier, J. Sznitman, S. Kassinosa, An efficient computational fluid-particle dynamics method to predict deposition in a simplified approximation of the deep lung. Eur. J. Pharm. Sci. **113**, 132–144 (2018)
41. Y. LeCun, Y. Bengio, G. Hinton, Deep learning. Nature **521**(7553), 436–444 (2015)
42. Y. LeCun, B. Boser, J.S. Denker, D. Henderson, R.E. Howard, W. Hubbard, L.D. Jackel, Backpropagation applied to handwritten zip code recognition. Neural Comput. **1**(4), 541–551 (1989)
43. Y. LeCun, L. Bottou, Y. Bengio, P. Haffner et al., Gradient-based learning applied to document recognition. Proc. IEEE **86**(11), 2278–2324 (1998)
44. T.-Y. Lin, P. Goyal, R. Girshick, K. He, P. Dollár, Focal loss for dense object detection, in *Proceedings of the IEEE International Conference on Computer Vision*, pp. 2980–2988 (2017)
45. A. Lintermann, J. Göbbert, K. Vogt, W. Koch, A. Hetzel, Rhinodiagnost-morphological and functional precision diagnostics of nasal cavities. InSiDE, Innov. Supercomput. Dtsch. **15**(2), 106–109 (2017)
46. G. Litjens, T. Kooi, B.E. Bejnordi, A.A.A. Setio, F. Ciompi, M. Ghafoorian, J.A. Van Der Laak, B. Van Ginneken, C.I. Sánchez, A survey on deep learning in medical image analysis. Med. Image Anal. **42**, 60–88 (2017)
47. P.W. Longest, G. Tian, R. Delvadia, M. Hindle, Development of a stochastic individual path (sip) model for predicting the deposition of pharmaceutical aerosols: Effects of turbulence, polydisperse aerosol size, and evaluation of multiple lung lobes. Aerosol Sci. Technol. **46**(12), 1271–1285 (2012)

48. P.W. Longest, G. Tian, N. Khajeh-Hosseini-Dalasm, M. Hindle, Validating whole-airway cfd predictions of dpi aerosol deposition at multiple flow rates. J. Aerosol Med. Pulm. Drug Deliv. **29**(6), 461–481 (2016)
49. W.E. Lorensen, H.E. Cline, Marching cubes: A high resolution 3d surface construction algorithm, in *ACM siggraph computer graphics*, vol. 21 (ACM, New York, 1987), pp. 163–169
50. F. Lu, F. Wu, P. Hu, Z. Peng, D. Kong, Automatic 3d liver location and segmentation via convolutional neural network and graph cut. Int. J. Comput. Assist. Radiol. Surg. **12**(2), 171–182 (2017)
51. B. Ma, K.R. Lutchen, An anatomically based hybrid computational model of the human lung and its application to low frequency oscillatory mechanics. Ann. Biomed. Eng. **34**(11), 1691–1704 (2006)
52. M. Malvè, A. Pérez del Palomar, S. Chandra, J. López-Villalobos, A. Mena, E. Finol, A. Ginel, M. Doblaré, Fsi analysis of a healthy and a stenotic human trachea under impedance-based boundary conditions. J. Biomech. Eng. **133**, 2 (2011)
53. K.-K. Maninis, J. Pont-Tuset, P. Arbeláez, L. Van Gool, Deep retinal image understanding, in *International conference on medical image computing and computer-assisted intervention* (Springer, Berlin, 2016), pp. 140–148
54. T.B. Martonen, Analytical model of hygroscopic particle behavior in human airways. Bull. Math. Biol. **44**(3), 425–442 (1982)
55. M. Mihaescu, E. Gutmark, R. Elluru, J.P. Willging, Large eddy simulation of the flow in a pediatric airway with subglottic stenosis, in *47th AIAA Aerospace Sciences Meeting including The New Horizons Forum and Aerospace Exposition*, p. 775 (2009)
56. F. Milletari, N. Navab, S.-A. Ahmadi, V-net: Fully convolutional neural networks for volumetric medical image segmentation. *2016 Fourth International Conference on 3D Vision (3DV)*, pp. 565–571 (2016)
57. M. Monjezi, R. Dastanpour, M.S. Saidi, A.R. SPishevari, Prediction of particle deposition in the respiratory track using 3dÂĂŞ1d modeling. Sci. Iran. **19**(6), 1479–1486 (2012)
58. A.I. Pack, Dynamic upper airway imaging during awake respiration in normal subjects and patients with sleep disordered breathing. Am. Rev. Respir. Dis. **148**, 1385–1400
59. A. Pandal-Blanco, R. Barrio-Perotti, R. Agujetas-Ortiz, A. Fernandez-Tena, Implementation of a specific boundary condition for a simplified symmetric single path cfd lung model with openfoam. Biomech. Model. Mechanobiol. **18**, 1759–1771 (2019)
60. C. Payer, D. Štern, H. Bischof, M. Urschler, Multi-label whole heart segmentation using cnns and anatomical label configurations, in *International Workshop on Statistical Atlases and Computational Models of the Heart* (Springer, Berlin, 2017), pp. 190–198
61. M.S. Pearce, J.A. Salotti, M.P. Little, K. McHugh, C. Lee, K.P. Kim, N.L. Howe, C.M. Ronckers, P. Rajaraman, A.W. Craft et al., Radiation exposure from ct scans in childhood and subsequent risk of leukaemia and brain tumours: a retrospective cohort study. Lancet **380**(9840), 499–505 (2012)
62. J. Pirnar, L. Dolenc-Grošelj, I. Fajdiga, I. Žun, Computational fluid-structure interaction simulation of airflow in the human upper airway. J. Biomech. **48**(13), 3685–3691 (2015)
63. F. Provost, Machine learning from imbalanced data sets 101, in *Proceedings of the AAAI 2000 workshop on imbalanced data sets*, vol. 68 (AAAI Press, Palo Alto, CA, 2000), pp. 1–3
64. H.R. Roth, L. Lu, N. Lay, A.P. Harrison, A. Farag, A. Sohn, R.M. Summers, Spatial aggregation of holistically-nested convolutional neural networks for automated pancreas localization and segmentation. Med. Image Anal. **45**, 94–107 (2018)
65. S. Ruder, An overview of gradient descent optimization algorithms. arXiv:1609.04747 (2016)
66. D. Rueckert, L.I. Sonoda, C. Hayes, D.L. Hill, M.O. Leach, D.J. Hawkes, Nonrigid registration using free-form deformations: Application to breast mr images. IEEE Trans. Med. Imaging **18**(8), 712–721 (1999)
67. T. Schlegl, S.M. Waldstein, W.-D. Vogl, U. Schmidt-Erfurth, G. Langs, Predicting semantic descriptions from medical images with convolutional neural networks, in *International Conference on Information Processing in Medical Imaging* (Springer, Berlin, 2015), pp. 437–448

68. R.J. Schwab, W.B. Gefter, A.I. Pack, E.A. Hoffman, Dynamic imaging of the upper airway during respiration in normal subjects. J. Appl. Physiol. **74**(4), 1504–1514 (1993)
69. M. Shakeri, S. Tsogkas, E. Ferrante, S. Lippe, S. Kadoury, N. Paragios, I. Kokkinos, Sub-cortical brain structure segmentation using f-cnn's, in *2016 IEEE 13th International Symposium on Biomedical Imaging (ISBI)* (IEEE, Prague, 2016), pp. 269–272
70. W.-C. Su, Y.S. Cheng, Estimation of carbon nanotubes deposition in a human respiratory tract replica. J. Aerosol Sci. **79**, 72–85 (2015)
71. W.-C. Su, B.K. Ku, P. Kulkarni, Y.S. Cheng, Deposition of graphene nanomaterial aerosols in human upper airways. J. Occup. Environ. Hyg. **13**(1), 48–59 (2016)
72. B. Suki, R.H. Habib, A.C. Jackson, Wave propagation, input impedance, and wall mechanics of the calf trachea from 16 to 1,600 hz. Am. Physiol. Soc. **75**(6), 2755–2766 (1993)
73. D.B. Taulbee, C.P. Yu, A theory of aerosol deposition in the humen respirtory tract. J. Appl. Physiol. **38**(1), 77–85 (1975)
74. M.H. Tawhai, A.J. Pullan, P.J. Hunter, Generation of an anatomically based three-dimensional model of the conducting airways. Ann. Biomed. Eng. **28**, 793–802 (2000)
75. A. Tena, J. Fernandez, E. Alvarez, P. Casan, D.K. Walters, Design of a numerical model of lung by means of a special boundary condition in the truncated branches. Int. J. Numer. Meth. Biomed. Eng. **33**(6), 1–9 (2016)
76. I.E. Vignon-Clementel, C.A. Figueroa, K.E. Jansen, C.A. Taylor, Outflow boundary conditions for three-dimensional finite element modeling of blood flow and pressure in arteries. Comput. Methods Appl. Mech. Eng. **195**, 3776–3796 (2006)
77. A. Voulodimos, N. Doulamis, A. Doulamis, E. Protopapadakis, Deep learning for computer vision: A brief review. Comput. Intell. Neurosci. **2018**, 7068349 (2018)
78. D.K. Walters, W.H. Luke, Computational fluid dynamics simulations of particle deposition in large-scale, multigenerational lung models. J. Biomech. Eng. **133**(1), 011003 (2011)
79. Y. Wang, J. Wang, Y. Liu, S. Yu, X. Sun, S. Li, S. Shen, W. Zhao, Fluid-structure interaction modeling of upper airways before and after nasal surgery for obstructive sleep apnea. Int. J. Numer. Methods Biomed. Eng. **28**(5), 528–546 (2012)
80. E. Weibel, *Morhometry of the human lung* (Academic Press, New York, 1963)
81. Z. Zhang, C. Kleinstreuer, C.S. Kim, Airflow and nanoparticle deposition in a 16-generation tracheobronchial airway model. Ann. Biomed. Eng. **36**(19), 2095–2110 (2008)
82. M. Zhao, T. Barber, P. Cistulli, K. Sutherland, G. Rosengarten, Predicting the treatment response of oral appliances for obstructive sleep apnea using computational fluid dynamics and fluid-structure interaction simulations, in *ASME 2013 International Mechanical Engineering Congress and Exposition*, American Society of Mechanical Engineers Digital Collection, New York (2013)
83. M. Zhao, T. Barber, P.A. Cistulli, K. Sutherland, G. Rosengarten, Simulation of upper airway occlusion without and with mandibular advancement in obstructive sleep apnea using fluid-structure interaction. J. Biomech. **46**(15), 2586–2592 (2013)
84. J.H. Zhu, H.P. Lee, K.M. Lim, S.J. Lee, L.S.L. Teo, D.Y. Wang, Passive movement of human soft palate during respiration: A simulation of 3d fluid/structure interaction. J. Biomech. **45**(11), 1992–2000 (2012)

Chapter 11
CFD Applications for Drug Delivery

Kendra Shrestha, Ross Walenga, Jinxiang Xi, Yidan Shang, Hana Salati, Jim Bartley, and David White

Abstract This chapter provides the latest research trends in applications of CFD for investigating nasal drug delivery. Contributions include research and outcomes from the regulatory body, US Food Drug Administration on recent projects. Additionally, the chapter showcases the application of multiphase flow for capturing nasal spray drug delivery for improving olfactory targeting for nose-to-brain drug delivery, and nasal irrigation targeting the sinuses.

11.1 Introduction

Nasal sprays are used to deliver drug for common problems such as nasal congestion and allergic rhinitis. It is also used as an alternative route for systemic therapy instead of injections and pills. However, it is a complex physical process which depends on numerous parameters such as nasal device design, drug liquid properties, and patient handling.

This chapter provides the latest research trends in applications of CFD for investigating nasal drug delivery. Since the inhaled air is mixed with the intended drug

K. Shrestha (✉) · Y. Shang
RMIT University, Melbourne, Australia
e-mail: kendra.shrestha@rmit.edu.au

Y. Shang
e-mail: yidan.shang@rmit.edu.au

R. Walenga
U.S. Food and Drug Administration, Silver Spring, USA

J. Xi
University of Massachusetts Lowell, Lowell, England

H. Salati · D. White
Auckland University of Technology, Auckland, New Zealand

J. Bartley
University of Auckland, Auckland, New Zealand

K. Inthavong et al. (eds.), *Clinical and Biomedical Engineering in the Human Nose*, Biological and Medical Physics, Biomedical Engineering,
https://doi.org/10.1007/978-981-15-6716-2_11

formulation, then a second phase (e.g. solid or liquid) is introduced with the inhaled air. Therefore, nasal drug delivery studies is inherently an application of multiphase flow where it involves: liquid-gas (liquid jet flow into nasal air cavity space during nasal irrigation); solid-gas (drug particle introduced into the nasal air cavity during nasal spray atomization), or even gas-gas (gas vapour mixed with inhaled air during vapourized drug formulations). This chapter showcases some of these flows that includes capturing nasal spray drug delivery for improving olfactory targeting for nose-to-brain drug delivery, and nasal irrigation targeting the sinuses.

11.2 CFD Modeling of Nasally Administered Drug Delivery in Regulatory Science Research

Ross Walenga

11.2.1 Generic Drug Regulatory Research Program

The Generic Drug User Fee Amendments (GDUFA) were signed into law in the United States of America as part of the Food and Drug Administration Safety and Innovation Act of 2012 [3] and was reauthorized in 2017. This legislation has allowed the U.S. Food and Drug Administration (FDA) to collect user fees from industry to provide funding for the generic drug review program at FDA, which includes allotments for staffing and research [4]. Under GDUFA, the Office of Generic Drugs (OGD) has established a regulatory science research program, where the purpose is to facilitate development of tools and datasets useful to industry for generic drug development as well as to provide FDA with enhanced tools for assessing bioequivalence (BE) [5]. This program includes several research projects that either focus on development of in silico tools that are useful for generic drug product development or use in silico methods such as CFD or physiologically-based pharmacokinetic (PBPK) modeling to answer questions pertinent to assessing BE for generic drug products.

To support FDA's current thinking on BE for nasal drug products, research projects have been initiated that relate to this class of products and are part of the regulatory science research program managed by OGD. Establishing BE between a test and reference nasal drug product for local action has been described in the FDA general guidance [1]. However, current FDA recommendations for establishing BE of nasal drug products are provided in product-specific guidances [2]. Assurance of equivalent local delivery between a test and reference nasal drug product, when formulated as a *solution*, can be accomplished through demonstration of (i) formulation sameness (i.e., qualitative (Q1) and quantitative (Q2)), (ii) comparable in vitro performance (i.e., in vitro BE studies), and (iii) device similarity (i.e., user interface).

In contrast, FDA's recommendations for establishing equivalence for nasal *suspension* drug products for local action also include in vivo pharmacokinetic (PK) and comparative clinical endpoint BE studies, in addition to the studies recommended for nasal solution drug products, to ensure equivalent systemic exposure and local drug delivery, respectively. Locally-acting nasal suspension drug products are intended for delivery to the site(s) of action in the nose, but drug deposited in the target regions of the nose may also enter the systemic circulation, so that determination of drug concentration in a biological fluid (e.g., plasma) is important for establishing BE between test and reference products due to possible systemic side effects [53]. Because in vitro performance is not a direct indicator of clinical efficacy, and measuring the amount of the drug in the blood or plasma is primarily recommended to address the potential systemic toxicity of the drug, a confirmatory comparative clinical endpoint study is recommended to provide direct evidence to assure equivalent local delivery, as this study is intended only to confirm the lack of important clinical differences between test and reference nasal spray suspension formulations [42]. However, comparative clinical endpoint BE studies can present challenges to potential generic applicants, since these studies can require a large study population to address higher variability and potentially lower sensitivity in detecting product differences that could impact regional deposition.

With the advancement of science and particle size characterization technology, FDA's current thinking with respect to BE assessment of locally-acting nasal spray suspension products is evolving. Novel technology such as Morphologically Directed Raman Spectroscopy (MDRS) may be able to distinguish the active pharmaceutical ingredient (API) from the suspended excipient particles and measure the API particle size distribution in test and reference nasal suspension products (which has the potential to influence the rate and extent of drug availability to nasal sites of action and to the systemic circulation) [1]. If so, potential applicants may submit comparative API PSD data as an alternative to the currently recommended comparative clinical endpoint BE study, in the context of a weight-of-evidence approach.

To facilitate development of generic locally acting nasal suspension drug products, FDA has been conducting several projects under its GDUFA-funded research program that focus on improving the FDA's understanding of the factors that may impact the local drug delivery to the sites of action within the nasal cavity. CFD is a useful tool for understanding the impact of device and formulation differences on in vitro metrics of interest and subsequently connecting these with in vivo nasal deposition. When CFD is combined with a PBPK model, it is possible to explore the impact of these differences on local and systemic PK metrics. Several GDUFA-funded studies have considered the use of CFD and in some cases PBPK to improve the tools available to industry for product development and to further develop BE methodology currently used at FDA.

11.2.2 General Model Development for CFD and PBPK Modeling of Nasal Drug Products

As of 2014, several nasal CFD-PBPK models had been developed to better understand local and systemic absorption of toxic vapors in animals, [6, 8, 9, 13, 25, 40] but there were no known efforts to develop a CFD-PBPK model for nasal suspension sprays. To meet this perceived need, the FDA funded a grant, 'Development of Hybrid CFD-PBPK Models for Absorption of Intranasal Corticosteroids' [21]. The purpose of the grant was to use CFD to predict regional nasal deposition of three intranasal corticosteroids (ICSs)—fluticasone propionate, budesonide, and mometasone furoate, which served as inputs to the nasal PBPK model that predicted local tissue and systemic plasma drug concentrations. Computed tomography (CT) scan data were used to generate two patient-specific geometric models, one healthy and one with allergic rhinitis, where rhinologists subdivided the models into anatomical relevant regions, as shown in Fig. 11.1 for the rhinitic model [37, 38]. Droplet size distribution data were collected using laser diffraction with actuation forces of 34.3 N, 56.9 N, and 84.3 N, which served as inputs to the three-dimensional, steady state, laminar CFD simulations, where a key result was that the actuation force had no significant effect on regional deposition fraction [34, 37]. Due to the inclusion of other excipients in nasal suspension spray formulations, translation of CFD regional deposition results to PBPK inputs requires a method for predicting API mass by droplet size, which was also developed as part of this grant [38]. The PBPK model that was developed includes considerations for dissolution, receptor binding, mucociliary clearance, and metabolism, and initial predictions of PK for fluticasone propionate administered via intravenous, oral, and nasal spray routes showed reasonably good agreement with in vivo PK data [33, 54]. The overall result of the grant was a viable CFD-PBPK model for prediction of local and systemic plasma concentration of locally-acting nasal suspension spray drug products that may be immediately useful or developed further if other enhancements are desired.

Another approach to predicting PK for nasal suspension spray drug products was presented through a grant titled 'Predictive Lung Deposition Models for Safety and Efficacy of Orally Inhaled Drug Products' [20]. A three-dimensional, steady state, Reynolds Stress Model turbulent CFD nasal model was used to predict nasal deposition, where the particle deposition locations were then transferred to a quasi-two-dimensional approximation of a nasal surface mucus layer [50]. Mucociliary transport of the particles in the mucus layer model was modeled using CFD while dissolution and absorption of the drug were also modeled simultaneously [50]. Following initial model development, the effect of differences in drug solubility and epithelium octanol-in-water partition coefficient on absorption were explored [49]. For the next step, a two-compartment PK model was coupled with the original CFD absorption model, where the initial prediction of mometasone furoate systemic plasma concentration demonstrated a good prediction of maximum plasma concentration when compared with in vivo PK data [49]. The CFD-PK model developed for this grant, as shown in Fig. 11.2, represented an enhancement in the ability of CFD to simul-

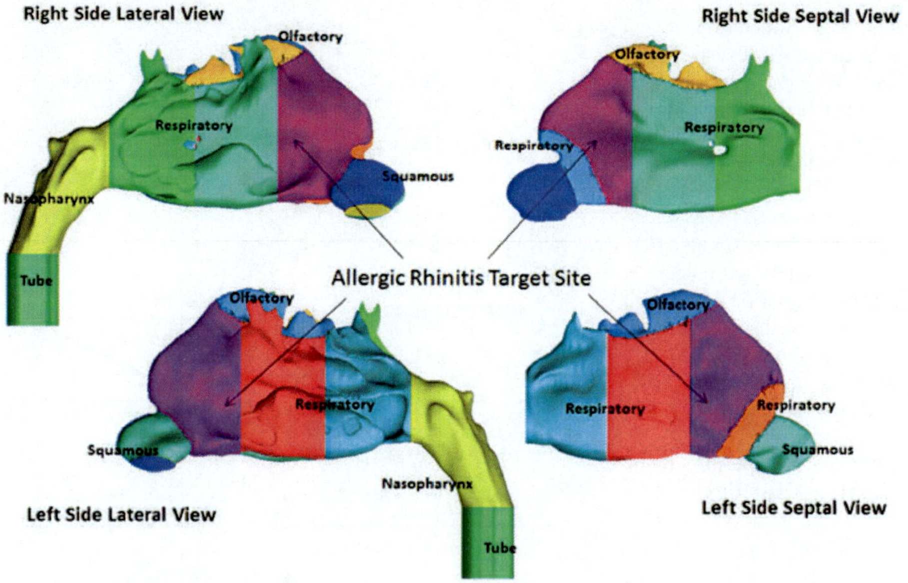

Fig. 11.1 Anatomical regions of rhinitic nasal model that was based on CT scan data. Image provided by Julia Kimbell

taneously model the effects of deposition, mucociliary clearance, dissolution, and absorption. A further extension of this concept is to model mucociliary clearance in a three-dimensional nasal surface model. A grant titled 'Nasal Mucociliary Clearance Affecting Local Drug-absorption in Subject-specific Geometries' [23] was funded to develop a three-dimensional nasal surface model, which will allow for predictions of local tissue concentration on a regional basis and is expected to be useful for drug products that target specific regions.

11.2.3 Application-Specific CFD Modeling of Nasal Drug Products

In addition to projects aimed at model development, two other GDUFA-funded projects illustrate how the FDA is using CFD to facilitate development and approval for generic nasal suspension spray drug products. This included the project 'Evaluating Relationships Between In Vitro Nasal Spray Characterization Test Metrics for Bioequivalence and Nasal Deposition In Silico and In Vitro' [22]. The project aims to develop a set of realistic nasal in vitro models that is capable of quantifying regional deposition variability from nasal suspension sprays, where the models are generated using rapid prototyping and drug delivery is evaluated using a high-performance liquid chromatography (HPLC) method. Once the models are complete, the effects of

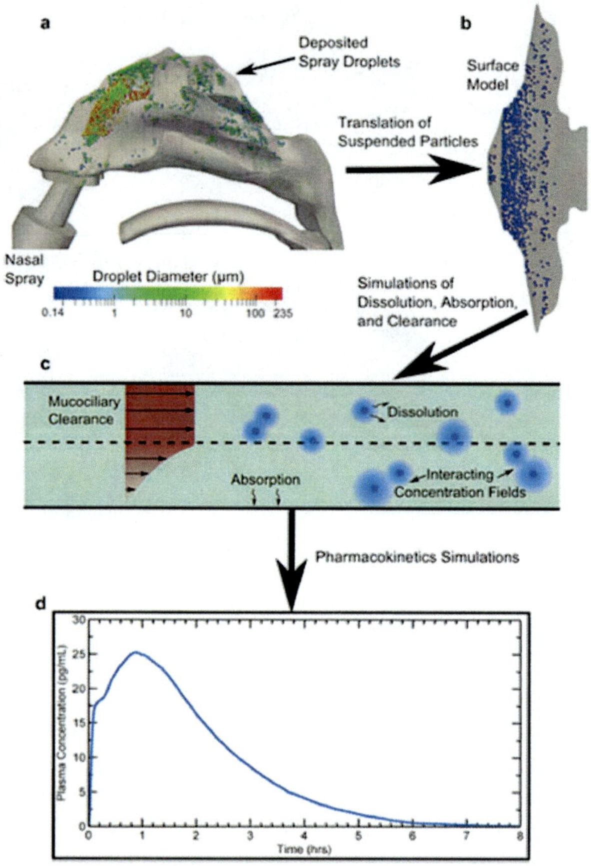

Fig. 11.2 Simulation scheme for nasal suspension spray drug delivery, where the steps are **a** deposition predictions using three-dimensional CFD, **b** translation of deposited particles to quasi-two-dimensional CFD simulation of nasal mucus layer, **c** prediction of mucus layer transit, dissolution, and absorption of deposited particles, and **d** two-compartment PK model of systemic absorption. Reprinted with permission of Elsevier. ©2016 Elsevier

differences in formulation, manufacturing, administration protocol, and pump design on regional nasal deposition will be measured using HPLC, where CFD simulations of the same experimental setups will be used to develop a better understanding of the mechanistic processes that connect product differences to changes in deposition results.

Another GDUFA-funded project is "Physiologically-based pharmacokinetic (PBPK) modeling of opioid abuse deterrent formulations (ADFs) to assess abuse deterrence potential via the nasal route," which is an internal FDA project that commenced in 2017 and is expected to be completed in the second half of 2020. Oxycodone hydrochloride extended release tablet is an abuse deterrent formulation for the abused nasal route of administration because it contains a polymer, polyethylene oxide, that is resistant to crushing and forms a sticky gel when it deposits in the nasal cavity. The primary goal of this project is to develop an in vitro dissolution method that is predictive of differences in systemic plasma concentration due to differences in drug product particle size due to manipulation, where a CFD-PBPK model based on the VCU model [48–50] is being used to predict systemic concentration. A separate internal study is developing the dissolution method and collecting data to serve as inputs for the CFD model, and in vivo data from another contract will be used to validate the model.

11.2.4 Conclusion

The primary utility of CFD in GDUFA-funded projects for nasally-administered drug products is its ability to connect in vitro metrics with in vivo performance. For generic drugs, this type of modeling is useful for taking the results of in vitro testing early in product development and understanding what changes may need to be made to the product prior to in vivo testing. Also, CFD may play a role in alternative BE approaches that rely more heavily on in vitro methods to establish BE. It is expected that CFD will continue to play an important role in GDUFA-funded research of nasally-administered drug products.

11.3 Improving Olfactory Targeting for Nose-to-Brain Drug Delivery

Jinxiang Xi

11.3.1 Introduction

The convoluted nasal passage filters most of the nasally inhaled aerosols and prevents effective drug delivery to the olfactory region or paranasal sinuses. Different methods have been explored to improve drug targeting to such secluded regions. These include pointed drug release, bi-directional intranasal delivery, electro-magnetic guidance, and pulsating flows. Both in vitro experiments and computational modeling were used to quantify the deposition rate to the olfactory region. A Sar-Gel based method was used to visualize the deposition distribution inside the nasal cavity. Results showed that point drug release and the bi-directional technique both yielded improved but limited deposition in the olfactory region. Aerosols with electric charges or magnetic properties can further improve olfactory targeting by applying an appropriate external electric or magnetic field in the nose. The olfactory deposition was sensitive to the voltage of the electrode close to the nose. For both the normal and bi-directional deliveries, electric field guidance resulted in a significant increase ($3 \approx 5$ times) in the olfactory deposition.

The capacity to deliver drugs to the olfactory region has many advantages over traditional therapies in the treatment of neurological disorders [28, 44]. Extremely low olfactory delivery using conventional nasal devices has deterred clinical applications of direct nose-to-brain drug delivery and forestalled the development of newly engineered neurological medications devised for the nasal route of administration [31]. Two factors can contribute to poor drug delivery to the olfactory region: (1) the convoluted nasal passage, which filters most of the inhaled aerosols well before they reach the olfactory region [66], and a lack of control over particle trajectories in the nose [44].

New delivery methods have been explored to increase the delivery efficiency to the olfactory region. A catheter was suggested by Wang et al. [61] to slide beneath the olfactory region to administer medications. However, this protocol has the risk to irritate the turbinate mucosa when inserting the catheter into the nasal passages. Gizurarson [51] evaluated the performance of a nasal spray with a small plume angle so that aerosols can reach the olfactory region. This technique requires a high exiting speed or high pressure to overcome the flow resistance. Hoekman and Ho [30] added a swirling flow in the nasal sprays and reported improved olfactory dosages in rats than the non-swirling flow. Data extrapolation from rat to human, however, is not straightforward due to the vast interspecies discrepancies [14, 68]. One such difference is the area of the olfactory epithelium, which covers 45% of the nasal surface in rats, but only 5.2% in humans [65]. Considering that the fate of a particle largely depends on its initial position (i.e., the positioning of nozzle), Si et al. [56] released aerosols into a specific point of the nostril and demonstrated increased olfactory deposition than the method that administers aerosols into the entire nostril. However, the overall olfactory delivery efficiency is still too low to be clinically significant. Examining all trails above, it is noted that after the aerosols enter the nose, their trajectories are solely determined by the aerodynamic force (drag) and inertia. Without further control, most of these aerosols will be passively filtered out by the labyrinthine nasal passages via either impaction or interception.

This section reviews the experimental and computational studies in our laboratory to find new delivery techniques to improve targeted drug delivery to the olfactory region and paranasal sinuses. These included point drug release, electric guidance, magnetic control, and pulsating aerosols, individually and as a combination.

11.3.2 Methods

An image-based nasal airway model was used to develop the hollow nose casts with a constant wall thickness [64]. The cast was separated into multiple sections to quantify subregional deposition fractions (Fig. 11.3a), and two delivery techniques were studied (normal and bi-directional), shown in Fig. 11.3. Step shaped grooves were built at the ends of each section for tight sealing and easy assembly [76, 77]. The influence of drug release position, electric field, magnetic control, and pulsating flow were studied on the efficiency of olfactory delivery [67].

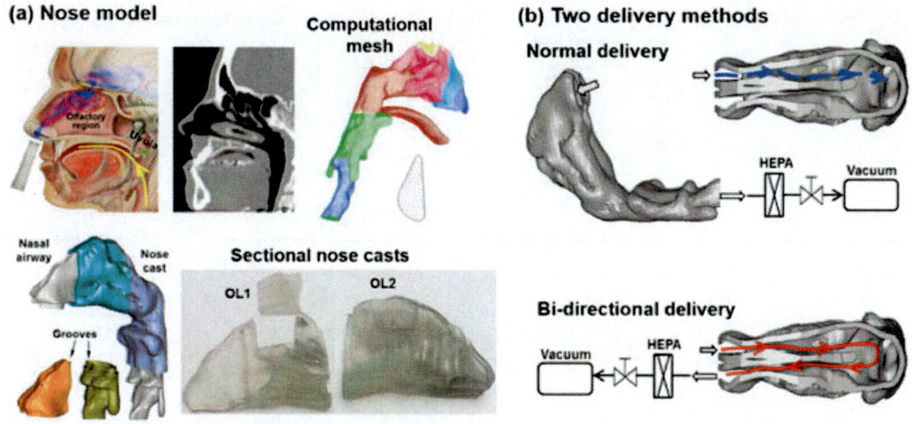

Fig. 11.3 Nose model and experimental methods: **a** computational and in vitro nasal airway models, and **b** normal and bi-directional delivery techniques

Electric Guidance

The inhalation device with electric guidance is envisioned to consist of two components: an aerosol generation/charging module and a head mounted brace [69]. Electrodes are embedded in the nasal mask, which is anchored by the brace relative to the patient nose (Fig. 11.4a). One conceptual design of the aerosol generation/charging device is depicted in Fig. 11.4b and has a nebulizer, an electric charger, and a nozzle for point release. Figure 11.4b shows a prototype manufactured by a 3D printer (Object30 Pro). Desirable parameters for such a design should include small particle size (<4 μm), low exiting velocity, and high electric charge for optimal electric guidance (28–30).

Numerical experiments of olfactory deposition were first conducted for scenarios with either point or entire nostril release with and without electric guidance. With a proper electric field, 3.7% of aerosols released from the entire nostril deposited in the olfactory region, in contrast to 0.06% in the absence of electric guidance (Fig. 11.4c). Point release further enhanced olfactory delivery by two orders of magnitude. The majority (90% in the optimal scenario) of inhaled aerosols were guided to the olfactory region (red ellipse) by avoiding being lost in the valve and turbinate (Fig. 11.4c).

Figure 11.4d compares experimental particle deposition patterns without and with electrostatic charges using powders. Similarly, most particles were trapped in the anterior nose in the absence of electric guidance and only a few particles reached the olfactory region. In contrast, olfactory delivery was significantly improved with electric guidance (blue dashed ellipse, Fig. 11.4d). A 5.2-fold enhancement was measured for normal (unidirectional) breathing and a 3.0-fold enhancement for bi-directional breathing (Fig. 11.4d).

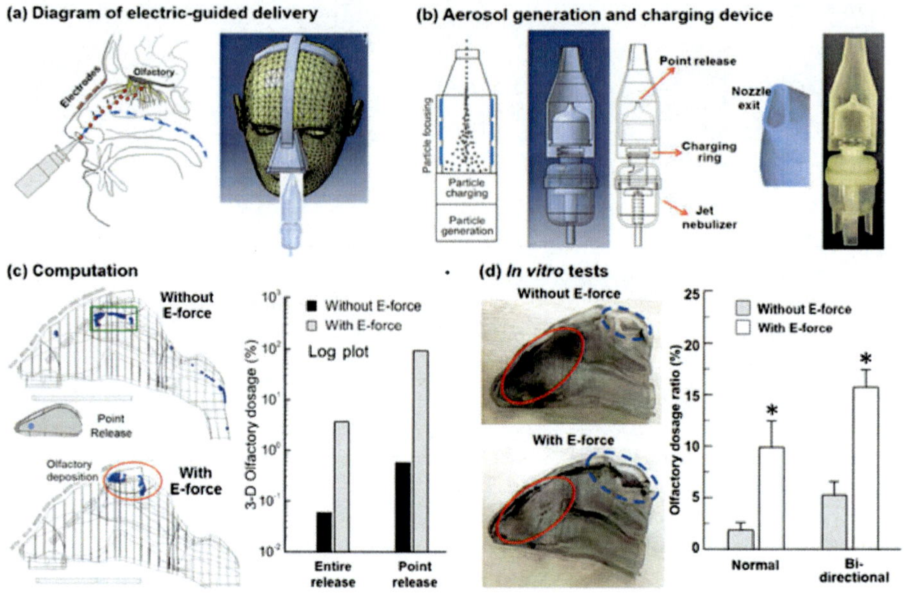

Fig. 11.4 Targeted olfactory delivery with electric guidance: **a** diagram of electric guided delivery, **b** aerosol generation and charging device, **c** computed olfactory deposition with/without electric guidance, and **d** in vitro experiments with/without electric guidance

Magnetic Guidance

The feasibility of magnetic guidance of particles was first evaluated in a two-plate channel (H: 10 mm, L: 150 mm) with permanent magnets on the outside edges [71]. Particle properties include a relative magnetic permeability of 700, a particle density of 1500 kg/m^3 as a mixture of drug and iron nanoparticles, and a particle size of 15 μm. In the absence of a magnetic force, particles follow parabolic paths due to gravitation (Fig. 11.5a). By choosing an appropriate intensity for the magnetic field, active controls of particle motion can be achieved, for instance, guiding particles to move horizontally instead of settling by counteracting the gravity (layout 1) or targeting particles to a specific site by applying a strong attraction force (layout 2). It is noted that the resultant magnet field was nonuniform, being stronger near the magnetic bars but quickly decreasing in strength away from them.

To test the magnetic control in olfactory targeting, magnet bars are put outside the nasal passage to produce a nonuniform magnetic field (vb). With the point release and an appropriate magnetic field, a deposition fraction of 45% was predicted (Fig. 11.5b, c). In contrast, only a small fraction (1.2%) of particles was predicted to deposit in the olfactory region when released from the entire nostril, even with an optimal magnetic field (Fig. 11.5b).

The effectiveness of magnetic control highly depends on the particle sizes. As shown in Fig. 11.5c, effective olfactory targeting occurs for particles in the range of 10–20μm only. Insignificant olfactory dosing occurs when the particle diameter is

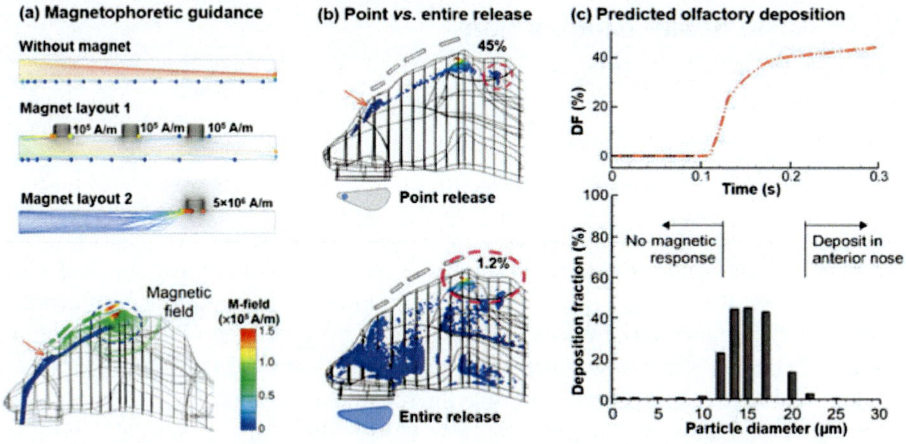

Fig. 11.5 Targeted olfactory delivery with magnetic guidance: **a** magnetophoretic guidance in a two-plate channel, **b** comparison of particle deposition patterns between point release and entire nostril release, and **c** numerically predicted olfactory deposition

smaller than 10μm or larger than 20μm. Particles smaller than 10μm have negligible responsiveness to the magnetic field, while particles larger than 20 μm will be lost in the anterior nose due to high inertia. As a reminder, the magnetic force on a particle increases with the third power of the particle diameter (d_p^3), as opposed to the first power of the magnetic field strength. In this study, the optimal olfactory deposition was predicted to be 13–17μm [71, 72].

11.3.3 Discussion

In this section, we reviewed a series of attempts of our lab to enhance drug targeting to the olfactory region or maxillary sinuses. Both experimental and numerical tests were performed to this objective by evaluating different mechanisms of particle control in the nose. Improved olfactory dosages were observed with strategies of drug point release, electric, and magnetic guidance, but at varying levels. It is practical to attain clinically significant doses by using a combination of the point release technique and the electric guidance of charged particles. In comparison to conventional inhalation devices, the drug point release approach significantly reduced the drug losses in the valve and turbinate region. However, even at its optimal delivery efficiency ($\approx$6.7%), the point release technique alone still falls short of the needs of the direct nose-to-brain drug delivery with effective outcomes and no noticeable adverse side-effects. Active controls of particle motion with an electromagnetic or acoustic field have been numerically demonstrated to dispense enhanced doses to the targeted sites. Complementary in vitro tests also demonstrated the potential of electric guidance; an olfactory delivery efficiency of 16% was measured with charged particles and bi-directional nasal delivery.

11.4 Nasal Spray Atomization

Kendra Shrestha

11.4.1 Nasal Spray Drug Delivery

Nasal sprays are used for common problems such as congestion, rhinitis, and allergies. However, it is well established that disproportional large numbers of droplets deposit in the anterior nasal cavity [11, 36, 39, 45, 58], with very little reaching the turbinates, due to geometric constraints of the nasal cavity. Improving its efficacy requires a holistic approach to overcome the restrictive geometry of the nasal cavity and spray atomization is one contributor to deliver the spray at optimal conditions to penetrate the nasal valve area. Studying near nozzle spray atomization is one approach to better understand the physical mechanisms of nasal spray drug delivery. This section presents experimental images of near nozzle atomization, and modelling techniques to account for some of the behaviour.

11.4.2 Experimental Visualisation

The atomization mechanism of the low-pressure nasal sprays has not been studied, although the same principles of atomization under high-pressure applications have been studied (e.g., engine sprays). This section describes experimental visualisation of the spray behaviour at the near nozzle region. A schematic of the experimental setup is shown in Fig. 11.6 which has a pneumatic actuator connected to a two-way solenoid valve, controlled by a programmable logic control (PLC) unit.

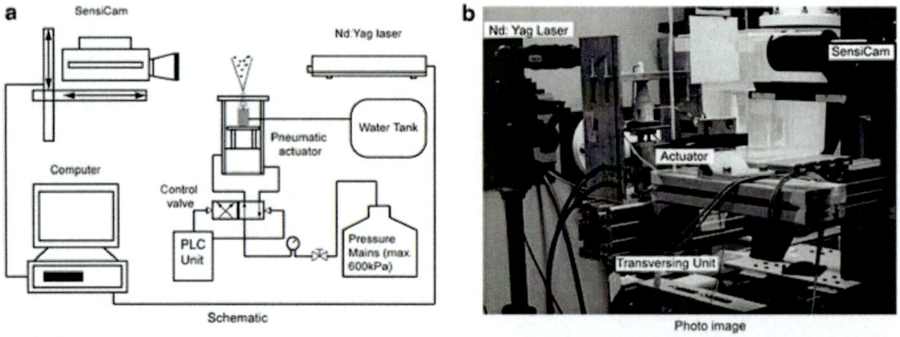

Fig. 11.6 **a** Schematic and **b** photo image of the experimental setup for PDIA and high-speed filming of spray atomization from nasal spray device. (Inthavong et al. [32]; reproduced with permission ©Mary-Ann Libert Publishers)

The spray bottle is fixed at its base onto the actuator, to avoid the undesired lateral motion during actuation. During actuation, the spray bottle moves up and down with the platform, while the spray nozzle remains in one position. This allows the system to capture spray images with a fixed nozzle position as a reference point. The strength of actuation force is controlled by a back pressure supplied by a connection to a pressure line. The pressure passing through was monitored and controlled by a pressure regulator set at 5 Bar. The actuation station was placed between a high-speed camera and a powerful 1000 W light source to produce shadowgraph images. Spray images were captured by Phantom V210 digital high-speed camera with 1280×800 CMOS sensor.

Figure 11.7 shows the spray development during the early pre-stable stage. As the water is ejected through the pressure swirl atomizer, a swirling liquid sheet is formed at the nozzle exit. The liquid sheet then disintegrates into ligaments because of instabilities caused by shearing with the surrounding air, which is characteristic of pressure swirl atomizers, implying the type of internal atomizer of the spray device. The spray develops into a tulip-like shape before expanding into to a full hollow cone shaped liquid sheet structure due to centrifugal force. Several forces are at play simultaneously including, deforming forces (inertial and centrifugal) and preserving forces (surface tension and viscosity) during the primary break-up. The images show that the liquid sheet is highly unstable due to its oscillating nature. The liquid sheet swirls vigorously and collapses into thin ligaments prior to its disintegration to droplets through the ligament break-up.

Figure 11.8 shows an instantaneous snapshot of the spray structure in the near nozzle region during the stable phase where image processing techniques were applied to highlight the different region. Three spray structures are identified: (i) a continuous swirling liquid sheet, where the interaction between the air and the liquid sheet is not well understood and is assumed that an aerodynamic instability in the form of Kelvin-Helmholtz waves grows on the sheet and causes the sheet to break up to ligaments demonstrated in structure. (ii) The ligaments are identified as liquid threads which is extended and stretched from the continuous liquid jet (blue colour). This distance is called sheet break-up length, and it is evident that this length is not constant over time due to instability. The final structure (iii) is the ligaments breaking into droplets. The breakup is dependent on the actuation pressure, ambient condition, nozzle geometry and liquid properties. The droplets are then determined by drag, collision, coalescence and secondary breakup.

11.4.3 Spray Droplet Characteristics

Droplet size distributions are typically measured using laser diffraction techniques. This is an optical method that measures the geometric size of droplets in flight, based on the principle that small particles will scatter light at large angles and large particles will scatter light at small angles. The laser measures a small region of the spray passing by the point, and therefore the measured results represent a droplet size

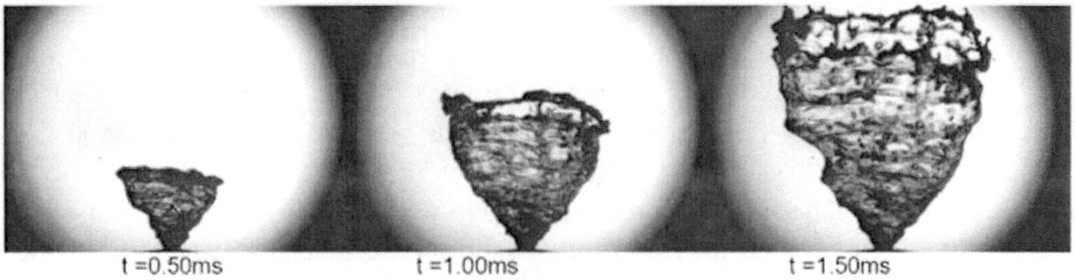

(a) Pre-Stable Stage

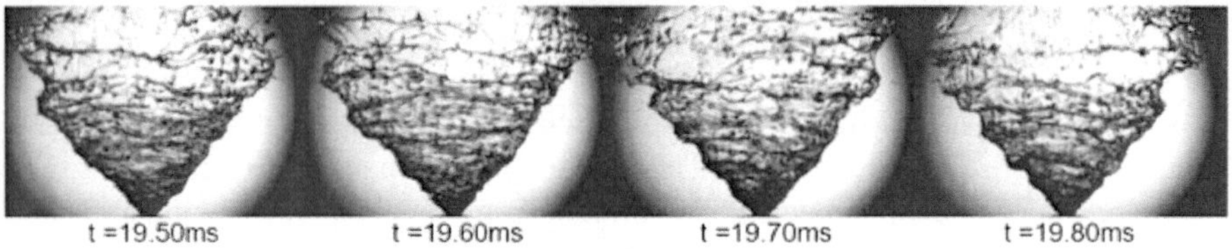

(b) Developed Stage

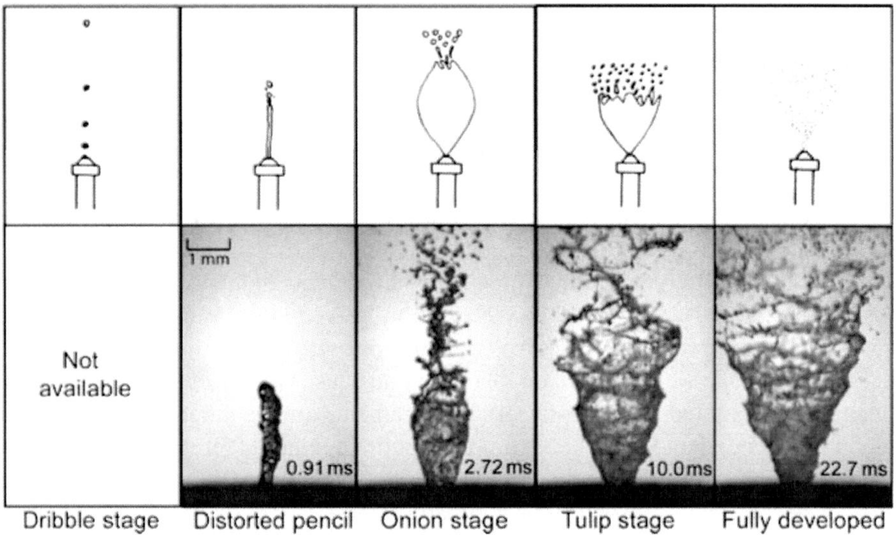

(c) Spray development shapes

Fig. 11.7 Spray plume development during **a** pre-stable stage and **b** developed stage. **c** Comparison of spray development stages, image taken from Inthavong et al. [32]; (reproduced with permission ©Mary-Ann Libert Publishers) of distorted pencil, onion stage, tulip stage, and fully developed.

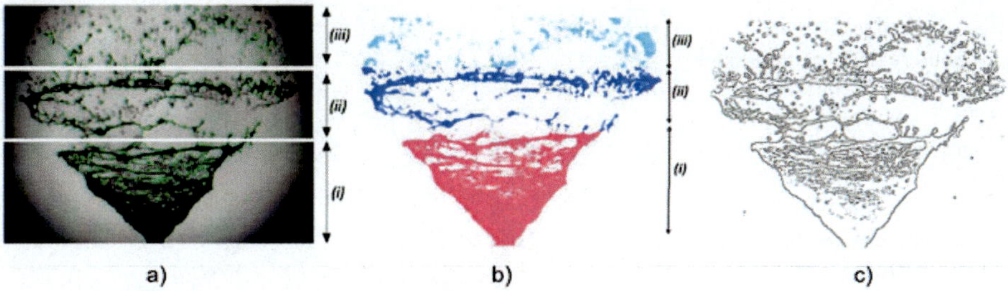

Fig. 11.8 Fluid dynamics structures during the stable stage of the spray atomization showing (i) Sheet formation (ii) Ligament formation and (iii) Droplet formation (**b**) Spray structures identified by different colors (**c**) Spray structures identified by edge detection displaying liquid boundary and features

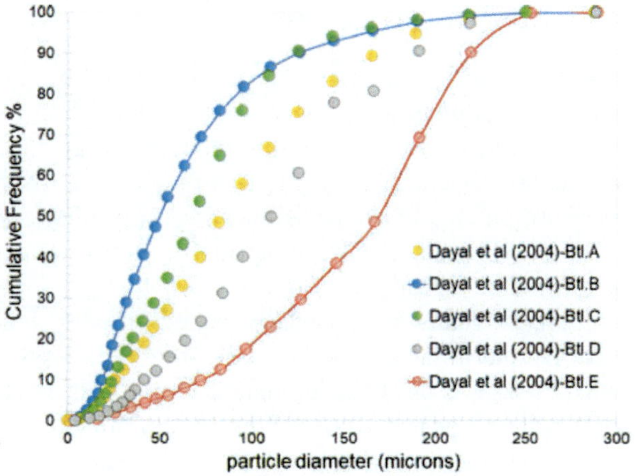

Fig. 11.9 Droplet size distribution data recompiled from results of Dayal et al. [17], combining the cumulative frequency distribution of droplets from atomization at 1.5 cm above the nozzle of five different spray bottles

distribution (DSD) at a specific location over time. Dayal et al. [17] measured the DSD from five bottles and concluded that nasal aerosol generation was dependent on a combination of actuation force, viscosity, rheological properties, surface tension, and device design. Figure 11.9 recasts Dayal's [17] data summarising the DSD produced at 3cm from the nozzle of five bottles. The results show how varied the DSD is due to the different spray parameters.

Cheng et al. [11] measured four bottes and found the DSD's fitted a log-normal distribution with mean values of 48 to 58μm. Foo et al. [24] reported the effects of various formulations in three different bottles and found mean DSD values of 37–44 μm for water, and this steadily increased to 40–54 μm for 30% glycerin mix, and 110–157 μm for 60% glycerin mix. Another technique is Phase Doppler Anenometry (PDA) which is a single point optical measuring technique which that also captures droplet velocity. Liu et al. [43] measured the DSD with phase doppler and found values between 30–42 μm at 3 cm above the nozzle. A comparison was made against

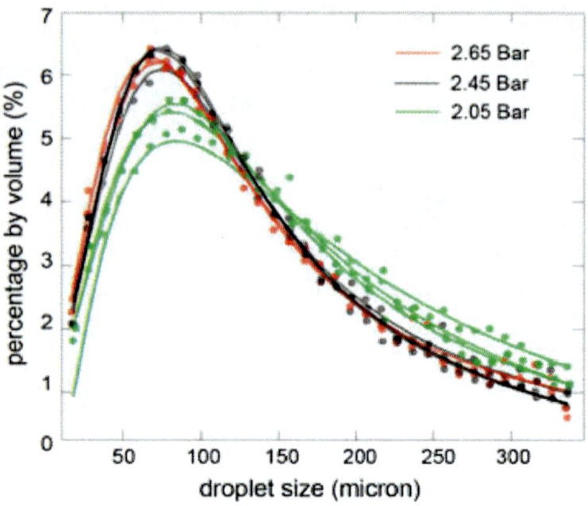

Fig. 11.10 Droplet size distribution data recompiled from results of Inthavong et al. [32], combining the volume percentage distribution of droplets from atomization for spray actuation pressures between 2.05–2.65 Bar

laser diffraction which found smaller values by 2–7% difference. Phase doppler measures droplets within a very small region (e.g., 80 μm in Liu et al. [43], while laser diffraction provides an ensemble method over a large measuring region based on the laser beam, typically 18 mm in diameter. This means that phase doppler can measure precise locations within the spray plume (e.g., centre of plume or outer plume), while laser light scattering provides greater statistical confidence over a larger plume region.

Inthavong et al. [32] used an imaged-based system through Oxford Laser's 'Particle Droplet Image Analyser' technique. This uses a backlit-imaging arrangement where a laser is used to illuminate a region that the droplets passes through. Images of droplets appear dark on a light background, and an automatic image processing algorithm determines the droplet size and velocity. Figure 11.10 compiles the DSD produced from different actuation pressure where a decreased pressure produced greater proportion of larger droplet sizes. The DSD mean values were between 47–50 μm.

Droplet velocities can be measured if two pieces of information can be gathered. In phase doppler, two coherent laser beams are needed to generate Doppler-equivalent frequency of the laser light scattered, while in an imaged based method, two short bursts of laser light illumination with high speed camera is required to capture two images of the spray droplets separated by very small time span. Liu et al. [43] found velocities ranging from 5 m/s to 20 m/s where the increase in velocity was correlated with increase in droplet diameter 10 to 20μm. Inthavong et al. [32] measured the droplet velocity at 7.5mm and 10.5mm above the nozzle shown in Fig. 11.10. Velocities ranged from 2–20/s at an actuation pressure of 2.05 Bar and 19–21/s at an actuation pressure of 2.65 Bar (Fig. 11.11).

The experimental measurements provide insight for setting up computational models of spray droplets for drug delivery studies. It was shown the droplet size distributions vary considerably due to different parameters, and that spray droplets in simulations should be within this wide range. The droplet velocities in the near nozzle region were on the order of 5–20/s where the highest velocities were corre-

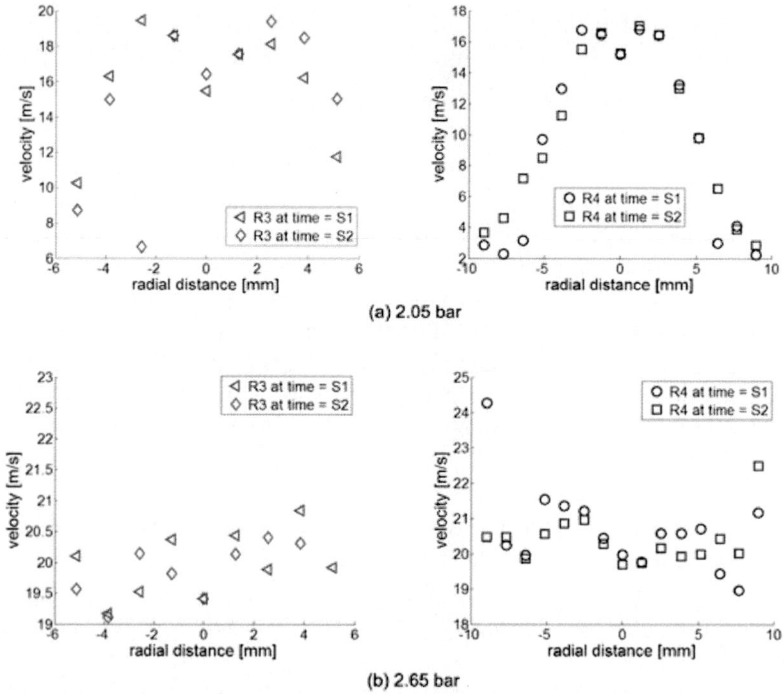

Fig. 11.11 Averaged droplet velocity across the rows R3 (7.5 mm from nozzle) and R4 (10.5 mm from nozzle), taken at **a** 2.05 bar for $t_1 = 126$ ms and, $t_2 = 168$ ms, **b** 2.65 bar, $t_1 = 88$ ms, $t_2 = 134$ ms after liquid has exit from nozzle

lated to larger sized droplets. It is expected that downstream of the nozzle, the droplet velocities reduce, due to drag.

The high velocities and large droplets near the nozzle create high inertia. Entry into the main nasal passage by penetrating through the narrowed nasal valve region becomes a challenge for nasal spray devices. New device designs are needed to overcome these barriers to reduce droplet deposition by inertial impaction in the anterior nasal cavity. The reduction of droplet inertia is most sensitive to droplet diameter, because the inertia is proportional to the square of droplet size. This reduction enhances the ability for droplets to trace or follow the airflow streamlines more readily, enhancing the transport through the nasal valve. However, if the droplets follow the airflow too faithfully,then the droplets may exit the nasal cavity completely.

11.4.4 Spray Models

Linear Instability Spray Atomization (LISA) model

The primary breakup of a spray is modelled through the LISA method that is described by Senecal et al. [55] and shown in Fig. 11.12. The model involves linear stability analysis which assumes that a two-dimensional, viscous, incompressible liquid sheet of thickness moving with a relative velocity through an inviscid, incompressible gas medium. A spectrum of infinitesimal disturbances is imposed onto the liquid sheet, initiating instability leading to breakup into ligaments and then droplets. The produced droplets are defined with an initial stochastic trajectory based on a dispersion angle and are injected within the dispersion angle in addition to the main spray angle to produce the total spray cone angle. Secondary droplet breakup is the breakup of parent droplets that are formed after primary breakup. The breakup criterion is determined by the gas Weber number of droplets.

Fung et al. [26] tuned the LISA model constants to match with experimental data to setup realistic spray conditions. It was found that the liquid sheet constant had a significant impact on the breakup length of liquid sheet that influences the liquid sheet thickness and the sheet ligament size. A liquid sheet constant value of 1 produced the best comparison with the experimental data. The dispersion angle provides the random dispersion caused by the instability-induced fluctuations on the liquid sheet surface. It defines the droplet's stochastic trajectories within a dispersion range in addition to its initialized spray cone angle. A dispersion angle of 0° implies no stochastic fluctuation of the liquid sheet and the injection angle does not vary. This leads to smaller droplets to drift toward the centerline position and accumulate within the spray core that is unrealistic and leads to an under-prediction of droplet size in the core region. For a large dispersion angle, the droplet size within the core increases. Exaggerated stochastic fluctuation of the liquid sheet will cause more droplets to be directed with a path closer toward the centerline and the atomized large droplets are more likely to reach the spray core. Fung et al. [26] found that a dispersion angle of 3° was appropriate in matching with the experimental result.

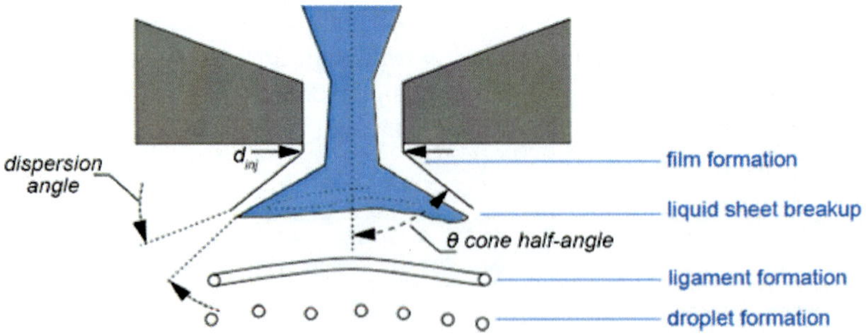

Fig. 11.12 Schematic depicting the LISA spray atomization model proposed in Senecal et al. [55]

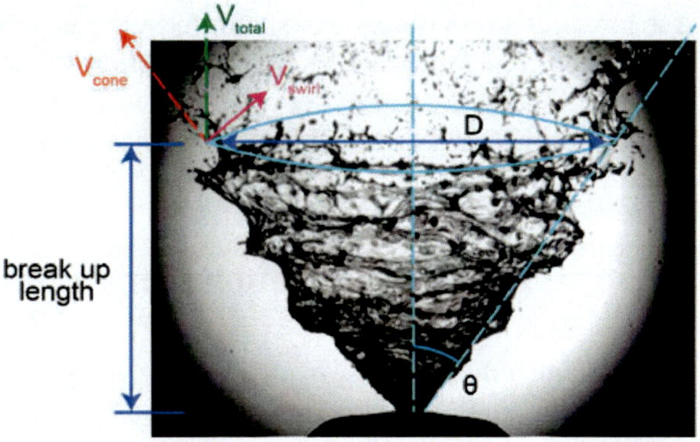

Fig. 11.13 **a** Schematic of cone injection depicting specific conditions relative to nasal spray atomization, where θ is the half-cone angle, D is the cone diameter at breakup

Hollow Cone Model

An alternative approach is to explicitly define the spray characteristics that defines the spray geometry. This includes, break-up length from the nozzle, break-up radius, spray angle, droplet velocities, and in the case of pressure-swirl atomizer types, a swirl fraction should be added. In this type of spray condition, particles are released from a circular 2D region above the nasal spray nozzle which represents a circular ring at the break-up length. The velocity magnitude and swirl fraction are specified for each droplet at the break-up length, which differs from the pressure swirl atomizer (LISA model) where injection pressure and dispersion angle are the input parameters. To account for the droplet to disperse radially and axially, the total spray velocity magnitude is given by

$$\vec{v}_{\text{total}} = \vec{v}_{\text{cone}} + \vec{v}_{\text{swirl}} \tag{11.1}$$

and the swirl component direction is defined using the right-hand rule about a cone axis. The resultant velocity moves the droplets towards the center of the spray direction whereas cone velocity maintains the spray cone angle and swirl velocity give tangential swirling motion to the particles. Figure 11.13 shows the representative hollow cone specified conditions relative to the liquid sheet breakup region.

11.4.5 Summary

Nasal spray atomization occurs at a much lower pressure than traditional spray systems such as engine or agricultural sprays. While similar principles of atomization are prevalent for both high and low pressure cases, studies for nasal sprays are relatively unknown. Most nasal sprays have adopted the pressure swirl atomizer which is what the well established atomization breakup model of the linear instability sheet atomization (LISA) model by Senecal et al. [55] was developed for. Its applicability for low pressure nasal sprays should be verified, since its initial development had targeted very high pressure engine combustion applications. Simpler models such as the hollow cone or even full cone spray could be used as an alternative if the initial conditions matched that produced by the nasal sprays at the breakup region.

11.5 Nasal Saline Irrigation Using a Neti-Pot

Hana Salati, Jim Bartley, David White

11.5.1 Introduction

Nasal saline irrigation is frequently utilised to manage allergic rhinitis [29], treat sinonasal infections [12, 41], and to aid recovery after nasal and sinus surgery [10, 18]. Nasal saline irrigation has been tested on cadavers, anatomical models and live patients. Visualization approaches using a variety of markers (e.g., fluorescein [7], food coloring [27], Technetium 99 m sulfur colloid [63], iodinated contrast solution [57]) have evaluated the final fluid distribution.

CFD can describe the fluid movement, pressure and shear forces, and time related distribution [75]. CFD simulations have also been used to investigate air-flow fields and drug deposition within the nasal cavities and paranasal sinuses [70]. The application of CFD to investigate saline irrigation through the nasal cavity and paranasal sinuses has been limited to a few studies [15, 16, 74]. This feasibility study compares CFD modelling results with blue-dyed saline applied to a simplified physical transparent nasal model using realistic saline flow conditions.

11.5.2 Method

To assess the feasibility of CFD to determine how saline travels through the nasal passages and maxillary sinuses with consideration to the nasal cycle, a simplified nasal model was constructed (Fig. 11.14). This model represented three main geometrical features of real nasal morphology: nasal passages, maxillary sinuses and ostia.

The nasal passages were modelled using two straight pipes and a 180° curved pipe. The pipe cross-section was chosen to be circular with different diameters on the two sides representing the patent and congested passages in the nasal cycle. The left and right passageways represent the patent and congested sides respectively. The dimension of each passage was taken from published MRI data as the average hydraulic diameters for each passage (Table 11.1) [62]. The maxillary sinuses were represented using cylinders of equivalent volume (Fig. 11.14). The maxillary sinuses

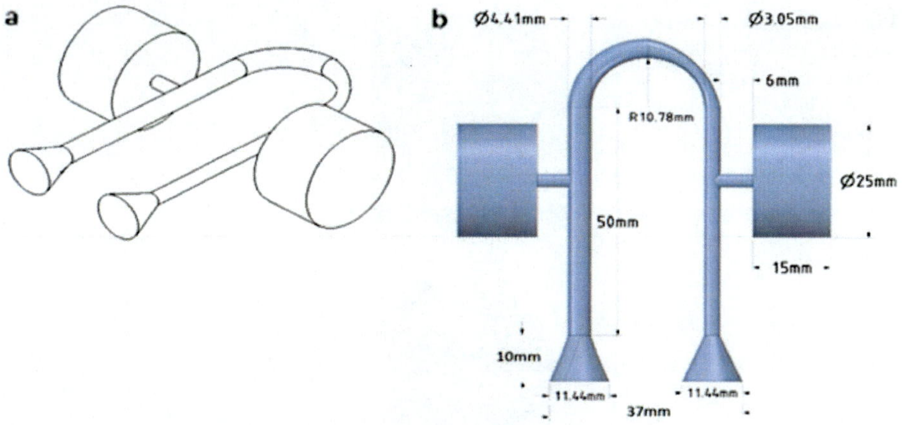

Fig. 11.14 Simplified model: **a** Isometric view **b** Top view

Table 11.1 Simplified model dimensions

Section	Dimension	Reference
Average hydraulic diameter-patent side (D1) (mm)	4.41	[15]
Average hydraulic diameter-congested side (D2) (mm)	3.05	[15]
Distance from nostril to ostium (mm)	44	[16]
Average length of the nasal passage (mm)	50	[17]
Average length of the ostia (mm)	6	[17]
Maxillary Volume (cm^3)	7.3	[18]
Nostril diameter (mm)	11.44	[19]
Nares Width (mm)	37	[20]

communicated with the nasal cavity through a short neck simulating the ostium. To represent the nasopharynx, the two passages were connected to each other via a 180° U-bend. The nares were represented as simple conical shapes. The two halves of the simplified nasal model were 3D printed using a PolyJet 3D printer using a clear acrylic material.

Physical and simulated irrigation volumes of 120 ml of saline solution per nostril were undertaken. This is representative of typical Neti pot volume that is recommended for either nostril by the device manufacturer. To investigate the effect of different head position on the saline flow field in the simplified model, two different head positions were considered (Fig. 11.15):

- Mygind position (lying with head back)
- 90° (tilting the head sideways at 90°)

Each head position was simulated for different inflow direction via each nostril to investigate the effect this has on saline distribution and flow characteristics.

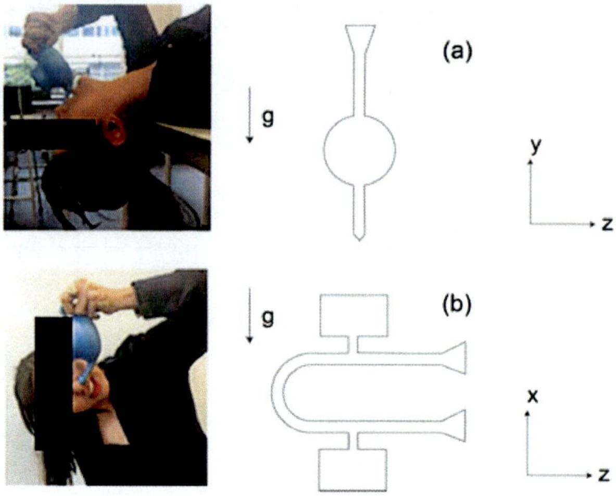

Fig. 11.15 Different head positions considered for simulation: **a** Mygind position **b** 90° lateral head position [52]

Unsteady 3D CFD simulations investigated the saline flow characteristics during irrigation. A multi-VOF (volume of fluid) model simulated the interaction between air and saline solution. The turbulent field was simulated using a realisable k-epsilon (k-ϵ) model. This model has been previously used in saline CFD visualization [74]. A tank and attached pipe, with the same dimensions as the Neti pot, modelled the delivery device in both the simulations and experiment. The pressure inlet boundary condition was assigned referencing the top of the tank and the pressure outlet was set at the other nostril to atmospheric pressure. An unstructured mesh, comprised of tetrahedral cells, was constructed to fill the simplified geometries [60]. During CFD simulation of the simplified nasal model, the monitored variables were almost constant when the cell quantity was over 800,000.

To control the saline irrigation entering the physical model, a ball valve was placed between the tank and physical model. In each case of physical model testing, a high-speed camera captured spatial-temporal video data of the inflow during irrigation (Fig. 11.16).

Saline penetration into the maxillary sinuses in both the computational and physical models was compared at three different time intervals of t=8 s, t=16 s, and t=24 s for all head positions and inflow side irrigations. A hole drilled into each side of the sinuses, covered by plasticine during irrigation, was used to assess the volume of solution that had entered. By stopping the experiment at the targeted time intervals the volume of fluid that had entered could be assessed. This entailed draining one sinus using a flexible tube connected to an oil suction gun. The solution within each nasal passage was then drained out and measured. The remaining solution in the other sinus was calculated by measuring the total mass of the model with solution remaining in one sinus, using a digital microscale, and subtracting from this the dry mass of the model. The difference between these two results being the mass of saline solution remaining in one sinus. This procedure was then repeated for each time increment and head position.

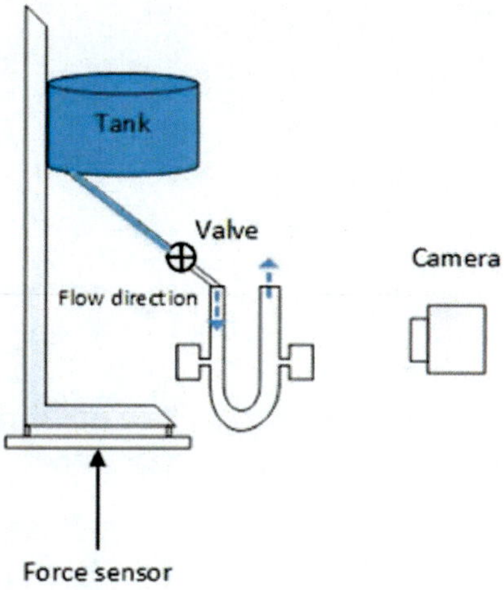

Fig. 11.16 Schematic view of experimental setup in [52]

11.5.3 Results

Visualization

Mygind Position Irrigated from Patent Side

The nasal cavities during the Mygind head position present as two vertical channels. During patent side irrigation the flow first moves down through the patent channel and in the same direction as gravity, and after changing direction, moves upwards through the congested channel towards the outlet against gravity. Figure 11.17 shows three images of the saline solution volume fraction at t = 1.5 s, t = 15 s, and t = 23 s. Each time interval was selected to demonstrate important flow features. These can be seen in the first few seconds as the liquid film flows on the patent side wall with flow detachment from the inner wall occurring at the U-bend. The flow then remains attached to the outer wall as it starts to fill the congested side. Midway through irrigation both nasal passageways are filled, and saline solution has entered both of sinuses.

Mygind Position Irrigated from Congested Side

Again selecting time intervals to demonstrate important flow features, it can be seen in the first few seconds that the greater flow resistance found in the congested side, due to its smaller hydraulic diameter, causes slower filling of this passage (Fig. 11.18) during the first few seconds when compared to irrigation into the patent side. At the later stages of irrigation, air is displaced out of both sinuses by passing through the upper side of each ostia while solution enters both sinuses by occupying the lower side of each ostia. This difference in inflow direction results in both sinuses receiving more solution than when irrigated from the patent side.

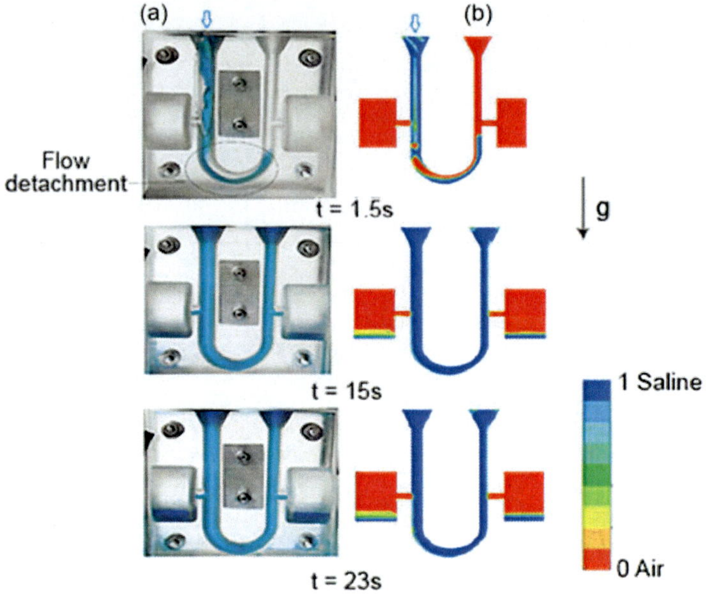

Fig. 11.17 Saline irrigation in the Mygind head position (**a**) physical model (**b**) computational model at different time intervals; t=1.5 s, t=15 s, and t=23 s with irrigation from the patent side

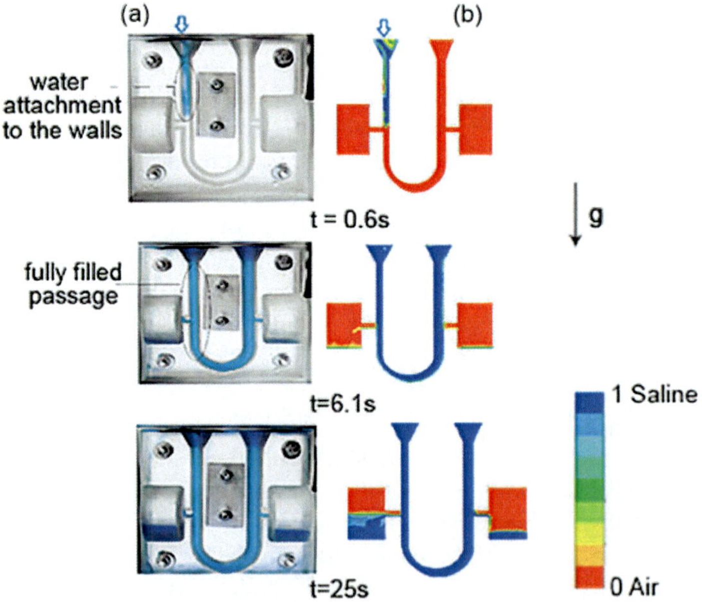

Fig. 11.18 Saline irrigation in the Mygind head position (**a**) physical model and (**b**) computational model when irrigated from the congested side at different time intervals t=0.6 s, t=6.1 s, and t=25 s. 90° head position irrigated from patent side

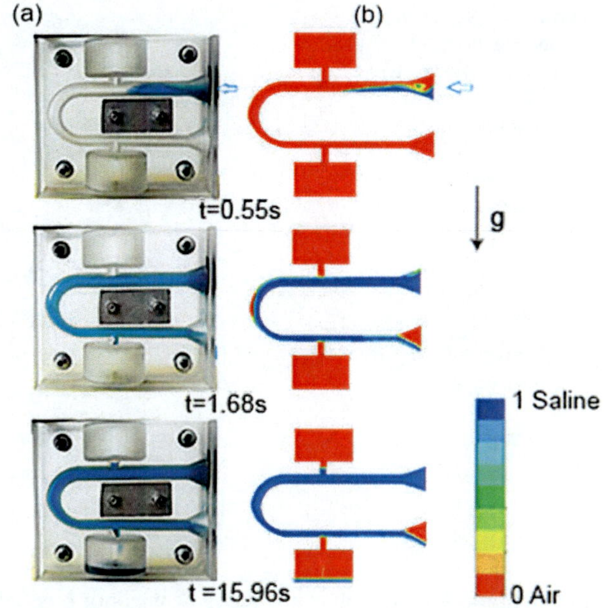

Fig. 11.19 Saline irrigation in the Mygind head position (**a**) physical model (**b**) computational model at different time intervals; t=1.5 s, t=15 s, and t=23 s with irrigation from the patent side

90° *Head Position Irrigated from Patent Side*

Both CFD and physical model results showed similar flow behaviour in the 90° head position when irrigated from either the patent (Fig. 11.19) or congested (Fig. 11.20) sides. While inflow through the ostium only occurred in the lower sinus cavity during patent side irrigation, the solution completely filled the conducting nasal cavity during irrigation. Despite the pressure head of irrigation inflow being greater than the height of the upper sinus, air trapped within this space prevented any solution from entering this cavity.

90° *Head Position Irrigated from the Congested Side*

During irrigation inflow into the congested side, the additional flow restriction imposed on this side of the nasal cavity causes flow separation to occur at the 180° U-bend. At this position, the centrifugal force acting on the turning flow causes the solution to detach from the nasal wall and form eddies and vortices (Fig. 11.20).

11.5.4 Sinus Penetration

Table 11.2 shows the amount of liquid remaining in each sinus of both the computational and physical models at different time intervals. Each experiment was performed four times, and the average results are shown in this table. There is good agreement in the amount of liquid that penetrated the sinuses between the computational and

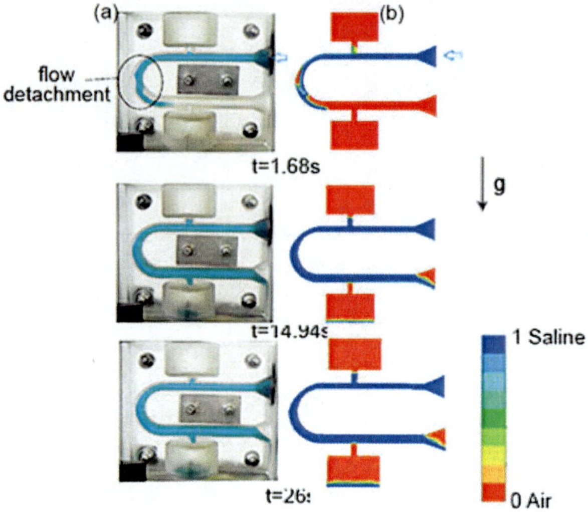

Fig. 11.20 Saline irrigation in the Mygind head position (**a**) physical model (**b**) computational model at different time intervals; t=1.5 s, t=15 s, and t=23 s with irrigation from the patent side

physical models. In some instances, it was not possible to measure the liquid penetration into the sinuses in the physical model where the liquid volume was very low. No saline penetration was observed in the 90° head position for the maxillary sinus located at the upper side where irrigation inflow occurred (Table 11.2). This was a result of trapped air in the sinus preventing solution inflow while in the 90° head position.

11.5.5 Discussion

In healthy humans, each nasal passageway alternates between being more congested or more patent in what is commonly termed as the ‘nasal cycle’ [19]. An almost fourfold change in unilateral resistance is associated with the nasal cycle [19], but up to eight-fold changes in unilateral nasal resistance have been reported [35, 46]. Depending on irrigation side, this increased resistance will be encountered either on the side irrigated or as the irrigation solution returns through the contralateral nasal passage. In a cadaveric state, both nasal passageways are larger than that found in vivo. In this situation, the significantly reduced nasal resistance could increase nasal flow and overestimate the extent of nasal saline irrigation. Previous CFD modelling using cadaver models has not considered the influence of the nasal cycle on nasal saline flow resistance [15, 16, 74]. A narrower nasal passage may predispose to air locking, where the saline irrigation may stop due to air trapping in the higher parts of the nose. Thus, the characteristics of fluid flow within each nasal passageway and paranasal sinus could vary depending on which side of the nose is irrigated. To be more representative of in vivo nasal irrigation, different circular cross-sectional

Table 11.2 Sinus penetration at different head positions and side directions. **A* is the sinus located at the patent side and B is the sinus located at the congested side

		Mygind patent	Mygind congested	90° patent	90° congested
Physical model (ml), standard deviation (t=8 s)	A*	0.25 ± 0.05			
	B*	0.30 ± 0.05	0.45 ± 0.07	0.85 ± 0.088	0
Numerical model (ml) (t=8 s)	A	0.208	0.008	0	0.114
	B	0.3	0.389	0.8	0
Physical model (ml), standard deviation (t=16 s)	A	0.49 ± 0.07	0.26 ± 0.056	0	0.25 ± 0.06
	B	0.6 ± 0.08	1.0 ± 0.097	1.55 ± 0.1	0
Numerical model (ml) (t=16 s)	A	0.43	0.2287	0	0.19
	B*	0.769	0.965	1.5	0
Physical model (ml), standard deviation (t=24 s)	A	0.72 ± 0.085	0.4 ± 0.065	0	0.28 ± 0.065
	B	1.0 ± 0.1	1.51 ± 0.1	2.03 ± 0.2	0
Numerical model (ml), (t=24 s)	A	0.64	0.46	0	0.2
	B	1.1	1.5	2.03	0

diameters were chosen to represent the patent and congested passages in the nasal cycle.

While nasal saline irrigation aids in the management of various sinonasal conditions, recommendations on the quantity and nature of the irrigation solution, the optimal head position, and the delivery devices vary [18, 47]. The cadaver and virtual models indicate that squeeze bottles and Neti pots, followed by syringes and electronic devices, provide the most effective forms of nasal irrigation [59]. In the literature, different terminologies describe similar head positions [18, 59]. The Mygind position is similar to the head down and forward or vertex down position which is optimal for post-operative sinus irrigation [59]. The 90° lateral head position was chosen to study whether lateral head position could influence irrigation solution to the maxillary sinuses. The methodologies and outcomes arising from this study of a simplified nasal model give confidence to the use of CFD measurements on progressively more complex models of the nasal cavity.

11.5.6 Conclusion

By comparing the flow features (sharp interfaces, flow detachment, and sinus penetration) captured in the images taken from the physical and computational models and sinus penetration, the multiphase and turbulence models were able to predict the important flow features. The experimental results generally agree with what the CFD calculations predicted, which verifies the methodology used. The methodologies and findings from this preliminary study have formed the basis for more complex models of the nasal cavities and CFD simulations which have been done by Salati [52].

11.6 Nasal Saline Irrigation from a Squeeze Bottle

Kendra Shrestha, Yidan Shang

11.6.1 Nasal Saline Irrigation

Nasal saline irrigation from a squeeze bottle provides additional pressure to deliver saline or drug formulation through the nasal cavity with greater momentum and force. To date, limited evidence from cadaver models have demonstrated the effect of irrigation penetration, distribution, and force within various paranasal sinuses. Computational fluid dynamics (CFD) represents a novel method of modeling air and fluid flows through the sinonasal tract that is reliable, quantifiable, reproducible, and allows multiple permutations to be rapidly assessed to determine the ideal combination of parameters. Despite this, only a few CFD studies of nasal irrigation have been performed. These include Zhao et al. [74] and Craig et al. [15, 16] who explored squeeze-bottle nasal irrigation and found the nose-to-ceiling head position was superior to the nose-to-floor position for irrigation to the sphenoid sinuses. This section presents new computational techniques to characterize nasal irrigation by identifying the critical fluid dynamics mechanisms that dominate the jet stream flowing through a human sinonasal cavity. The techniques provide quantitative analysis for determining sinonasal surface coverage of nasal irrigation, residence times across the mucosal surfaces, and surface wall shear stress.

11.6.2 Model Setup

A realistic nasal cavity geometry was reconstructed from a high-resolution CT scan of a healthy 25-year-old, Asian female (161 cm height, 53 kg mass, without nasal septum deviation, turbinate hypertrophy or sinusitis) at resting conditions (Fig. 11.21). The CT data were acquired using a Siemens Dual Source CT scanner with image

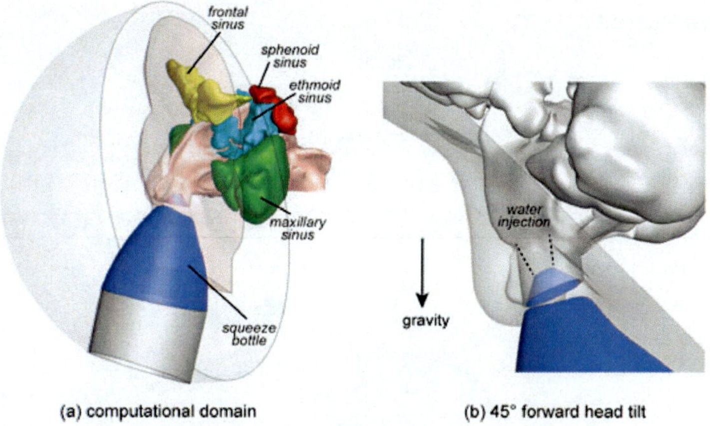

Fig. 11.21 CFD geometry model of the nasal cavity and squeeze bottle

parameters of 0.39 × 0.39 mm pixel size, 512 × 512 pixels image dimensions and 0.5 mm slice thickness. The nasal cavity and the entire paranasal sinuses including the maxillary, ethmoid, frontal and sphenoid sinuses (see Fig. 11.22) were extracted using 3D-Slicer's segmentation tool by a computer engineer in conjunction with an ENT surgeon and has been used in a previous study [73]. The model contained 1.1 million poly-hexcore cells (equivalent to 4.6 million tetrahedral cells) with five prism layers (first layer height of 0.07mm, and total height of prism layer of 0.64 mm) using ANSYS-Fluent (R2019v1) meshing.

A liquid jet was introduced as a mass flow inlet at the squeeze bottle nozzle which had a 5mm diameter. All other surfaces were set to a wall boundary condition. Figure 11.21 demonstrates the nasal geometry with a squeeze bottle (featuring a tapered nozzle) inserted into the left nostril until the nozzle occluded the nostril opening which was approximately 10 mm. Two cases are demonstrated which compared the jet flow with the head at zero tilt and a 45° forward head tilt with the nozzle jet aligned vertically.

In general, consensus on administration technique is a 45° forward head tilt, although there is no evidence-based recommendation for this approach as the most efficient nasal irrigation delivery. Salati [52] evaluated four different head positions for a gravity-fed Neti-pot nasal irrigation. These were supine (lying with head back), 90° (tilting the head sideways at 90°), head back (head is oriented 45° upward from the ground), and head forward (head is inclined downwards at 45° to the ground). Instructions that accompany the sinus rinse bottle (Flo®, ENT Technologies, Hawthorn, Victoria, Australia) suggest the head angled forward at 45° (head over sink).

The nozzle jet diameter was 5 mm based on a commercially available nasal irrigation bottle. The liquid jet was introduced into the airway from zero to a flow rate of 35 mL/sec (approximately 1.79 m/s) held for 2 s (see Fig. 11.23). This gave a total volume of 70 mL based on preliminary testing of short physical squeezes of a squeeze bottle, where less than half the volume remained. The squeeze duration represented

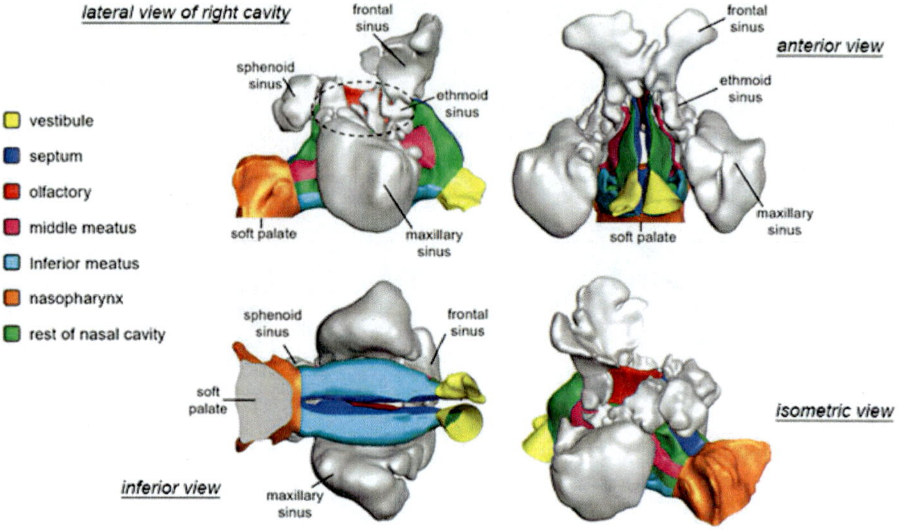

Fig. 11.22 Components of the nasal cavity used to study the nasal irrigation flow

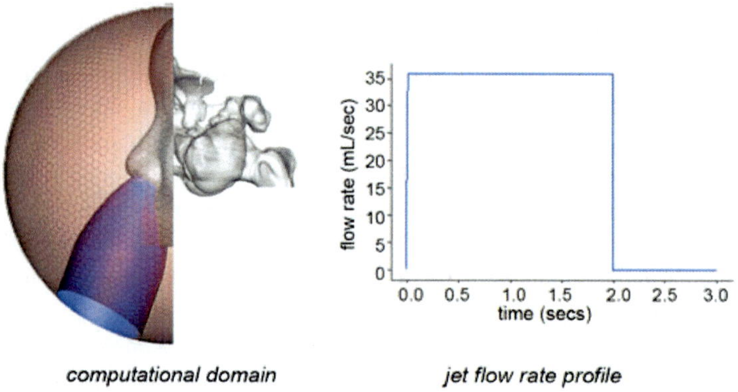

Fig. 11.23 Components of the nasal cavity used to study the nasal irrigation flow

a rapid squeeze with the hand. The initial ramp-up and ramp-down times were both set at 0.01 s, selected to preclude ramp periods as a physical test variable.

The CFD simulation uses the volume of fluid (VOF) method which tracks the interface between the liquid jet and air. The VOF model requires the two fluids to be predominantly separated, and that they are not interpenetrating. The flow can therefore be considered as an interfacial multiphase flow. From Chap. 6 (meshing), the computational meshed nasal airway model contains millions of cells. Within each computational cell, the simulation determines a volume fraction for either fluid (e.g., air or water) where the sum of both fluids within any cell is equal to one. Figure 11.24 demonstrates the interface between two fluids and the volume fractions for each individual cell.

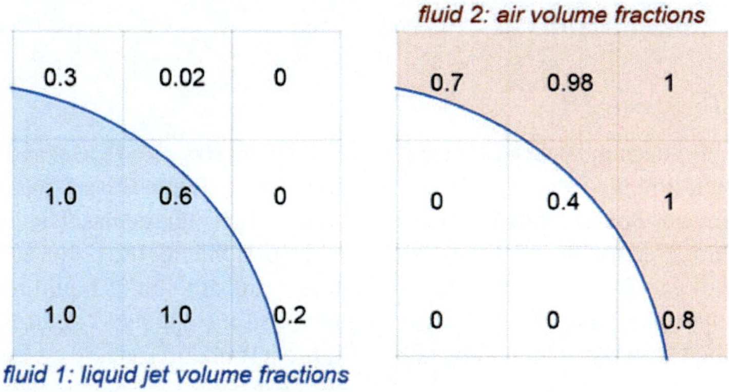

Fig. 11.24 Volume of fluid (VOF) interface where the computational cell displays the volume fraction of each fluid

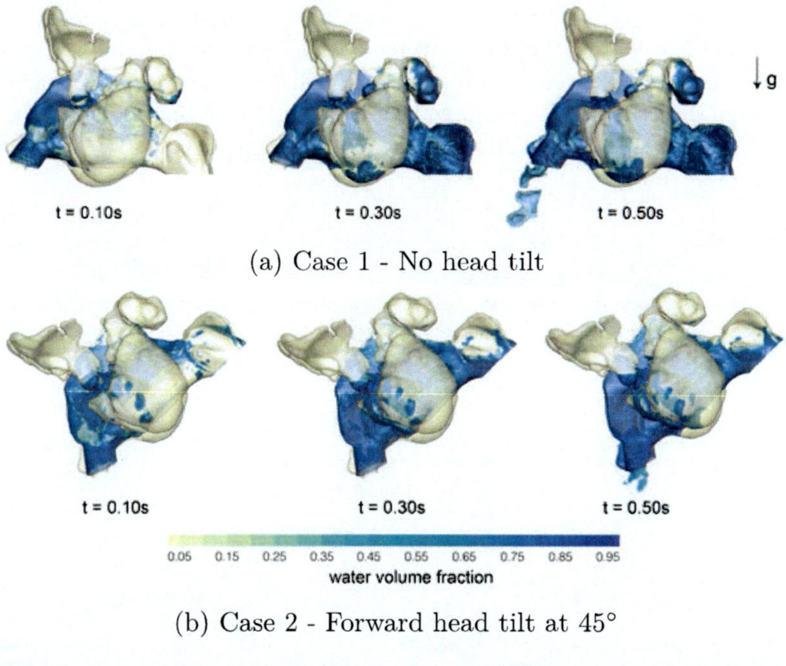

Fig. 11.25 Liquid jet (water) volume fraction penetrating through the nasal cavity for **a** Case 1—no head tilt; **b** Case 2—forward head tilt at 45°

11.6.3 Comparisons

Jet Flow Distribution

The jet flow entering the nasal cavity at times of $t = 0.10\,\text{s}$, $t = 0.30\,\text{s}$, and $t = 0.50\,\text{s}$ are shown in Fig. 11.25 where Case 1 has no head tilt, while Case 2 has a 45° head tilt. The images demonstrate the water penetrating through the nasal cavity in terms of its volume fraction during the immediate time period after the bottle squeeze. The influence of gravity is evident with a reduction in the amount of liquid reaching the most the nasopharynx (most posterior region), and the sphenoid sinuses when the head is tilted forward. The jet appears to fill the left maxillary sinuses better for no head tilt.

Liquid Jet (Water) Volume Fraction Coverage

The volume fraction for the liquid jet was averaged across different surfaces (Figs. 11.26 and 11.27), which represents the amount of surface coverage. Due to the large amount of data, the results were separated into the left and right cavities, and the main nasal passages and the sinuses. Consistently high-volume fraction was found along the vestibule, septum, and turbinate surfaces. Surface coverage in the sinuses was significantly less, although there was a high peak up to 0.7 found for the left sphenoid.

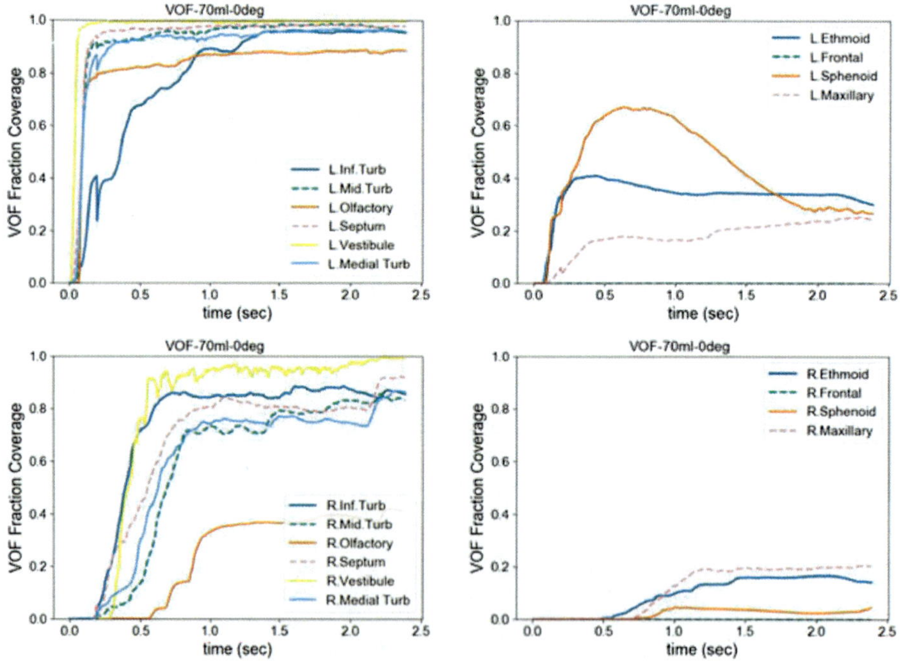

Fig. 11.26 Surface coverage of liquid jet using the averaged volume fraction for Case 1—no head tilt

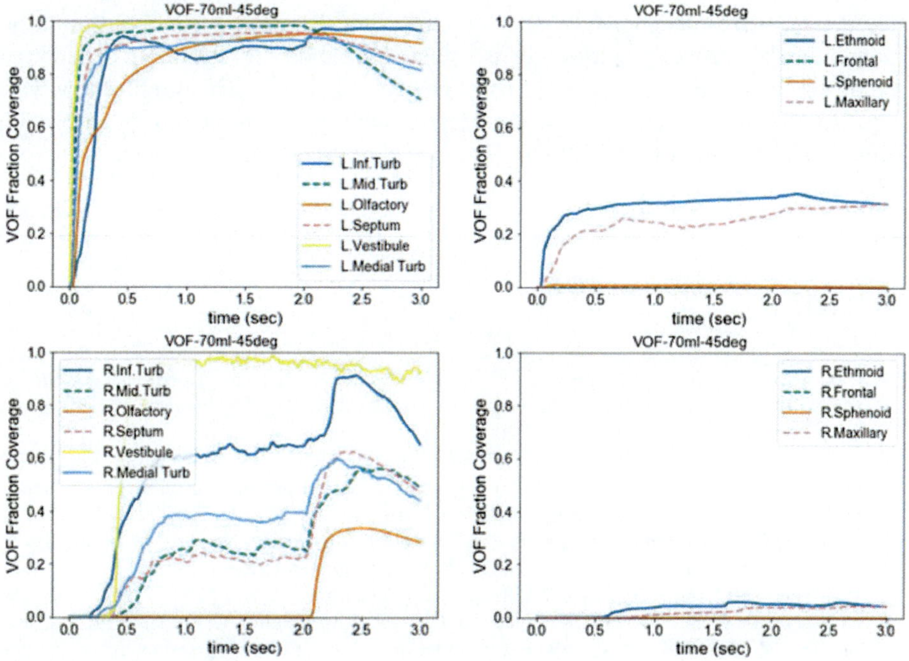

Fig. 11.27 Surface coverage of liquid jet using the averaged volume fraction for Case 2—forward head tilt of 45°

The influence of a forward head tilt is not evident in the main passages and the maxillary, and ethmoid sinuses of the left cavity where the jet first enters. However, the surface coverage found in the sphenoid disappeared for the forward head tilt case. After the jet has turned from the nasopharynx, a forward head tilt causes a delay in the liquid reaching right cavity surfaces (after $t = 2\,\mathrm{s}$). There is also a reduction in the penetration into the right sinuses.

The results suggest that the liquid jet momentum needs to overcome gravity force, and when the head is tilted forward the liquid takes longer to reach the nasopharynx. This causes a reduction in momentum of the liquid, diminishing the possible beneficial effects of high shear to remove any debris on the mucus. The lower liquid momentum also leads to slower movement within the right nasal cavity.

11.6.4 Summary

Nasal irrigation from a squeeze bottle was modelled using the VOF method where two scenarios were presented to demonstrate the current computational modelling capability. The VOF technique is an ideal method to track the liquid-air interface, thereby determining how much each surface of the nasal cavity is covered by the liquid jet. The results showed that a forward head tilt reduced the initial liquid jet momentum because of gravity, and this hindered the ability for the liquid to penetrate the posterior regions, and the right nasal cavity sufficiently. This preliminary work suggests that a forward head tilt impedes the ability for the liquid jet to provide sufficient irrigation to the nasal cavity, in terms of both surface coverage, and shearing effects.

References

1. Center for Drug Evaluation and Research, FDA. Bioavailability and Bioequivalence Studies for Nasal Aerosols and Nasal Sprays for Local Action (2003). https://www.fda.gov/media/70867/download. Accessed 12 Aug. 2009
2. U.S. Food and Drug Administration, Product-Specific Guidances for Generic Drug Development (2019). https://www.fda.gov/drugs/guidances-drugs/product-specific-guidances-generic-drug-development. Accessed 10 Sep. 2019
3. Food and Drug Administration Safety and Innovation Act, Pub. L. No. 112–144, Title iii
4. U.S. Food and Drug Administration. GDUFA II Commitment Letter
5. U.S. Food and Drug Administration. Science & Research
6. M. Andersen, R. Sarangapani, R. Gentry, H. Clewell, T. Covington, C. Frederick, Application of a hybrid CFD-PBPK nasal dosimetry model in an inhalation risk assessment: an example with acrylic acid. Toxicol. Sci. **57**(2), 312–325 (2000)
7. B.S. Bleier, I. Debnath, R.J. Harvey, R.J. Schlosser, Temporospatial quantification of fluorescein-labeled sinonasal irrigation delivery. Int. Forum Allergy Rhinol. **1**(5), 361–5 (2011)
8. M. Bush, C. Frederick, J. Kimbell, J. Ultman, A CFD-PBPK hybrid model for simulating gas and vapor uptake in the rat nose. Toxicol. Appl. Pharmacol. **150**(1), 133–145 (1998)
9. J. Campbell, M. Andersen, H. Clewell, A hybrid CFD-PBPK model for naphthalene in rat and human with ivive for nasal tissue metabolism and cross-species dosimetry. Inhal. Toxicol. **26**(6), 333–344 (2014)
10. X. Chen, S. Feng, L. Chang, X. Lai, X. Chen, X. Li, G. Zhang, The effects of nasal irrigation with various solutions after endoscopic sinus surgery: Systematic review and meta-analysis. J. Laryngol. Otol. **132**(8), 673–679 (2018)
11. Y. Cheng, T. Holmes, J. Gao, R. Guilmette, S. Li, Y. Surakitbanharn, C. Rowlings, Characterization of nasal spray pumps and deposition pattern in a replica of the human nasal airway. J. Aerosol Med. **14**(2), 267–280 (2001)
12. L.Y. Chong, K. Head, C. Hopkins, C. Philpott, S. Glew, G. Scadding, M.J. Burton, A.G. Schilder, Saline irrigation for chronic rhinosinusitis. Cochrane Database Syst. Rev. **4**, CD011995 (2016)
13. R. Corley, K. Minard, S. Kabilan, D. Epstein, A. Kuprat, J. Harkema, J. Kimbell, M. Gargas, J. Kinzell, Magnetic resonance imaging and computational fluid dynamics (CFD) simulations of rabbit nasal airflows for the development of hybrid CFD/PBPK models. Inhal. Toxicol. **21**(6), 512–518 (2009)
14. R.A. Corley, S. Kabilan, A.P. Kuprat, J.P. Carson, K.R. Minard, R.E. Jacob, C. Timchalk, R. Glenny, S. Pipavath, T. Cox, C.D. Wallis, R.F. Larson, M.V. Fanucchi, E.M. Postlethwait, D.R. Einstein, Comparative computational modeling of airflows and vapor dosimetry in the respiratory tracts of rat, monkey, and human. Toxicol. Sci. **128**(2), 500–516 (2012)
15. J.R. Craig, J.N. Palmer, K. Zhao, Computational fluid dynamic modeling of nose-to-ceiling head positioning for sphenoid sinus irrigation. Int. Forum Allergy Rhinol. **7**(5), 474–479 (2017)
16. J.R. Craig, K. Zhao, N. Doan, S. Khalili, J.Y. Lee, N.D. Adappa, J.N. Palmer, Cadaveric validation study of computational fluid dynamics model of sinus irrigations before and after sinus surgery. Int. Forum Allergy Rhinol. **6**(4), 423–8 (2016)

17. P. Dayal, M. Shaik, M. Singh, Evaluation of different parameters that affect droplet-size distribution from nasal sprays using the malvern spraytec. J. Pharm. Sci. **93**(7), 1725–1742 (2004)
18. S.H. De Paiva Leite, R.G. Douglas, How does sinus surgery affect topical irrigation distribution? Curr. Opin. Otolaryngol. Head Neck Surg. **26**(1), 21–26 (2018)
19. R. Eccles, Nasal airflow in health and disease. Acta Otolaryngol. **120**(5), 580–95 (2000)
20. FDA, Predictive lung deposition models for safety and efficacy of orally inhaled drug products (Worth Longest, P), in *Generic Drug User Fee Amendments Grant 1U01FD004570* (2012)
21. FDA, Development of hybrid CFD-PBPK models for absorption of intranasal corticosteroids (Schroeter, J), in *Generic Drug User Fee Amendments Grant 1U01FD005201* (2014)
22. FDA, Evaluating relationships between in vitro nasal spray characterization test metrics for bioequivalence and nasal deposition in silico and in vitro (Golshahi, L). *HHSF223201810144C* (2018)
23. FDA, Nasal mucociliary clearance affecting local drug-absorption in subject-specific geometries (Kleinstreuer, C), in *Generic Drug User Fee Amendments Grant 1U01FD006537* (2018)
24. M. Foo, Y. Cheng, W. Su, M. Donovan, The influence of spray properties on intranasal deposition. J. Aerosol Med. **20**)(4), 495–508 (2007)
25. C. Frederick, P. Gentry, M. Bush, L. Lomax, K. Black, L. Finch, J. Kimbell, K. Morgan, R. Subramaniam, J. Morris, J. Ultman, A hybrid computational fluid dynamics and physiologically based pharmacokinetic model for comparison of predicted tissue concentrations of acrylic acid and other vapors in the rat and human nasal cavities following inhalation exposure. Inhal. Toxicol. **13**(5), 359–376 (2001)
26. M.C. Fung, K. Inthavong, W. Yang, J. Tu, CFD modeling of spray atomization for a nasal spray device. Aerosol Sci. Technol. **46**(11), 1219–1226 (2012)
27. A. Grobler, E.K. Weitzel, A. Buele, C. Jardeleza, Y.C. Cheong, J. Field, P.J. Wormald, Pre- and postoperative sinus penetration of nasal irrigation. Laryngoscope **118**(11), 2078–81 (2008)
28. L. Hanson, W. Frey, Strategies for intranasal delivery of therapeutics for the prevention and treatment of neuroaids. J. Neuroimmune Pharmacol. **2**(1), 81–86 (2007)
29. K. Head, K. Snidvongs, S. Glew, G. Scadding, A.G. Schilder, C. Philpott, C. Hopkins, Saline irrigation for allergic rhinitis. Cochrane Database Syst. Rev. **6**, CD012597 (2018)
30. J.D. Hoekman, R.J.Y. Ho, Effects of localized hydrophilic mannitol and hydrophobic nelfinavir administration targeted to olfactory epithelium on brain distribution. AAPS Pharmscitech. **12**(2), 534–543 (2011)
31. J.D. Hoekman, R.J.Y. Ho, Effects of localized hydrophilic mannitol and hydrophobic nelfinavir administration targeted to olfactory epithelium on brain distribution. AAPS Pharmscitech. **12**(2), 534–543 (2011)
32. K. Inthavong, M.C. Fung, W. Yang, J. Tu, Measurements of droplet size distribution and analysis of nasal spray atomization from different actuation pressure. J. Aerosol Med. Pulm Drug Delivery **28**(1), 59–67(2014)
33. J.D. Schroeter, D. Stricklin, J.S. Kimbell, R.R. Delvadia, X. Zhang, A physiologically-based pharmacokinetic model framework to estimate systemic bioavailability of fluticasone propionate nasal spray, in *Respiratory Drug Delivery*, April 17–21, Scottsdale, AZ, USA (2016)
34. J.D. Schroeter, J.S. Kimbell, B. Saluja, R.R. Delvadia, E.L. Vallorz, P. Sheth, The impact of actuation force on droplet size distribution and spray duration of three commercially available nasal sprays, in *Respiratory Drug Delivery*, April 17–21, Scottsdale, AZ, USA (2016)
35. G. Jo, S. Chung, Y. Na, Numerical study of the effect of the nasal cycle on unilateral nasal resistance. Respir. Physiol. Neurobiol. **219**, 58–68 (2015)
36. J.T. Kelly, B. Asgharian, J.S. Kimbell, B.A. Wong, Particle deposition in human nasal airway replicas manufactured by different methods. Part I: Inertial regime particles. Aerosol Sci. Technol. **38**, 1063–1071 (2004)
37. J.S. Kimbell, J.D. Schroeter, P. Sheth, G. Tian, R.R. Delvadia, B. Saluja, R. Walenga, Effect of actuation force on simulated regional nasal spray deposition in a healthy nasal cavity. *Respiratory Drug Delivery*, April 17–21, Scottsdale, AZ, USA (2016)

38. J. Kimbell, J. Schroeter, G. Tian, R. Walenga, A. Babiskin, R. Delvadia, Estimating size-specific numbers of active pharmaceutical ingredient particles in the regional deposition of a nasal spray. J. Aerosol Med. Pulm. Drug Deliv. **30**(3), 18–19 (2017)
39. J.S. Kimbell, R.A. Segal, B. Asgharian, B.A. Wong, J.D. Schroeter, J.P. Southall, C.J. Dickens, G. Brace, F.J. Miller, Characterization of deposition from nasal spray devices using a computational fluid dynamics model of the human nasal passages. J. Aerosol Med. **20**(1), 59–74 (2007)
40. J. Kimbell, R. Subramaniam, Use of computational fluid dynamics models for dosimetry of inhaled gases in the nasal passages. Inhal. Toxicol. **13**(5), 325–334 (2001)
41. D. King, B. Mitchell, C.P. Williams, G.K. Spurling, Saline nasal irrigation for acute upper respiratory tract infections. Cochrane Database Syst. Rev. **4**, CD006821 (2015)
42. B. Li, F. Jin, S. Lee, T. Bai, B. Chowdhury, H. Caramenico, D. Conner, Bioequivalence for locally acting nasal spray and nasal aerosol products: standard development and generic approval. AAPS J. **15**(3), 875–883 (2013)
43. X. Liu, W.H. Doub, C. Guo, Evaluation of droplet velocity and size from nasal spray devices using phase doppler anemometry (pda). Int. J. Pharm. **388**, 82–87 (2010)
44. A. Mistry, S. Stolnik, L. Illum, Nanoparticles for direct nose-to-brain delivery of drugs. Int. J. Pharm. **379**(1), 146–157 (2009)
45. S.P. Newman, F. Moren, S.W. Clarke, Deposition pattern of nasal sprays in man. Rhinology **26**, 111–120 (1988)
46. R. Patel, G. Garcia, D. Frank-Ito, J. Kimbell, J. Rhee, Simulating the nasal cycle with computational fluid dynamics. Otolaryngol. Head Neck Surg. **152**(2), 353–60 (2015)
47. N. Principi, S. Esposito, Nasal irrigation: an imprecisely defined medical procedure. Int. J. Environ. Res. Public Health **14**, 5 (2017)
48. A. Rygg, M. Hindle, P. Longest, Absorption and clearance of pharmaceutical aerosols in the human nose: effects of nasal spray suspension particle size and properties. Pharm. Res. **33**(4), 909–921 (2016)
49. A. Rygg, M. Hindle, P. Longest, Linking suspension nasal spray drug deposition patterns to pharmacokinetic profiles: a proof-of-concept study using computational fluid dynamics. J. Pharm. Sci. **105**(6), 1995–2004 (2016)
50. A. Rygg, P. Longest, Absorption and clearance of pharmaceutical aerosols in the human nose: development of a CFD model. J. Aerosol Med. Pulm. Drug Deliv. **29**(5), 416–431 (2016)
51. S. Gizurarson, Method for administration of active substances to the olfactory region. in.: Patents ca 2298596
52. Salati, H., *Investigation Into Nasal Saline Irrigation Within a Healthy Human Nose*. Thesis (2019)
53. B. Saluja, B. Li, S. Lee, *Bioequivalence for Orally Inhaled and Nasal Drug Products* (Springer, New York, 2014), pp. 369–394
54. J. Schroeter, J. Kimbell, R. Walenga, A. Babiskin, R. Delvadia, A CFD-PBPK model to simulate nasal absorption and systemic bioavailability of intranasal fluticasone propionate. J. Aerosol Med. Pulm. Drug Deliv. **30**(3), 13–14 (2017)
55. P.K. Senecal, D.P. Schmidt, I. Nouar, C.J. Rutland, R.D. Reitz, M.L. Corradini, Modeling high-speed viscous liquid sheet atomization. Int. J. Multiph. Flow **25**(6–7), 1073–1097 (1999)
56. X. Si, J. Xi, J. Kim, Y. Zhou, H. Zhong, Modeling of release position and ventilation effects on olfactory aerosol drug delivery. Respir. Physiol. Neurobiol. **186**(1), 22–32 (2013)
57. K. Snidvongs, P. Chaowanapanja, S. Aeumjaturapat, S. Chusakul, P. Praweswararat, Does nasal irrigation enter paranasal sinuses in chronic rhinosinusitis? Am. J. Rhinol. **22**, 483–86 (2008)
58. J.D. Suman, B.L. Laube, T. Lin, G. Brouet, R. Dalby, Validity of in vitro tests on aqueous spray pumps as surrogates for nasal deposition. Pharm. Res. **19**(1), 1–6, Springer (2002)
59. W.W. Thomas, R.J. Harvey, L. Rudmik, P.H. Hwang, R.J. Schlosser, Distribution of topical agents to the paranasal sinuses: an evidence-based review with recommendations. Int. Forum Allergy Rhinol. **3**(9), 691–703 (2013)
60. J. Tu, K. Inthavong, G. Ahmad, *Computational Fluid and Particle Dynamics in the Human Respiratory System* (Springer, Berlin, 2013)

61. J. Wang, J. Bentz, R. Anderson, Nasal device for delivery to the olfactory region
62. D. White, A. Al-Jumaily, J. Bartley, J. Lu, Correlation of nasal morphology to air-conditioning and clearance function. Respir. Physiol. Neurobiol. **179**(2–3), 137–41 (2011)
63. P. Wormald, T. Cain, L. Oates, L. Hawke, I. Wong, A comparative study of three methods of nasal irrigation. Laryngoscope **114**(2), 2224–7 (2004)
64. J. Xi, A. Berlinski, Y. Zhou, B. Greenberg, X. Ou, Breathing resistance and ultrafine particle deposition in nasal-laryngeal airways of a newborn, an infant, a child, and an adult. Ann. Biomed. Eng. **40**(12), 2579–2595 (2012)
65. J. Xi, J. Kim, X.A. Si, R.A. Corley, Y. Zhou, Modeling of inertial depositions in scaled models of rat and human nasal airways: towards in vitro regional dosimetry in small animals. J. Aerosol Sci. **99**, 78–93 (2015)
66. J. Xi, P.W. Longest, Characterization of submicrometer aerosol deposition in extrathoracic airways during nasal exhalation. Aerosol Sci. Technol. **43**(8), 808–827 (2009)
67. J. Xi, X. Si, W. Longest, Electrostatic charge effects on pharmaceutical aerosol deposition in human Nasal–Laryngeal airways. Pharmaceutics **6**(1), 26–35 (2014)
68. J. Xi, X.A. Si, J. Kim, Y. Zhang, R.E. Jacob, S. Kabilan, R.A. Corley, Anatomical details of the rabbit nasal passages and their implications in breathing, air conditioning, and olfaction. Anat. Rec. **299**(7), 853–868 (2016)
69. J. Xi, J. Yuan, M. Alshaiba, D. Cheng, Z. Firlit, A. Johnson, A. Nolan, W. Su, Design and testing of electric-guided delivery of charged particles to the olfactory region: experimental and numerical studies. Curr. Drug Deliv. **10** (2015)
70. J. Xi, J.E. Yuan, Y. Zhang, D. Nevorski, Z. Wang, Y. Zhou, Visualization and quantification of nasal and olfactory deposition in a sectional adult nasal airway cast. Pharm. Res. **33**(6), 1527–41 (2016)
71. J. Xi, Z. Zhang, X. Si, Improving intranasal delivery of neurological nanomedicine to the olfactory region using magnetophoretic guidance of microsphere carriers. Int. J. Nanomed. **10**, 1211–1222 (2015)
72. J. Xi, Z. Zhang, X.A. Si, J. Yang, W. Deng, Optimization of magnetophoretic-guided drug delivery to the olfactory region in a human nose model. Biomech. Model. Mechanobiol. **15**(4), 877–891 (2016)
73. Y. Zhang, Y. Shang, K. Inthavong, Z. Tong, B. Sun, K. Zhu, A. Yu, G. Zheng, Computational investigation of dust mite allergens in a realistic human nasal cavity. Inhal. Toxicol. **31**(6), 224–235 (2019). PMID: 31431101
74. K. Zhao, J. Craig, N. Cohen, N. Adappa, S. Khalili, J. Palmer, Sinus irrigations before and after surgery - visualization through computational fluid dynamics simulations. Laryngoscope **126**(3), E90–E96 (2016)
75. K. Zhao, P. Dalton, The way the wind blows: implications of modeling nasal airflow. Curr. Allergy Asthma Rep. 7(2), 117–25
76. Y. Zhou, M. Guo, J. Xi, H. Irshad, Y.-S. Cheng, Nasal deposition in infants and children. J. Aerosol Med. Pulm. Drug Deliv. **27**(2), 110–116 (2014)
77. Y. Zhou, J. Xi, J. Simpson, H. Irshad, Y.-S. Cheng, Aerosol deposition in a nasopharyngolaryngeal replica of a 5-year-old child. Aerosol Sci. Technol. **47**(3), 275–282 (2013)

Chapter 12
Future Topics, Challenges

Kiao Inthavong

Abstract Rapidly developing computational techniques and technologies have led to increased research activity in computational modelling of nasal airflow. The research has reached a critical crossroad where novel techniques and ideas will be needed to develop and shape the future trends in nasal airway modelling, including advanced multiphysics approaches with real breathing conditions. This chapter also discusses some recent developments including whole respiratory airway and lung models, and how big data and artificial intelligence can be integrated into current capability to advance the field.

12.1 Whole Respiratory Models

The respiratory airway consists of the upper and lower respiratory tract which consists of the respiratory organs that form a path to conduct the inhaled air into the deep lung region, i.e. from the nose to the bronchioles (Fig. 12.1). Clinicians often specialise in separate regions of the respiratory airway (typically upper or lower respiratory tracts) and increasingly, it is being recognised that a complication or disease in the upper respiratory tract could be related to some issue that is coming from the lower respiratory tract.

Studies of the nasal airway often end with the outlet defined at the pharynx, mainly due to the limitation of CT/MR scans. This meant that the airway geometry was limited to the upper respiratory airway which is suitable for analysis involving the inhalation phase e.g., inhaled air, particles, and drug delivery but not necessarily suitable for the exhalation phase. This is because of the requirement of defining a boundary condition at the inlet and outlet boundaries. During inhalation, the upstream airflow emanates from the surrounding environment and enters the nasal cavity through the nostrils, and many studies prior to 2010 would truncate the models at the nostrils

K. Inthavong (✉)
Mechanical & Automotive Engineering, School of Engineering, RMIT University, Melbourne, Australia
e-mail: kiao.inthavong@rmit.edu.au

K. Inthavong et al. (eds.), *Clinical and Biomedical Engineering in the Human Nose*, Biological and Medical Physics, Biomedical Engineering,
https://doi.org/10.1007/978-981-15-6716-2_12

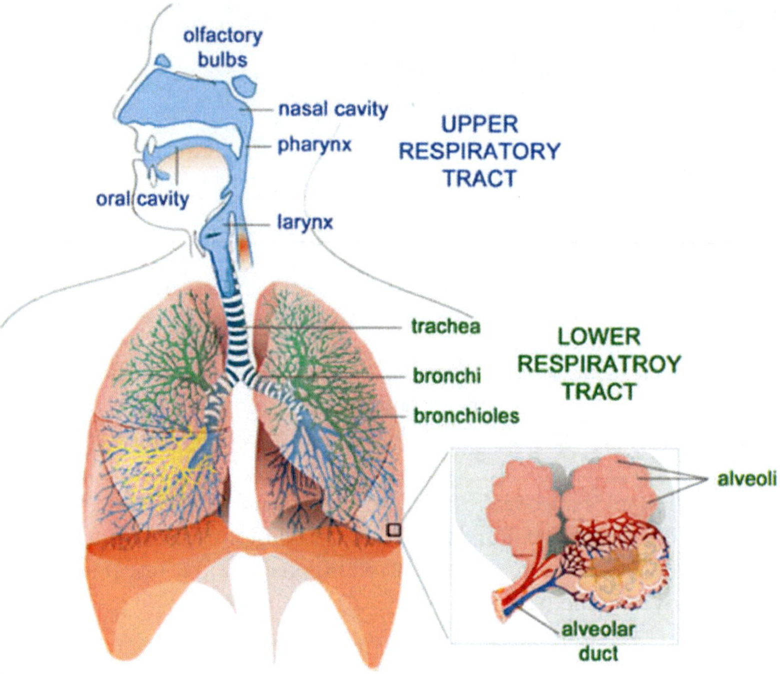

Fig. 12.1 Schematic of the respiratory system displayed by the upper and lower respiratory tract region

and define a uniform or plug-flow velocity profile at the boundary. Later, studies demonstrated that the outer face and surrounding environment was needed to allow the modelled air flow to naturally enter the nostrils thereby removing the incorrect definition of uniform or plug-flow velocity profiles at a truncated nostril inlet [5, 17].

During exhalation, the upstream airflow emanates from the lungs, and enters the nasal cavity via the nasopharynx. The upstream conditions for exhalation is vastly different to that experienced during inhalation. Now, the air originating from the lungs passes through the complex larynx geometry and performs a sharp 90° bend before entering the nasal cavity, and a prescribed uniform or plug-flow velocity profile at the nasopharynx boundary is incorrect. This suggests that (similar to the inhalation phase) extending the geometry beyond the larynx may be needed to account for the upstream disturbances in the larynx during exhalation. However, obtaining the whole respiratory tract from scanned images is rarely possible, because of limited images obtained from MR/CT scans. Future studies may need to determine the extent and level of disturbances and perhaps create an artificial boundary profile that is suitable as a replacement for the entire lower respiratory tract. To overcome this, studies that have modelled the entire respiratory tract have done so by artificially connecting different anatomical regions from multiple models [16]. Future advancements in imaging modalities such as low dose CT may allow the entire respiratory system to be captured.

12.2 Dynamic Meshing

Moving meshes can provide dynamic modelling of physiological functions that are transient in nature. Its use in respiratory modelling is evident in capturing changes in the lung shape [13] or sleep apnea applications [3, 20, 24], and in Sect. 10.2.1 demonstrated its use to mimic the muscular wall motions during breathing of the upper respiratory airway. The large deformations found in the 4D-cine imaging by Bates et al. [1] suggest that the wall motion is not solely a response to aerodynamic forces of the inhaled/exhaled air but is contributed by activation of the airway muscles from neuro-physiological responses of the breathing cycle. This creates significant challenges because the impact of the aerodynamic force on the wall motion may be masked by the airway wall muscular force, which is extremely difficult to measure in-vivo. Future work to account for both airway muscle and aerodynamic forces will be a new and challenging problem.

Nevertheless, dynamic meshing application to soft palate flutter during snoring and sleep apnea will continue to break new ground through the coupled fluid and structural equations. There have been ongoing studies that have demonstrated a coupled FSI approach [3, 4, 9, 11, 20, 24]. The large deformations (i.e. movement) of the structural walls, leads to occlusion or near occlusion of the airway which causes significant skewness in finite volume computation mesh elements. The modelling becomes highly susceptible to numerical solution divergence. Novel techniques to manage the deformation and couple the structural flutter with the fluid dynamics equations are needed to avoid these problems.

Future applications of dynamic meshing will also be needed to investigate different nasal airway collapse. This includes the nasal valve which is susceptible to collapse, where it experiences the highest resistance to flow because of its narrowed geometry. The high resistance leads to a drop in intraluminal pressure increasing the potential for collapse especially in cases where there is a deviated septum [21], or an over-projecting nose and narrow nostrils or wide columella. Furthermore, the ageing process can naturally weaken the nasal valve structures increasing the the likelihood of nasal valve collapse. In modelling the structural deformation of the behaviour of the nose, the elasticity of the nose is needed to facilitate the deformation. Efforts to detect nasal valve collapse [2] concluded that while the skin in patients with nasal valve collapse appeared to maintain its elasticity, it demonstrates an increased ability to stretch and lengthen compared with nasal skin of those without valve collapse. The elasticity of the nose varies in patients and in particular with age where the ageing process weakens the region which make up the nasal valve. This suggests future work in this field will be important in societies that have high percentage of ageing population.

For aerodynamic induced deformation, the fluid-structure-interaction method will be needed, but for muscular induced wall movement, then a dynamic mesh method would be used. Take for example the changes in lung shape and volume that occur during inhalation and exhalation. When the muscles of the diaphragm move downwards during inhalation, the rib cage expands creating more room for the lungs to expand. The volume of the lung increases which decreases the pressure relative to the outside air. This pressure difference drives the movement of air for inhalation. The lung motion is created with a moving or dynamic grid which expand and contract without deformation between the part of the grid that is attached to the static regions which has been demonstrated in Feng et al. [6] and Mead-Hunter et al. [13].

12.3 Machine Learning and Big Data

A drawback of CFD results is the intensive requirement of computational time and storage, particularly for transient flows. The prospect of using *'big data'* and machine learning techniques such as Artificial Neural Network (ANN) may provide rapid results as an alternative to CFD models. Its accuracy is reliant on the input CFD data used as training samples, and the algorithm used for machine learning to describe the fluid dynamics behaviour.

An application of machine learning that will greatly benefit clinicians is the automated segmentation from CT/MR scans. Currently it takes approximately 8–10 hours for scans to be segmented [15] and labelled by components/regions in detail for simulations. Through Deep Learning algorithms each region of the nasal cavity could be determined and its performance would improve as more data is submitted. One challenge in the algorithm is to overcome artifacts and noise which can be difficult to separate materials of similar density (see Chap. 5) and this may be achieved through Convolutional Neural Network (CNN) architecture that focus on shape recognition.

Studies have demonstrated the application of machine learning to segmentation, where Men et al. [14] proposed a Deep Deconvolutional Neural Network (DNN) for segmentation of nasopharyngeal cancer from CT images using data from 184 patients and another 46 patients for assessing the DNN performance; Xiong et al. [22] applied a fully convolutional network (FCN), to segment the paranasal sinuses; Souadih et al. [19] used a 3D CNN to automatically segment the sphenoid sinus and; Iwamoto et al. [8] applied a FCN as an automatic approach to segment the paranasal sinuses from CT images. These studies demonstrate a critical capability—that of rapid turnaround time of modelling which is essential to move *in-silico* modelling to truly become a real time point of care technology.

Similarly, the techniques could be applied to fluid flow predictions including predicting collapsible sites in the upper airways through machine learning [23], and nasal patency following surgery. Indeed, its application to surgery has huge potential. Clinicians are confronted with difficult decisions to make in functional nasal airway surgery, principally relying on surgeons' anecdotal experience, and on individual patient symptoms, medical history, and physical examination. Unfortunately, these methods lack robust scientific evidence or data to support the decisions. With increased availability of evidence-based medicine and patient data/outcomes, machine learning could suggest the best surgical technique simply from scanned CT/MR data.

Optimised and personalised health care will also benefit from the availability of massive data (e.g. *Big Data* concept) leading to the next revolution in health care in the so-called Digital Twin. This term refers to the digital representation of a physical product such as the human anatomy. With the development of the *Internet of Things* (IoT), digital twins are becoming a potential across many industries, including healthcare, and it is expected that with machine learning, this will become a reality. However, its success is anticipated to rely on large data sets and well established results to ensure that the computational analysis will provide reliable results.

12.4 Advanced Physiology

Nasal physiology is multi-functional with simultaneous process occurring during each respiration cycle. There have been significant progress made to address these functions, and future efforts will continue to advance the current state of the art, in particular there will be a focus on a multi-physics approach including the fluid-structure-interactions and dynamic meshing discussed in Sect. 12.2.

Multiphase flows will play a big role in widening the clinical applications that would involve inhalation toxicology, drug delivery, olfaction, and air conditioning. Many studies adopt a discrete particle phase to represent a disperse phase of inhaled particles [12] or sprayed drug delivery [7, 18]. More physics that will enhance the modelling capability includes two-way coupled fluid-particle interactions, and four-way coupling (to include turbulence), treating the secondary phase as droplets to allow deformation and heat and mass transfer effects for breakup, evaporation, and condensation, and chemical reaction modelling of droplets with its environment. When the particle or droplet deposits onto the nasal walls, its interactions can be captured by by accounting for the mucus layer. The nose's mucociliary function serves as a defense mechanism that provides a pathway for eventual dissolution into the mucus sublayers. Kirch et al. [10] provided an early study on the effect of mucociliary clearance on nanoparticle deposition, while Shang et al. [16] provided new methods for modelling the mucus motion and its influence on clearance mechanisms. An alternative technique (already in use) is the Physiologically Based PharmacoKinetic (PBPK) model for predicting the absorption dosimetry analysis where dosimetry flux from particle deposition rates onto the boundary walls in CFD can be used as boundary values for PBPK models.

A wider scope of multiphase flows will become more common because of its wide application. This includes higher resolved spray breakup models such as the Eulerian Volume of Fluid (VOF) model that treats the liquid sheet ejecting from a spray nozzle as a continuous fluid and allowed to break up into droplets. A coupled VOF and Discrete-Particle-Model [11] can be used to investigate in detail the particle behaviour for drug delivery in the nasal cavity. The VOF method has also been used to investigate nasal irrigation applications where a liquid flow entering the nasal cavity by a jet stream demonstrated in Chap. 11.

If these multi-functional physiology can be captured simultaneously, then this can be combined with real-world data to create the *Digital Twin* concept. A digital twin is a replicate computational model of the real thing, and in this case the human nasal cavity, designed for monitoring, simulation, and prediction of future outcomes from a range of forthcoming inputs or scenarios. The digital twin concept comes from industrial 'Internet of Things' which integrate devices such as sensors that consume real data to anticipate maintenance, impending failure, and optimise performance. The success of this bold endeavour will contribute towards 'personalised medicine', an emerging practice in healthcare that treats patients as individuals and not just apply 'standard of care' treatments. The challenge will be to capture biological and physiological data and in real time that would be integrate with computational analysis. Nevertheless, there is huge potential and value of using the available data for predicting the health outcomes.

The future of computational modelling of the human nasal cavity is extremely strong and vibrant. The availability of computational power for everyday users provides huge capability for modelling and understanding new concepts and insights into the functions and physiology of the nose. The contributions from authors of this book from all over the world demonstrate the vibrant research community in this field, which undoubtedly will lead to advances in the current capabilities, new discoveries, and help shape clinicians' practice in advancing the standard of care of patients to a more personalised and targeted approach.

References

1. A.J. Bates, A. Schuh, K. McConnell, B.M. Williams, J.M. Lanier, M.M. Willmering, J.C. Woods, R.J. Fleck, C.L. Dumoulin, R.S. Amin, A novel method to generate dynamic boundary conditions for airway cfd by mapping upper airway movement with non-rigid registration of dynamic and static mri. Int. J. Numer. Methods Biomed. Eng. **34**(12), e3144 (2018), e3144 CNM-Mar-18-0089.R1
2. J.P. Bonaparte, R. Campbell, Assessment of pliability and elasticity of the external nasal skin in patients with unilateral nasal valve collapse A static biomechanical evaluation. JAMA Facial Plastic Surg. **20**(6), 475–479 (2018)
3. K.K. Chang, K.B. Kim, M.W. McQuilling, R. Movahed, Fluid structure interaction simulations of the upper airway in obstructive sleep apnea patients before and after maxillomandibular advancement surgery. Amer. J. Orthodon. Dentofacial Orthopedics **153**(6), 895–904 (2018)
4. G.C. Cheng, R.P. Koomullil, Y. Ito, A.M. Shih, S. Sittitavornwong, P.D. Waite, Assessment of surgical effects on patients with obstructive sleep apnea syndrome using computational fluid dynamics simulations. Math. Comput. Simul. (2014)
5. D.J. Doorly, D.J. Taylor, R.C. Schroter, Mechanics of airflow in the human nasal airways. Respir. Physiol. Neurobiol. **163**(1–3), 100–110 (2008)

6. Y. Feng, J. Zhao, C. Kleinstreuer, Q. Wang, J. Wang, D.H. Wu, J. Lin, An in silico inter-subject variability study of extra-thoracic morphology effects on inhaled particle transport and deposition. J. Aerosol Sci. (2018)
7. K. Inthavong, P. Das, N. Singh, J. Sznitman, In silico approaches to respiratory nasal flows: a review. J. Biomech. (2019)
8. Y. Iwamoto, K. Xiong, T. Kitamura, X.H. Han, N. Matsushiro, H. Nishimura, Y.W. Chen, Automatic segmentation of the paranasal sinus from computer tomography images using a probabilistic atlas and a fully convolutional network, in *Proceedings of the Annual International Conference of the IEEE Engineering in Medicine and Biology Society, EMBS* (2019)
9. S.J. Jeong, W.S. Kim, S.J. Sung, Numerical investigation on the flow characteristics and aerodynamic force of the upper airway of patient with obstructive sleep apnea using computational fluid dynamics. Med. Eng. Phys. (2007)
10. J. Kirch, M. Guenther, U.F. Schaefer, M. Schneider, C.-M. Lehr, Computational fluid dynamics of nanoparticle disposition in the airways: mucus interactions and mucociliary clearance. Comput. Visual. Sci. **14**(7), 301–308 (2011)
11. Y. Liu, J. Mitchell, Y. Chen, W. Yim, W. Chu, R.C. Wang, Study of the upper airway of obstructive sleep apnea patient using fluid structure interaction. Respir. Physiol. Neurobiol. (2018)
12. N. Lu Phuong, N. Dang Khoa, K. Inthavong, K. Ito, Particle and inhalation exposure in human and monkey computational airway models. Inhalation Toxicol. (2018)
13. R. Mead-Hunter, A.J. King, A.N. Larcombe, B.J. Mullins, The influence of moving walls on respiratory aerosol deposition modelling. J. Aerosol Sci. **64**, 48–59 (2013)
14. K. Men, X. Chen, Y. Zhang, T. Zhang, J. Dai, J. Yi, Y. Li, Deep deconvolutional neural network for target segmentation of nasopharyngeal cancer in planning computed tomography images. Front. Oncol. (2017)
15. S. Pirner, K. Tingelhoff, I. Wagner, R. Westphal, M. Rilk, F.M. Wahl, F. Bootz, K.W. Eichhorn, CT-based manual segmentation and evaluation of paranasal sinuses. Eur. Arch. Oto-Rhino-Laryngology (2009)
16. Y. Shang, J. Dong, L. Tian, K. Inthavong, J. Tu, Detailed computational analysis of flow dynamics in an extended respiratory airway model. Clin. Biomech. **61**, 105–111 (2019)
17. Y.D. Shang, K. Inthavong, J.Y. Tu, Detailed micro-particle deposition patterns in the human nasal cavity influenced by the breathing zone. Comput. Fluids. (2015)
18. J. Siu, K. Shrestha, K. Inthavong, Y. Shang, R. Douglas, Particle deposition in the paranasal sinuses following endoscopic sinus surgery. Comput. Biol. Med. **116**, 103573 (2020)
19. K. Souadih, D. Ben Salem, A. Belaid, Automatic segmentation of the sphenoid sinus in CT-scans volume with DeepMedics 3D CNN architecture. Med. Technol. J. (2019)
20. Y. Wang, J. Wang, Y. Liu, S. Yu, X. Sun, S. Li, S. Shen, W. Zhao, Fluid-structure interaction modeling of upper airways before and after nasal surgery for obstructive sleep apnea. Int. J. Numer. Methods Biomedi. Eng. (2012)
21. M. Wittkopf, J. Wittkopf, W.R. Ries, The diagnosis and treatment of nasal valve collapse (2008)
22. K. Xiong, T. Kitamura, Y. Iwamoto, X.H. Han, N. Matsushiro, H. Nishimura, Y.W. Chen, Semi-automatic segmentation of paranasal sinus from CT images using fully convolutional networks, in *2018 IEEE 7th Global Conference on Consumer Electronics, GCCE 2018* (2018)
23. S.H. Yeom, J.S. Na, H.-D. Jung, H.-J. Cho, Y.J. Choi, J.S. Lee, Computational analysis of airflow dynamics for predicting collapsible sites in the upper airways: machine learning approach. J. Appl. Physiol. **127**(4), 959–973 (2019). PMID: 31318618
24. M. Zhao, T. Barber, P.A. Cistulli, K. Sutherland, G. Rosengarten, Simulation of upper airway occlusion without and with mandibular advancement in obstructive sleep apnea using fluid-structure interaction. J. Biomech. (2013)

Appendix A
List of Useful Computational Software

A.1 Dicom Viewers

The listed DICOM viewers have similar functionality

3DimViewer	www.3dim-laboratory.cz/en/software/3dimviewer
Acculite	www.accuimage.com
Agnosco	www.e-dicom.com
Aliza DICOM Viewer	www.aliza-dicom-viewer.com
AMIDE	www.amide.sourceforge.net
Athena DICOM	www.medicalharbour.com/en/athena-dicom-viewer
Cobra DICOM	www.exxim-cc.com/free_DICOMviewer.html
Dicompyler	www.dicompyler.com
DICOMWorks	www.dicomworks.com
JiveX	www.visus.com/en/downloads/jivex-dicom-viewer.html
Irfanview	www.irfanview.com
Mango	www.ric.uthscsa.edu/mango
MicroDicom	www.microdicom.com
Onis DICOM viewer	www.onis-viewer.com
Open DICOM viewer	www.sourceforge.net/projects/opendicomviewer
OsiriX (MAC)	www.osirix-viewer.com
Sante DICOM Viewer	www.santesoft.com
Weasis	www.nroduit.github.io
xMedCon	www.xmedcon.sourceforge.net//Main/HomePage
XNView	www.xnview.com

K. Inthavong et al. (eds.), *Clinical and Biomedical Engineering in the Human Nose*, Biological and Medical Physics, Biomedical Engineering,
https://doi.org/10.1007/978-981-15-6716-2

A.2 Open Source Medical Imaging and Segmentation

3D Slicer	www.slicer.org
CVIPTools	www.ee.siue.edu/CVIPtools
Fiji/ImageJ	www.pacific.mpi-cbg.de/wiki/index.php/Fiji
GemIdent	www.gemident.com
ITK-SNAP	www.itksnap.org
Nasal-GEOM	https://dx.doi.org/10.17632/d23m5ykyw2.1

A.3 Commercial Medical Imaging and Segmentation

3D Doctor	www.ablesw.com/3d-doctor
Amira	www.visageimaging.com/amira.html
Analyse	www.analyzedirect.com
Mimics	www.materialise.com
SliceOmatic	www.tomovision.com
TurtleSeg	www.oxipita.com
Vida Diagnostics	www.vidadiagnostics.com

A.4 Open Source Computer Aided Design Software

FreeCAD	www.sourceforge.net/projects/free-cad
Open CASCADE	www.opencascade.org
BRL CAD	www.brlcad.org
OpenSCAD	www.openscad.org

A.5 Commercial Computer Aided Design Software

Geomagic	www.geomagic.com
CATIA	www.3ds.com
Autodesk	www.autodesk.com
Solidworks	www.solidworks.com
PRO/Engineer	www.ptc.com
IronCAD	www.ironcad.com

A.6 Open Source CFD Packages

dolfyn	www.dolfyn.net
FEniCS	www.fenicsproject.org
NEK5000	www.nektar.info
OpenFOAM	www.openfoam.com
OpenFVM	www.openfvm.sourceforge.net
OpenLB	www.openlb.net
Palabos	www.gitlab.com/unigespc/palabos

A.7 Commercial CFD Packages

ANSYS	Includes ICEM meshing, CFX, Fluent, CFDPost, www.ansys.com
Siemens	Includes STAR-CCM, www.plm.automation.siemens.com
Autodesk CFD	www.autodesk.com/products/cfd
Intelligent Light	www.ilight.com
Fieldview	www.fieldviewcfd.com
Flow 3D	www.flow3d.com
Numeca	www.numeca.be
Phoenics	www.cham.co.uk

A.8 Third Party Post-Processing Software

ParaView	www.paraview.org
GNU Plot	www.gnuplot.info
OpenDX	www.opendx.org
Tecplot Visualisation	www.tecplot.com
Plot3D Interactive	www.openchannelfoundation.org

Index

K. Inthavong et al. (eds.), *Clinical and Biomedical Engineering in the Human Nose*, Biological and Medical Physics, Biomedical Engineering,
https://doi.org/10.1007/978-981-15-6716-2